AF756784

Handbook of Artificial Intelligence Techniques in Photovoltaic Systems

Handbook of Artificial Intelligence Techniques in Photovoltaic Systems

Modeling, Control, Optimization, Forecasting and Fault Diagnosis

Adel Mellit and Soteris Kalogirou

ELSEVIER

ACADEMIC PRESS

An imprint of Elsevier

Academic Press is an imprint of Elsevier
125 London Wall, London EC2Y 5AS, United Kingdom
525 B Street, Suite 1650, San Diego, CA 92101, United States
50 Hampshire Street, 5th Floor, Cambridge, MA 02139, United States
The Boulevard, Langford Lane, Kidlington, Oxford OX5 1GB, United Kingdom

Copyright © 2022 Elsevier Ltd. All rights reserved.

MATLABs is a trademark of The MathWorks, Inc. and is used with permission. The MathWorks does not warrant the accuracy of the text or exercises in this book. This book's use or discussion of MATLABs software or related products does not constitute endorsement or sponsorship by The MathWorks of a particular pedagogical approach or particular use of the MATLABs software.

No part of this publication may be reproduced or transmitted in any form or by any means, electronic or mechanical, including photocopying, recording, or any information storage and retrieval system, without permission in writing from the publisher. Details on how to seek permission, further information about the Publisher's permissions policies and our arrangements with organizations such as the Copyright Clearance Center and the Copyright Licensing Agency, can be found at our website: www.elsevier.com/permissions.

This book and the individual contributions contained in it are protected under copyright by the Publisher (other than as may be noted herein).

Notices
Knowledge and best practice in this field are constantly changing. As new research and experience broaden our understanding, changes in research methods, professional practices, or medical treatment may become necessary.

Practitioners and researchers must always rely on their own experience and knowledge in evaluating and using any information, methods, compounds, or experiments described herein. In using such information or methods they should be mindful of their own safety and the safety of others, including parties for whom they have a professional responsibility.

To the fullest extent of the law, neither the Publisher nor the authors, contributors, or editors, assume any liability for any injury and/or damage to persons or property as a matter of products liability, negligence or otherwise, or from any use or operation of any methods, products, instructions, or ideas contained in the material herein.

ISBN: 978-0-12-820641-6

For information on all Academic Press publications
visit our website at https://www.elsevier.com/books-and-journals

Publisher: Charlotte Cockle
Acquisitions Editor: Edward Payne
Editorial Project Manager: Aleksandra Packowska
Production Project Manager: Anitha Sivaraj
Cover Designer: Christian J. Bilbow

Typeset by STRAIVE, India

Working together
to grow libraries in
developing countries

ELSEVIER Book Aid International

www.elsevier.com • www.bookaid.org

Contents

Supplementary materials

The ancillary material containing codes written in MATLAB/Simulink is available at https://www.elsevier.com/books-and-journals/book-companion/9780128206416

Authors' biographies

Adel Mellit is Professor at the Faculty of Sciences and Technology, Jijel University, Algeria. He received his MS and PhD in electronics from the University of Sciences and Technology Houari Boumediene (USTHB), Algiers in 2002 and 2006, respectively. His research interests focus on the application of artificial intelligence techniques in photovoltaic systems and microgrids (control, fault diagnosis, optimization, and real-time applications). Dr. Adel Mellit has authored or coauthored more than 150 papers in international peer-reviewed journals (mostly with Elsevier), papers in conference proceedings (mostly with the IEEE) mainly on photovoltaic systems, six book chapters, and two books. He is the director of the Renewable Energy Laboratory at the Jijel University, Algeria, and is an associate member at the ICTP Trieste, Italy. He is serving on the editorial board of the *Renewable Energy* and is Editor of the *IEEE Journal of Photovoltaic* and of *Energy* (Elsevier Ltd). https://orcid.org/0000-0001-5458-3502.

Soteris Kalogirou is Professor at the Department of Mechanical Engineering and Materials Sciences and Engineering of the Cyprus University of Technology, Limassol, Cyprus. He is currently the Dean of the School of Engineering and Technology. In addition to PhD, he holds the title of DSc. He is a fellow of the European Academy of Sciences and a founding member of the Cyprus Academy of Sciences, Letters, and Arts. For more than 35 years, he is actively involved in research in the area of solar energy. He is also involved in research related to the use of artificial intelligence for the modeling, performance prediction, and fault finding of renewable energy systems. He has a large number of publications in books, book chapters, international scientific journals, and refereed conference proceedings. He is the editor-in-chief of *Renewable Energy* and deputy editor-in-chief of *Energy*. He is the editor of the book *Artificial Intelligence in Energy and Renewable Energy Systems*, published by Nova Science Inc.; the coeditor of the book *Soft Computing in Green and Renewable Energy Systems*, published by Springer; the editor of the book *McEvoy's Handbook of Photovoltaics*, published by Academic Press of Elsevier; and the author of the books *Solar Energy Engineering: Processes and Systems* and *Thermal Solar Desalination: Methods and Systems*, published by Academic Press of Elsevier. https://orcid.org/0000-0002-4497-0602.

Preface

This book is the result of many years of collaboration between two researchers in the field of artificial intelligence (AI) techniques and solar energy systems. The idea behind writing the book is to give to readers the first book on the application of AI techniques, including machine learning, and deep learning in photovoltaic systems.

Recently, the use and the application of AI techniques in renewable energy systems (RESs) have increased considerably, showing substantial advances in solving some complex problems. However, in some cases, the application of AI techniques is considered a magic solution by some researchers, which is not correct. Therefore, the main objective of writing this book is to demonstrate to the readers a comprehensive way and detailed methods on how, when, and why to use and apply AI techniques in solving various photovoltaic problems, including solar radiation forecasting, optimization and management, control and maximization of output power, fault diagnosis, power forecasting, power sharing, and smart monitoring systems using the Internet of things (IoT) technique.

The book provides readers with a comprehensive overview of the application of AI techniques in photovoltaic systems. We believe that this comprehensive illustration of the AI techniques and their application in photovoltaic systems will be a valuable key guidance for scientists, researchers, and industry sectors working in this area, to apply these techniques effectively. The book can also be used as advanced background reading by students who are studying postgraduate courses and who are motivated and interested in the application of AI techniques in photovoltaics. In addition, readers will be able to get an idea about different photovoltaic configurations, including grid-connected, hybrid, and micro-grids. Understating the main challenging issues in the implementation of AI in photovoltaic applications is another objective.

The most valuable features and content in the book are (1) understanding how to apply AI techniques; (2) showing the advantages of AI to solve some challenging problems in photovoltaics (autonomous, grid-connected, hybrid, and micro-grids), such as fault detection, identification, and localization of faults; advanced management and control process; output power forecasting; and smart monitoring systems; (3) showing the most important aspects for the implementation of intelligent methods in real time using some reconfigurable circuits such as field programmable gate arrays (FPGAs), microcontrollers (Arduino), and Minco-processor (Raspberry); and (4) designing smart prototypes based on embedded technologies.

All examples, including codes presented in this book, are written in Python language and Matlab®/Simulink. Readers should have basic skills in Matlab/Simulink and programming language (Python and C), as well as general knowledge about AI techniques, including machine learning and deep learning algorithms.

The book contains eight chapters:

Chapter 1 consists of two parts: the first one provides a short introduction to solar radiation, which is the input power for solar energy systems, including photovoltaic systems. The main equations and models used for modeling and simulation of solar radiation components are presented in this section. Some examples of these applications are provided using Python and Matlab languages. The second part of this chapter covers a brief introduction to photovoltaics, including solar cell conversion, photovoltaic technologies, latest solar cell efficiency, photovoltaic configuration, and type of photovoltaic systems (e.g., stand-alone, grid-connected, and hybrid photovoltaic systems). Examples for modeling and simulation of photovoltaic systems, from solar cell to complete photovoltaic systems, using Matlab/Simulink are also presented.

Chapter 2 provides an introduction to AI techniques, including machine learning, ensemble learning, and deep learning. Several basic and comprehensive examples of different machine learning algorithms (such as support vector machine, *k*-nearest neighbors, linear regression, logistic regression, decision tree, random forest, *k*-means, fuzzy c-means, principal component analysis, hierarchical clustering, and neural networks), ensemble learning algorithms (such as Bagging, Extra tree, AdaBoost, histogram-based gradient boosting and gradient boost, voting, and stacking), and deep learning algorithms (such as convolutional neural networks, long short-term memory, and other hybrid configurations) are given. The examples presented have been developed using the Python programming language and offer a clear idea to readers on how to apply and use these algorithms for solving photovoltaic problems. This chapter is not intended to provide the theoretical development of these algorithms; however, it is focused on the application of these algorithms to solve certain photovoltaic systems problems.

Chapter 3 discusses four applications. The first application is about forecasting one step ahead of the daily global horizontal irradiation (GHI), using machine learning methods. The second application is related to one step ahead forecasting of in-plane solar irradiance using deep learning neural networks. The third application concerns the investigation of deep learning natural networks for multistep ahead forecasting of in-plane solar irradiance. In the last application, emphasis is given to the use of neural networks for one step ahead of daily and hourly GHI from meteorological parameters. A total of 16 examples are provided in order to help readers understand the application of machine learning and deep learning in this topic. Four databases are used to develop the different models.

Chapter 4 covers numerous examples of the application of machine learning and deep learning algorithms for short-term photovoltaic output power forecasting. The first application is on forecasting one step ahead photovoltaic output power using machine learning algorithms; in this application, only historical data were used. The second application focuses on the use of artificial neural networks (ANNs) to forecast photovoltaic output power using measured photovoltaic power and other meteorological parameters. The third application concerns the investigation of ensemble learning methods for one step ahead forecasting. The fourth application deals with the utilization of deep learning natural networks for one step and multistep ahead forecasting of output photovoltaic power. The investigated algorithms include deep neural networks such as long short-term memory (LSTM), bidirectional LSTM, gated recurrent unit (GRU), convolutional neural networks (CNNs), and hybrid configurations. The fifth application shows how to qualify the variation in the predicted results by calculating the confidence interval and the prediction interval, which is very important for analyzing the uncertainties of the forecasters.

Chapter 5 deals with the optimization of photovoltaic power, particularly under shading abnormal conditions that are responsible for the creation of various problems such as hotspot, shading effects, dust accumulation, and aging, among others. This chapter covers the application of optimization methods based on AI techniques for maximizing the photovoltaic output power under uniform and non-uniform solar conditions. The material presented is mainly focused on the application of fuzzy logic, neural networks, neuro-fuzzy, particle swarm optimization, and various hybrid combinations of these techniques. There are mainly two ways for maximizing the PV output power: by tracking the maximum power from a photovoltaic array using maximum power tracking algorithms and by photovoltaic array reconfiguration (static or dynamic techniques), which are globally used to prevent shading effects. Eight examples are presented in detail.

Chapter 6 is intended to address various applications of machine learning and deep learning methods for fault diagnosis of photovoltaic arrays. The machine learning methods investigated in this chapter are *k*-nearest neighbor, support vector machine, decision tree, random forest, principal component analysis, and ensemble learning. A total of 15 examples with eight listings are presented in detail. In this chapter readers will learn how to prepare a dataset, perform the main relevant features extraction, reduce the dimensionality of a dataset, and understand how to apply various types of machine learning methods, such as deep convolutional neural networks to detect and classify the major defects that affect the reliability and safety of photovoltaic arrays.

Chapter 7 addresses three issues: control of hybrid-grid connected PV systems, energy management systems, and power sharing of micro-grids (distributed generation). The six examples presented in this chapter have been developed and simulated using Matlab/Simulink and Python language. We have shown how to apply neural networks, fuzzy logic, and metaheuristic algorithms to address the following topics: (1) control of grid-connected and hybrid PV systems, (2) energy management of a micro-grid (MG) with an electrical vehicle (EV) station, and (3) power sharing in islanded MGs.

Chapter 8 provides a real-time implementation of some methods described in the previous chapters, in microcontrollers (Arduino), microprocessor (Raspberry Pi), and reconfigurable circuits such as field programmable gate array (FPGA). We verify experimentally the ability of these methods for real-time applications. The idea is to design smart embedded systems using IoI and AI techniques. Five applications are presented: (1) smart PV monitoring systems using the IoT technique and programmable devices (Arduino and Raspberry Pi), (2) FPGA-based implementation of an intelligent off-grid photovoltaic system emulator, (3) implementation of intelligent

Maximum Power Point Tracking (MPPT) algorithms using an FPGA board, (4) co-simulation of an intelligent fault-diagnosis method for a photovoltaic array using the Xilinx System Generator FPGA, and (5) implementation of PV fault diagnosis techniques in the Raspberry device using machine learning and deep learning algorithms.

Acknowledgments

The editors thank Dr. Alessandro Massi Pavan and Dr. Vanni Lughi from Trieste University (Italy), Prof. Mohamed Benghanem from Medinah University (KSA), Dr. Boubakeur Sahbi from Jeddah University (KSA), Dr. Kellil Nouamen from UDES, Bousmail (Algeria), and Dr. Abdessalam Belaout from Setif University (Algeria) for providing the datasets used in this book.

The first author Dr. Adel Mellit is grateful to his ex-PhD students Dr. Nedjwa Chettibi, Dr. Wafa Chine from Jijel University (Algeria), and Dr. Adnane Messai from Birine COMENA (Algeria) for their help in writing examples.

Solar radiation and photovoltaic systems: Modeling and simulation

1

Chapter outline

1.1 Introduction

Among the various renewable energy sources, solar energy utilization, through the use of photovoltaic (PV), is one of the most important sources of electrical energy production around the world. Over the past few years, photovoltaic installations have experienced rapid growth worldwide. At the end of 2020, the cumulative installed capacity exceeded 760.4GW [1]. Fig. 1.1 shows the evolution of annual PV installations around the world. With reference to IRENA [2], photovoltaics (PV) is the leading energy technology in terms of annual growth rate (over 35% in 2010–19).

The contribution of PV to decarbonizing the energy mix is progressing, with PV savings for as much as 720 million tons of CO_{2eq}. At the end of 2019, PV contributed to the reduction of the global CO_2 emissions by 1.7% or 2.2% of the energy-related emissions and 5.3% of the electricity-related emissions as compared to a world without PV [3]. The role played by PV in the reduction of the CO_2 emissions from electricity is continuously increasing [1].

According to the latest cost data from the IRENA, the global weighted-average levelized cost of electricity (LCOE) of utility-scale solar photovoltaics fell by 82% between 2010 and 2019 [2]. Currently, power purchase agreements (PPAs) have been announced for large-scale PV systems at below 2 USD cents/kWh/m^2. The share of monocrystalline technology is now about 66%, achieved by refining the architecture and the manufacturing of PV modules [4]. As reported in Ref. [1] for several countries, the PV contribution to the electricity demand has passed the 5% mark. PV is now the most competitive electricity source in some market segments. The availability of this cheap electricity is starting to allow the breakthrough of green fuels [1].

Handbook of Artificial Intelligence Techniques in Photovoltaic Systems. https://doi.org/10.1016/B978-0-12-820641-6.00001-6
Copyright © 2022 Elsevier Ltd. All rights reserved.

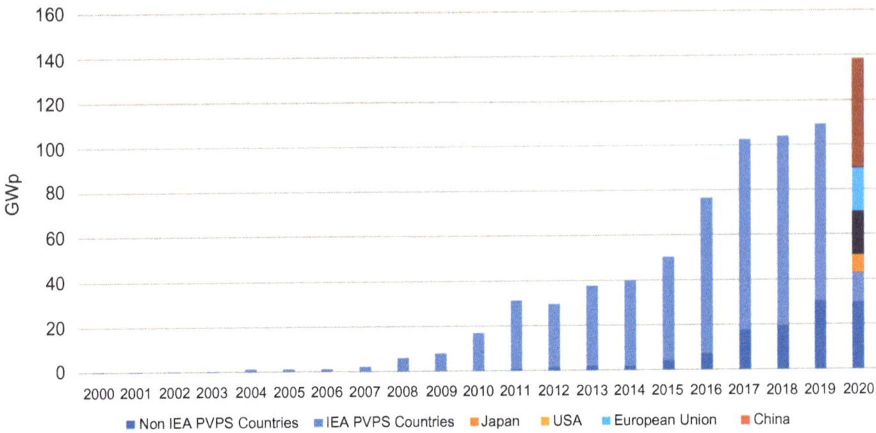

Fig. 1.1 Evolution of annual photovoltaic installations (IEA) [1].

Photovoltaic applications include many areas: (1) solar farms (e.g., large PV systems), (2) remote locations (sites with no access to grid, such as electrification of Sahara regions, mountains, rural locations, etc.), (3) stand-alone power (e.g., water pumping, telecommunication, parking meters, radio transmitter, etc.), (4) space (e.g., satellite), (5) transportation (e.g., electrical vehicles and boats), (6) buildings (e.g., PV on the roof, integrating transparent PV window, parking structures, etc.), (7) military sector (e.g., drones and other equipment), and (8) supply power to electronic devices.

In this chapter, a short introduction to solar radiation is initially given, which plays a very important role for solar energy systems, including photovoltaic systems. The main equations and models used for modeling and simulation of solar radiation components are presented in this section. Some examples of these applications are provided using Python and Matlab® languages. Next, we present a brief introduction to photovoltaics, including solar cell conversion, PV technologies, latest solar cell efficiency, PV configuration, and type of photovoltaic systems (e.g., stand-alone, grid-connected, and hybrid photovoltaic systems). Examples for modeling and simulation of photovoltaic systems, from solar cell to PV systems, using Matlab/Simulink are also presented.

1.2 Solar radiation

Solar radiation is the energy radiated from the Sun in the form of electromagnetic waves, including visible and ultraviolet light and infrared radiation. Fig. 1.2 shows the extraterrestrial solar spectrum (ETS) at the mean Sun-Earth distance [5].

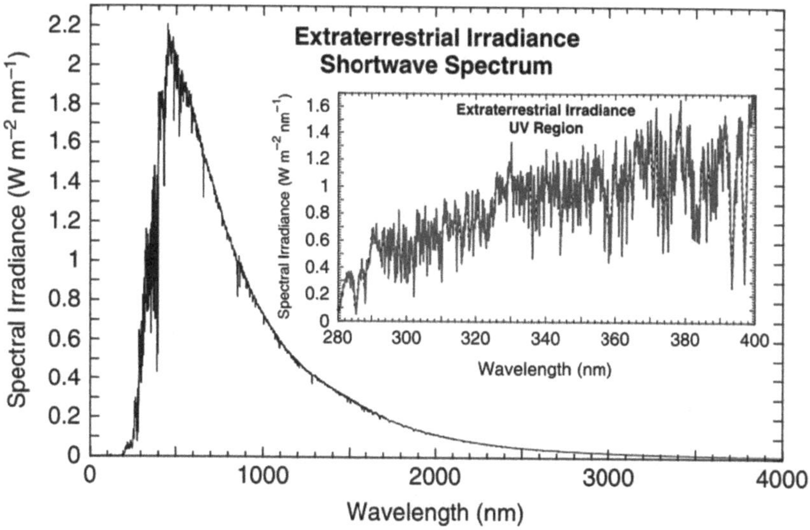

Fig. 1.2 Extraterrestrial solar spectrum in the shortwave at low resolution (0.5 – 5 nm).

Example 1.1: Solar spectrum

The following code (**Listing 1.1**) shows an example of how to plot the solar spectrum [6] using the open-source PVLib-Python library [7] with the use of the *Spectrum. spectrl2* function.

Listing 1.1: Solar spectrum modeling

```
lat=36;lon=5 ;tilt=36;azimuth=180
pressure=9 ;water_vapor_content=0.5
tau500=0.1;ozone=0.31 ;albedo=0.2
times=pd.date_range('2020-03-2012:00', freq='h', periods=4, tz='Etc/
  GMT+2')
solpos=solarposition.get_solarposition(times, lat, lon)
aoi=irradiance.aoi(tilt, azimuth, solpos.apparent_zenith, solpos.
  azimuth)
spectra=spectrum.spectrl2(
    apparent_zenith=solpos.apparent_zenith,
    aoi=aoi,
    surface_tilt=tilt,
    ground_albedo=albedo,
    surface_pressure=pressure,
    relative_airmass=relative_airmass,
    precipitable_water=water_vapor_content,
    ozone=ozone,
    aerosol_turbidity_500nm=tau500
)
```

So to execute the above code, first the PVLib library should be installed and then the required functions such as spectrum, solar position, irradiance, and atmosphere should be imported; see the following command line.

```
from pvlib import spectrum, solarposition, irradiance, atmosphere
```

There is a need to also import Pandas and Matplotlib libraries to plot the curve using the lines below:

```
import pandas as pd
import matplotlib.pyplot as plt
```

The plot of the modeled spectral irradiance for different air mass (AM) and zenith values is shown in Fig. 1.3.

Terrestrial applications of renewable energy utilizing solar radiation generally rely on radiation, or photons, referred to as "optical radiation," with a spectral range of about 300–4000nm [5]. Solar irradiance is the intensity of solar energy impacting an imaginary unit surface. Expressed in W/m^2, it is an instantaneous value of energy and does not represent accumulated energy over a period of time. The pyranometer is the main instrument used for measuring solar irradiance by meteorologists. Reference solar cells are also used for measuring global solar irradiance denoted by G [8]. As an example, Fig. 1.4A and B shows solar irradiance measurement using a pyranometer and a reference cell made from Si (polycrystalline).

As can be observed from Fig. 1.4C, acceptable correction exists between data measured by a pyranometer or a reference solar cell.

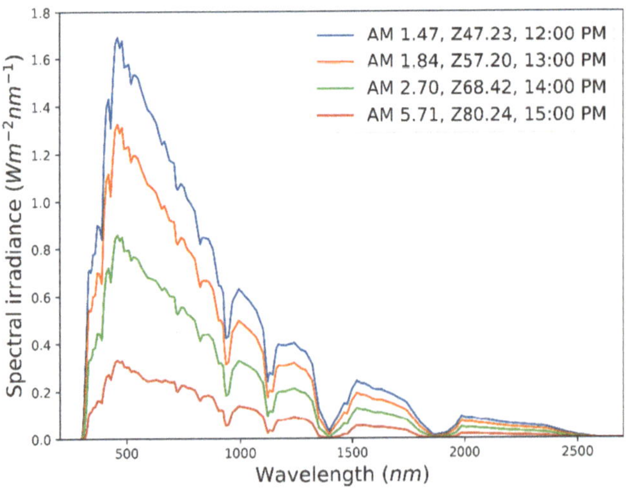

Fig. 1.3 Plot of spectral irradiance for different values of AM and zenith (latitude = 36°, longitude = 5°, and tilt angle = 36°).

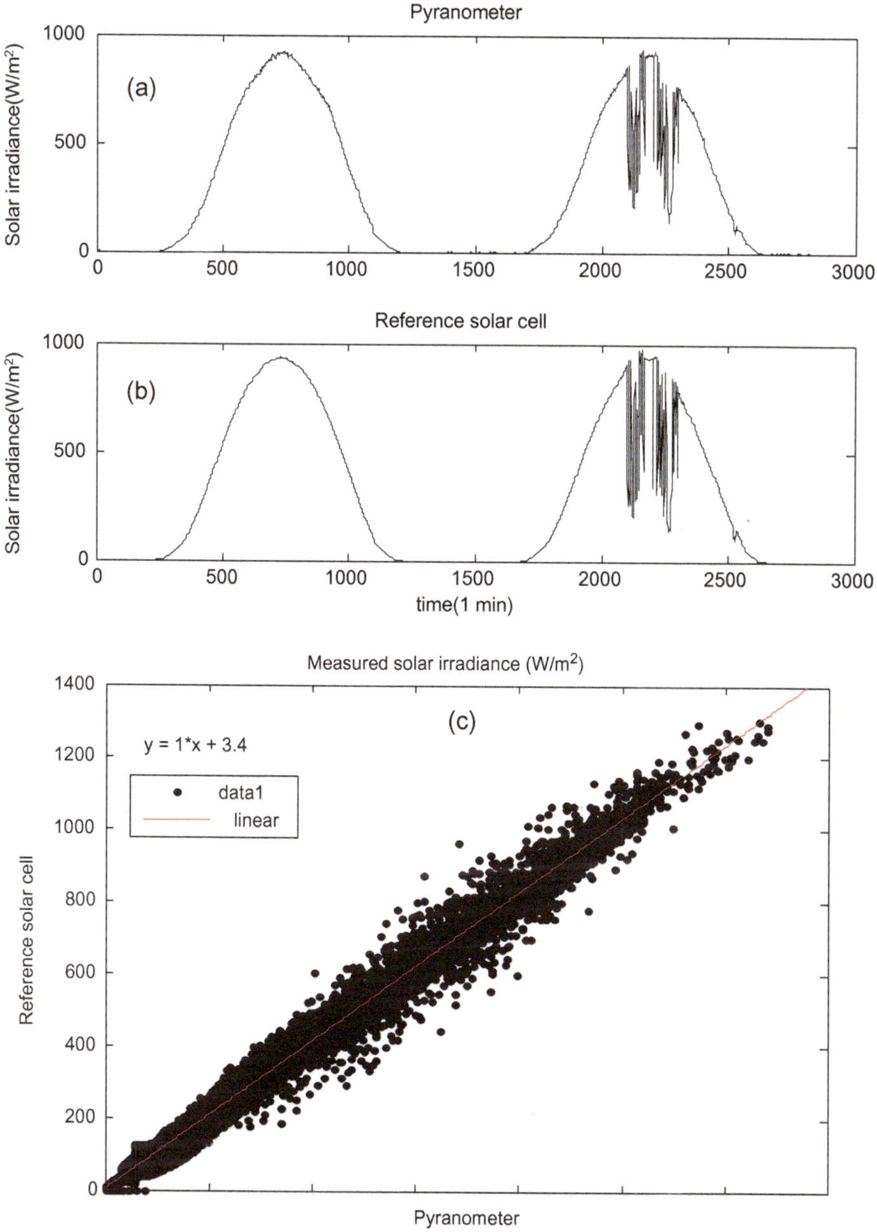

Fig. 1.4 (A) Measured solar irradiance by (B) a pyranometer and a reference solar cell. Geographical coordinate: latitude: 45.6536°, longitude: 13.7784, 45° 39′ 13″ North, 13° 46′ 42″ East, and altitude 2 m, corresponding to 2 days (July 1 and 2, 2012). (C) Scatterplot of measured *G* by a pyranometer and a reference solar cell.

Insolation is the solar energy that reaches the earth surface over the course of the day, and is expressed as kWh/m^2/day [9]. Fig. 1.5 shows an example of the insolation measured for a period of 1 year, 2020.

The daily extraterrestrial radiation on a horizontal surface (G_o) is given as follows [10]:

$$G_o = \frac{24}{\pi}\left[1 + 0.033\cos\left(\frac{2\pi d}{365}\right)\right](\cos(\varphi)\cos(\delta)\sin(\omega_s) + \omega_s\sin(\varphi)\sin(\delta)) \quad (1.1)$$

where d, φ, and δ are the day of the year, the latitude, and the declination, respectively. ω_s, the sunset hour angle for the horizontal surface ω_s, is given by

$$\omega_s = \cos^{-1}(-\tan(\varphi)\tan(\delta)) \quad (1.2)$$

The length of the day, S_o, is the time duration between sunrise and sunset, and it is one of the extraterrestrial variables of the classical Angstrom (1924) equation that could be estimated from

$$\cos(S_o) = -\tan(\varphi)\tan(\delta) \quad (1.3)$$

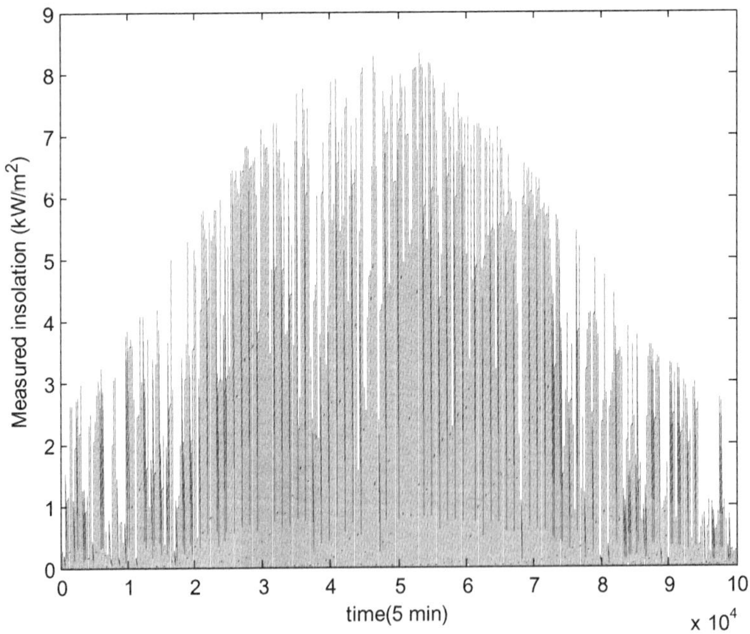

Fig. 1.5 Measured insolation 1 year (2020), geographical coordinate: latitude: 45.6536°, longitude: 13.7784, 45° 39′ 13″ North, 13° 46′ 42″ East, and altitude 2 m.

Example 1.2: Extraterrestrial radiation and the length of the day

G_o and S_o can be calculated based on the following Matlab function (**Listing 1.2**).

%Listing 1.2: Extraterrestrial radiation and sunshine duration (length of the day)

```
function [s0,g0]=h0s0(lat,beta,Nj);
lat=pi*lat/180;beta=pi*beta/180;s0=[];h0=[];%lat:latitude,   beta:
  declination
for j=1:Nj,%Nj:number of day
    I0=1367*(1+0.033*cos(pi*0.984*j/180));
    delta=(pi/180)*23.45*sin(pi*0.98*(284+j)/180);
    ws=acos(-tan(lat-beta)*tan(delta));
    g0=[g0,(24/pi)*I0*cos(lat-beta)*cos(delta(j))*(sin(ws)-ws*cos(ws))];
    s0=[s0,(24/pi)*acos(-tan(lat)*tan(delta))];
end
```

The daily clearness index K_t is defined as the ratio of the daily global radiation on a horizontal surface to the daily extraterrestrial radiation on a horizontal surface:

$$K_t = \frac{G}{G_o} \tag{1.4}$$

As an example, Fig. 1.6 shows a plot of extraterrestrial radiation (G_o) and the calculated K_t.

Fig. 1.7 shows the global horizontal irradiation (GHI) distribution around the world.

A large number of models were developed in the literature for the estimation of GHI [11–15]. Most of them are based on the basic formula of Angstrom [16] and its improved versions [17,18]. Generally, these can be classified into correlation models using S_o and G_o [11], models based on meteorological data such as air temperature [19] and cloud variation [20], and models based on image satellites [21].

1.3 Photovoltaics

1.3.1 Photovoltaic effect

The photovoltaic effect was discovered for the first time by E. Becquerel in 1839, using an electrochemical cell [22]. The process of conversion of light to electricity is called the *photovoltaic effect*. It simply means the production of DC current from sunlight [23] as depicted in Fig. 1.8. A basic structure of a solar cell comprises two layers of semiconductors (P and N); the junction between P and N acts as a diode allowing electrons to move from N to P. So, when photons with sufficient energy hit the

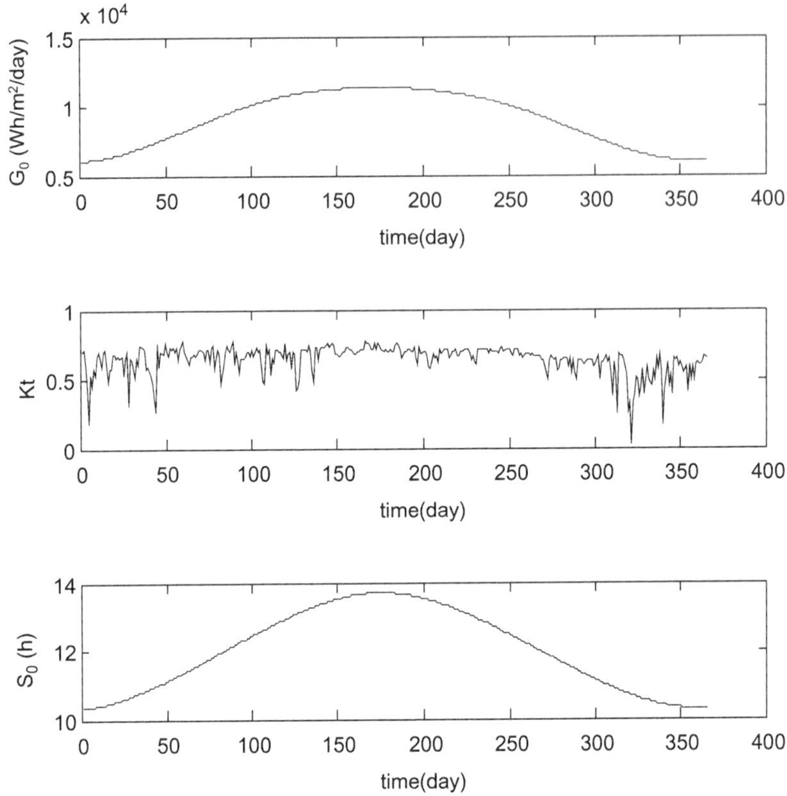

Fig. 1.6 The daily G_o, K_t, and S_o variation for a year, for geographical coordinates (27°23′11″, 41°38′49″).

cell, they create a movement (from N to P only) causing excess electrons in the N layer and a shortage in the P layer.

1.3.2 Photovoltaic technologies

There are mainly three photovoltaic technologies: (1) crystalline silicon or first-generation solar cells, (2) thin films or second-generation solar cells, and (3) emerging materials or third-generation solar cells [23].

1.3.2.1 Crystalline silicon

The first generation consists of wafer-based cells, which are made of crystalline silicon (Si). This includes materials such as monocrystalline Si and polycrystalline Si. The best wafer-based cell efficiency for Si-based technologies reached 26.7% [24]. Remarkable technological improvements have been achieved in the case of PV modules, which reached 24.4%. Si-wafer-based PV technologies account for about 95% of the total production in 2019 [25]. Fig. 1.9 shows a photo of monocrystalline and polycrystalline solar cells.

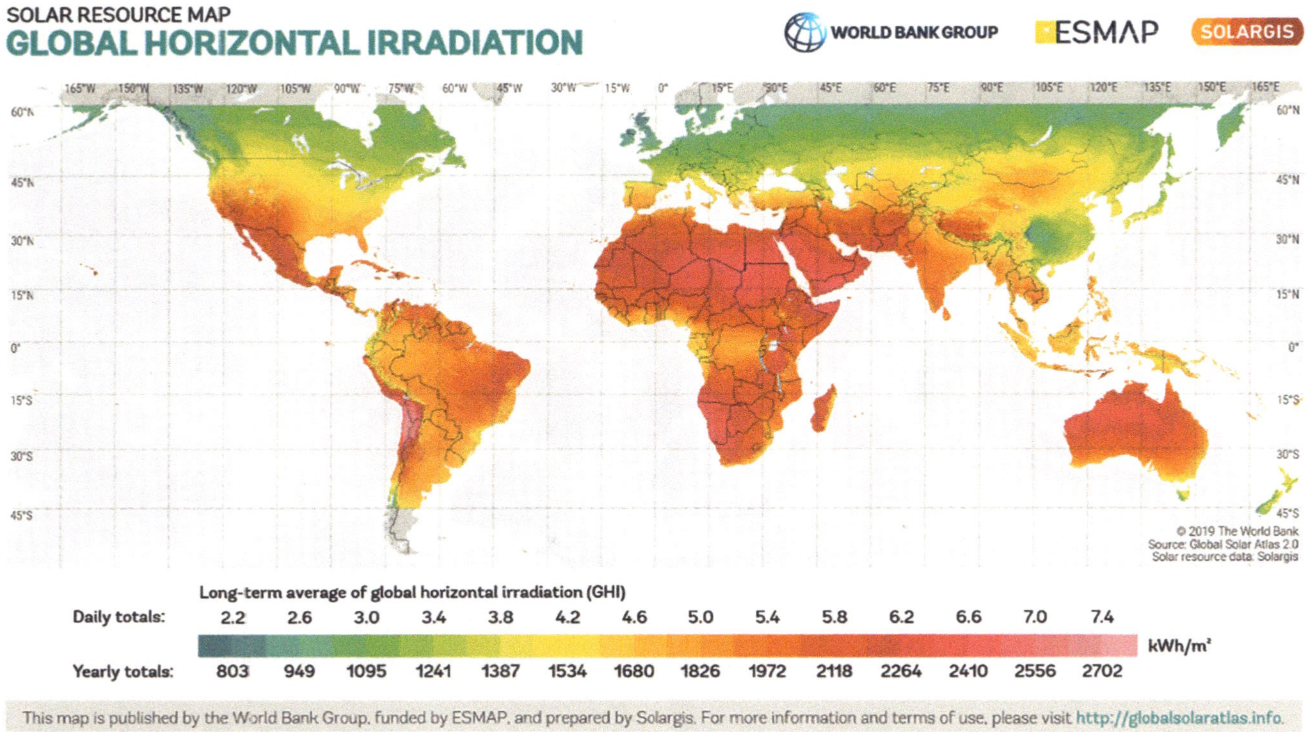

Fig. 1.7 GHI distribution around the world.

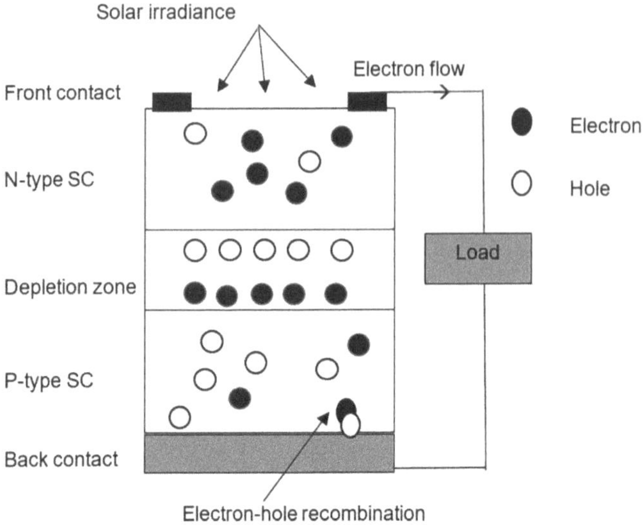

Fig. 1.8 The photovoltaic effect.

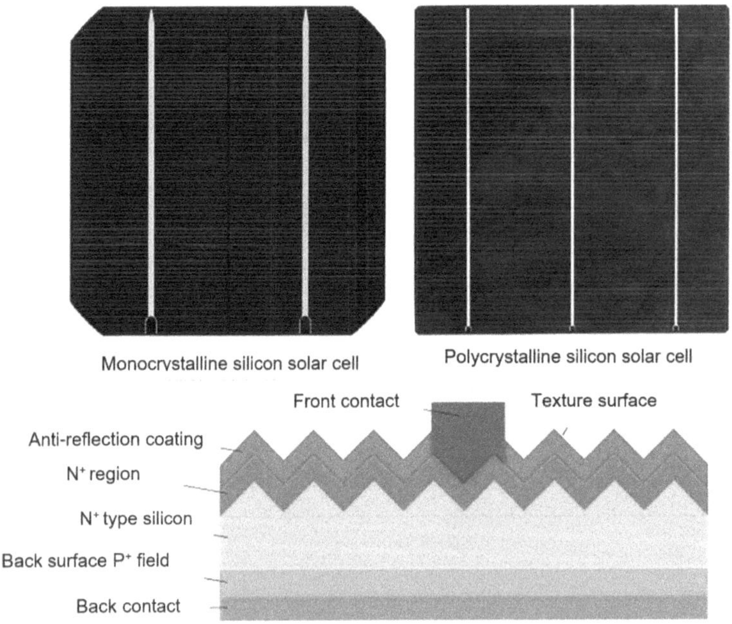

Fig. 1.9 A typical structure of silicon solar cell: monocrystalline and polycrystalline.

1.3.2.2 Thin films

The second-generation solar cells, named thin films, include copper indium gallium selenide (CIGS), cadmium telluride (CdTe), amorphous silicon (a-Si), and amorphous silicon: hydrogenated (a-Si:H) cells. This technology can be easily implemented as the fabrication process is simpler, compared to wafer-based Si. However, their efficiency is still lower compared to the first-generation cells. The best thin-film solar cells reached 23.35% (CIGS) [24]. This technology is the second most commercialized around the world after wafer-based Si; it contributes to the PV market by about 7% [23]. Producing CdTe and a Si:H solar cells is easier than CIGS, while, the stability of CIGS- and CdTe-based cells is higher than that of a-Si:H cells [26]. Fig. 1.10 shows an example of a typical structure of GIGS solar cell [27].

1.3.2.3 Emerging technologies

The third generation consists of emerging technologies and includes organic, dye-sensitized solar cells (DSSC), quantum dot (QD), perovskite, and multi-junction cells. Emerging technologies are considered as a low-cost PV technology. From the various types belonging to this generation, perovskites (thin films) present the highest efficiency of 25.2% [25]. A typical structure of multi-junction solar cells is shown in Fig. 1.11.

1.3.3 Modeling and simulation

The effective use of PV modules requires reliable modeling methods, aiming at predicting the behavior of a PV system at conditions different from those characterized and presented in the manufacturer's datasheets. Thus, accurate models can help for

(a) studying the behavior of solar cells
(b) monitoring a PV system performance
(c) analysis of losses in solar PV systems

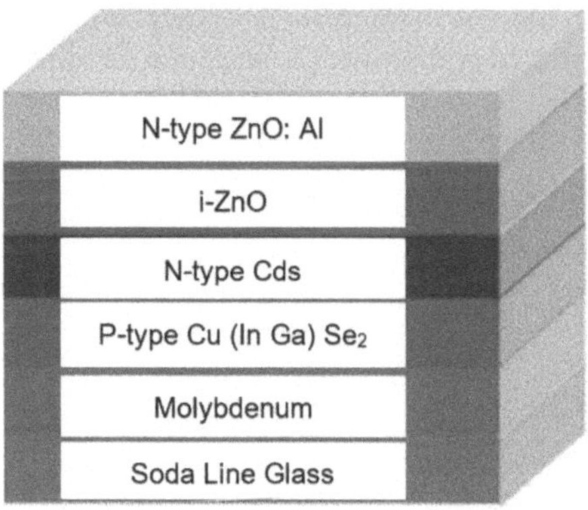

Fig. 1.10 A typical structure of CIGS solar cell.

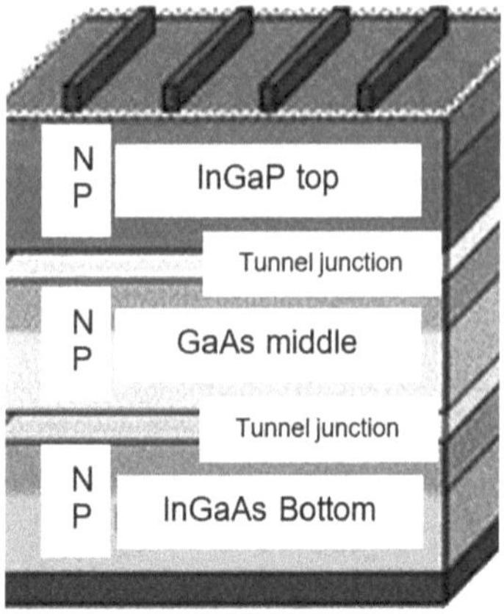

Fig. 1.11 A typical structure of multi-junction solar cell.

(d) forecasting the produced PV output power
(e) development and testing of maximum power point tracking algorithms
(f) fault diagnosis of PV systems

1.3.3.1 Solar cell modeling

The single diode model shown in Fig. 1.12 is the most used equivalent circuit for solar cells (called also five parameters model: R_{Sh}, R_S, a, I_{Ph}, and I_0) [28].

The circuit shown in Fig. 1.12 can be described by the following equation:

$$I = I_{Ph} - I_0\left[\exp^{a\frac{q(V+IR_S)}{BT_C}} - 1\right] - \frac{V - IR_S}{R_{Sh}} \tag{1.5}$$

where I_{Ph} is the light-generated current, i.e., the short-circuit current neglecting the parasitic resistances, I_0 is the dark saturation current due to recombination, q is the electron charge, R_S is a series resistance, B is the Boltzmann's constant, T_C is the solar cell temperature (in K), R_{Sh} is a shunt resistance, and a is an ideality factor. The light-generated current is directly proportional to the solar irradiance.

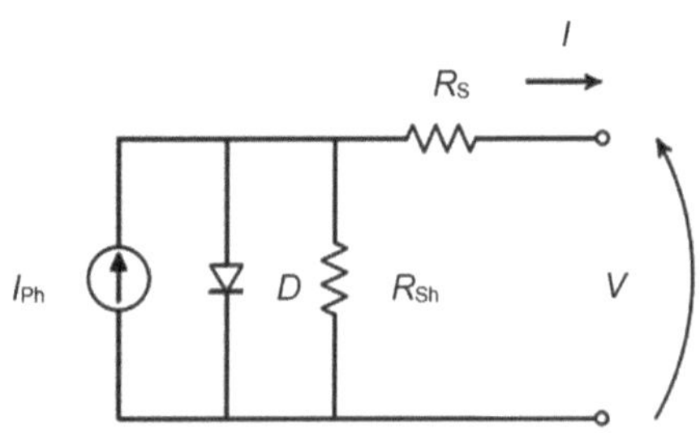

Fig. 1.12 Solar cell equivalent circuit—five-parameters model.

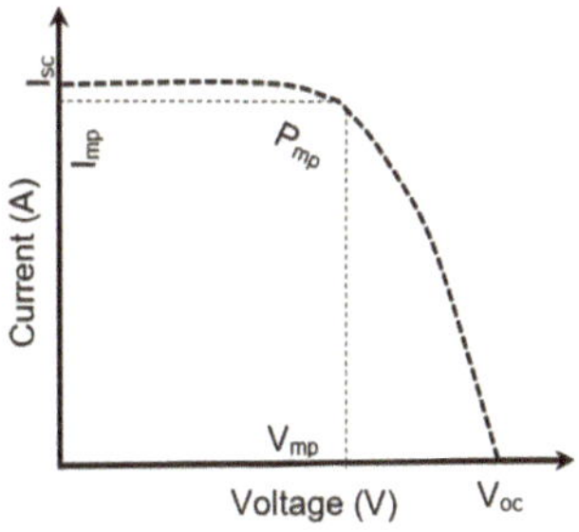

Fig. 1.13 Typical *I-V* curve of a solar cell.

Fig. 1.13 shows an example of a typical *I-V* curve of a solar cell. Each solar cell is characterized by the produced peak power (P_{m_p}), the fill factor (FF), and the efficiency (η). So [28]

$$FF = \frac{I_{mp} \times V_{mp}}{I_{sc} \times V_{oc}} \text{ and } \eta = \frac{I_{mp} \times V_{mp}}{S \times G} \tag{1.6}$$

where I_{mp} and V_{mp} are the maximum current and voltage, S is the effective surface of the solar cell, and G is the incident solar irradiance.

Latest solar cell efficiencies can be found in Ref. [29]. Fig. 1.14 shows the best researched solar cells reported by NREL [30]. Multi-junction solar cell technology reached 47.1% (four-junction or more).

Numerous solar cell models are available in the literature and they can be classified into two types: implicit and explicit models.

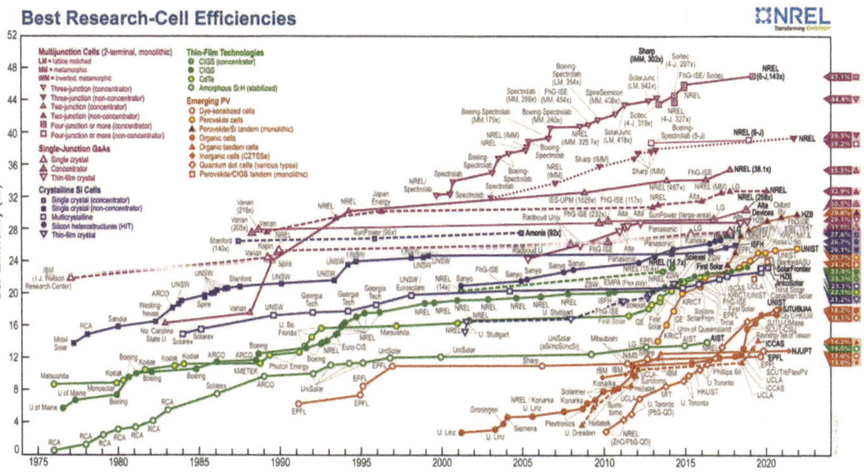

Fig. 1.14 Best research—cell efficiencies (NREL).

(a) Implicit models

Implicit models are represented by $I=f(I, V)$, and numerical methods are required to identify the parameters (I_{Ph}, I_0, R_S, R_{Sh}, and a) [31–33]. By identifying these parameters, the *I-V* curves could be plotted easily for a given condition of G and T.

(b) Explicit models

Explicit models are represented by a simple relation between I and V, $I=f(V)$ [34–37]. Parameters from the datasheet (P_{m_p}, I_{mp}, V_{mp}, I_{sc}, and V_{oc}) are enough to draw the *I-V* curves at a given condition (G and T_C).

More detail about the main equations for modeling of the one-diode model (implicit model) is reported in Appendix A.

1.3.3.2 PV module modeling

A PV module consists of a number of solar cells connected in series and in parallel configuration. For example, a BP solar module consists of 72 cellules; four branches of 18 solar cells are connected in parallel with four bypass diodes for each branch.

Example 1.3: *I-V* curve using Matlab function (one diode model)

Fig. 1.15 depicts the plotted *I-V* curve for a BP solar 120 W_p module, at STC conditions ($G=1000\,W/m^2$, $T_C=25°C$). The following Matlab function is used to calculate the current $I=f(V, G, \text{and } T_C)$. In this function, the Newton method is used to the estimate the parameters. The code used is as follows (**Listing 1.3**).

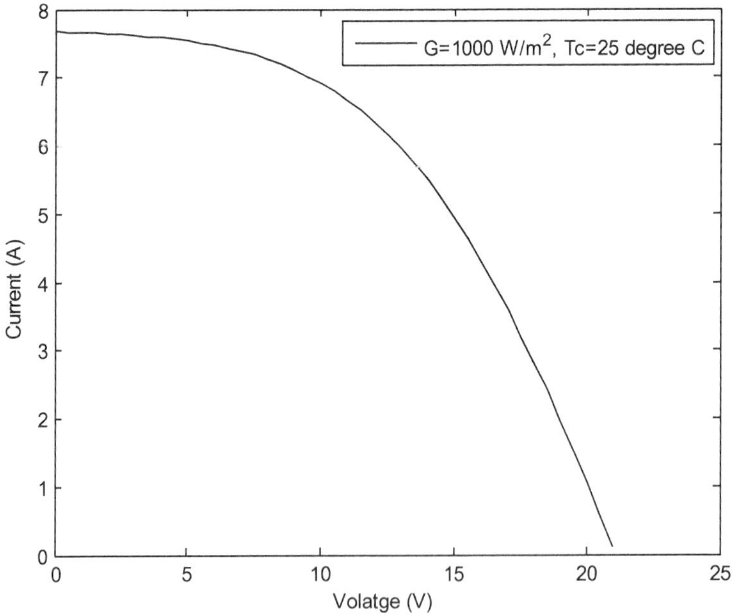

Fig. 1.15 *I-V* curve of a PV module (BP solar of 120W_p) at STC conditions.

Listing 1.3: I-V curve modeling using Matlab

```
function Ia = one_diode_model(Va,G,TaC);
k = 1.381e-23; q = 1.602e-19; n =1.6506;Eg = 1.12; Ns = 72; TrK = 298;
Voc_TrK = 21.1/Ns; Isc_TrK = 7.7;N=100;alpha = 0.65e-3; TaK = 273 +
  TaC;
Vc = Va / Ns;
Isc = Isc_TrK * (1 + (alpha * (TaK - TrK)));
Iph = G * Isc;
Vt_TrK = n * k * TrK / q;
b = Eg * q /(n * k);
Ir_TrK = Isc_TrK / (exp(Voc_TrK / Vt_TrK) -1);
Ir = Ir_TrK * (TaK / TrK)o(3/n) * exp(-b * (1 / TaK -1 / TrK));
dVdI_Voc = -1.0/Ns;
Xv = Ir_TrK / Vt_TrK * exp(Voc_TrK / Vt_TrK);
Rs = - dVdI_Voc - 1/Xv;
Vt_Ta = n * k * TaK / q;
% Solve for Ia by Newton's method: Ia2 = Ia1 - f(Ia1)/f'(Ia1)
I=zeros(size(Vc));
for j=1:N;
I = I - (Iph - Ia - Ir .* ( exp((Vc + Ia .* Rs) ./ Vt_Ta) -1))...
./ (-1 - Ir * (Rs ./ Vt_Ta) .* exp((Vc + Ia .* Rs) ./ Vt_Ta));
end
```

Example 1.4: *I-V* curve using PVLib open source [38] (implicit model)

In this example, `pvsystem.singlediode` and `pvsystem.calcparams_desoto` functions available in PVlib [39] are used to plot the *I-V* curve shown in Fig. 1.16. See the following **Listing 1.4**.

Listing 1.4: I-V curve modeling using PVLIB library /Python

```
from pvlib import pvsystem
import pandas as pd
import matplotlib.pyplot as plt
# Example module parameters for the Canadian Solar CS5P-220M:
parameters={
    'Name': 'Canadian Solar CS5P-220M',
    'BIPV': 'N',
    'Date': '10/5/2009',
    'T_NOCT': 42.4,
    'A_c': 1.7,
    'N_s': 96,
    'I_sc_ref': 5.1,
    'V_oc_ref': 59.4,
    'I_mp_ref': 4.69,
    'V_mp_ref': 46.9,
    'alpha_sc': 0.004539,
```

```
    'beta_oc': -0.22216,
    'a_ref': 2.6373,
    'I_L_ref': 5.114,
    'I_o_ref': 8.196e-10,
    'R_s': 1.065,
    'R_sh_ref': 381.68,
    'Adjust': 8.7,
    'gamma_r': -0.476,
    'Version': 'MM106',
    'PTC': 200.1,
    'Technology': 'Mono-c-Si',
}
conditions=pd.DataFrame([(1000, 55)], columns=['Geff', 'Tcell'])
# adjust the reference parameters according to the operating
# conditions using the De Soto model:
IL, I0, Rs, Rsh, nNsVth=pvsystem.calcparams_desoto(
    conditions['Geff'],
    conditions['Tcell'],
    alpha_sc=parameters['alpha_sc'],
    a_ref=parameters['a_ref'],
    I_L_ref=parameters['I_L_ref'],
    I_o_ref=parameters['I_o_ref'],
    R_sh_ref=parameters['R_sh_ref'],
    R_s=parameters['R_s'],
    EgRef=1.121,
    dEgdT=-0.0002677
)
# plug the parameters into the SDE and solve for IV curves:
curve_info=pvsystem.singlediode(
    photocurrent=IL,
    saturation_current=I0,
    resistance_series=Rs,
    resistance_shunt=Rsh,
    nNsVth=nNsVth,
    ivcurve_pnts=100,
    method='lambertw'
)
# plot the calculated curves:
#plt.figure()
for i, case in conditions.iterrows():
    plt.plot(curve_info['v'][i], curve_info['i'][i], color='k')
    v_mp=curve_info['v_mp'][i]
    i_mp=curve_info['i_mp'][i]
plt.xlabel('Module voltage (V)',fontsize=16)
plt.ylabel('Module current (A)',fontsize=16)
plt.title(parameters['Name'])
plt.show()
```

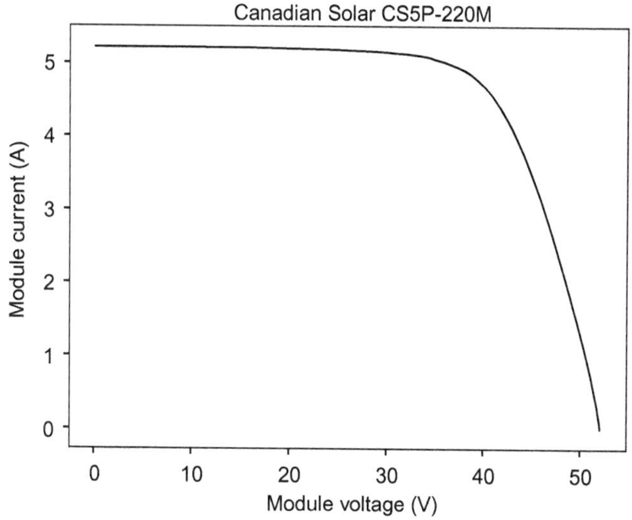

Fig. 1.16 *I-V* curve plot of a PV module (Canadian solar CSSP-220M) using a Desoto model developed in PVlib at STC conditions.

Example 1.5: Parameter extraction using the Newton-Raphson method

PV module parameter extraction using a one-diode or two-diode model was solved based on iterative techniques, including metaheuristic algorithms (genetic algorithm, particle swarm optimization, etc.). In this example, we present a basic method to estimate the main parameters of the two diode model: I_{Ph}, I_0, a_1, a_2, R_S and R_{Sh}. Assumptions have been made to simplify the method. The following code (**Listing 1.5**) allows the computing of the parameters based on the Newton-Raphson algorithm.

#Listing 1.5: PV module Parameters extraction

```
import numpy as np
#InformationmodelfortheMSX-120solararraydatasheet
Isc_ref=3.87;Voc_ref=42;Vmp=33.7;Imp=3.56;Pmax_e=120;Ns=72;
kv=-0.08;ki=Isc_ref*(0.065/100);
G0=1000; T0=25+273.15; T=T0; k=1.3806503e-23;
q=1.60217646e-19; Vt0=Ns*k*T0/ q;     Vt=k*T*Ns/q;
a1=1;a2=1.3;p=2.3;
Rs=0; Rs_max=(Voc_ref-Vmp)/Imp;Rp_min=Vmp/(Isc_ref-Imp)-Rs_max;
Rp=Rp_min;tol=0.001;
error1=float('inf');
I01=Isc_ref/(np.exp(Voc_ref/(((a1+a2)/p)*Vt))-1);
I0=I01;Isc=Isc_ref;Iph=Isc_ref;
V=[];
for i in range(Voc_ref):
    V.append(i)
print(V)
```

```
I=[];II=[];N=100;
while(error1>tol):
  Rp=(Vmp+(Imp*Rs))/(Iph-I0*(np.exp((Vmp+(Imp*Rs))/Vt)+np.exp((Vmp
 +(Imp*Rs))/((p-1)*Vt))+2)-(Pmax_e/Vmp));#Calculate Rp
 I=Isc_ref;
 II=[];
 for n in V:
   for j in range(N):
     f=Iph-I-I0*((np.exp((V[n])+(I*Rs))/(Vt))+(np.exp((V[n]
 +(I*Rs))/((p-1)*Vt)))+2)-((V[n]+(I*Rs))/Rp)
     dg=-1-I0*(((Rs/(Vt))*np.exp((V[n]+(I*Rs))/(Vt))+((Rs/((p-1)
 *Vt))*np.exp((V[n]+(I*Rs))/((p-1)*Vt))))-(Rs/Rp),
     I=I-(f/df)
     II=II.append(I)
     P[n]=V(n)*II(n)
 Pmax_m=max(P)
 error1=(Pmax_m-Pmax_e)
 Rs=Rs+0.01;Rs=Rs; Rp=Rp
print('a1=',a1,'a2=',a2,'Io=',I0,'Io1=',Io1,'Rs=',Rs,'Rsh=',Rp,'Iph=',Iph)
```

The results are shown in Fig. 1.17.

Example 1.6: *I-V* curve based on an explicit model [40] using Matlab

The explicit model developed in Ref. [40] is used to plot the *I-V* curve for the QPro PV module (Q-cells). The main code with functions is written in Matlab (as shown in **Listing 1.6**). Fig. 1.18 shows the modeled *I-V* curve at STC conditions.

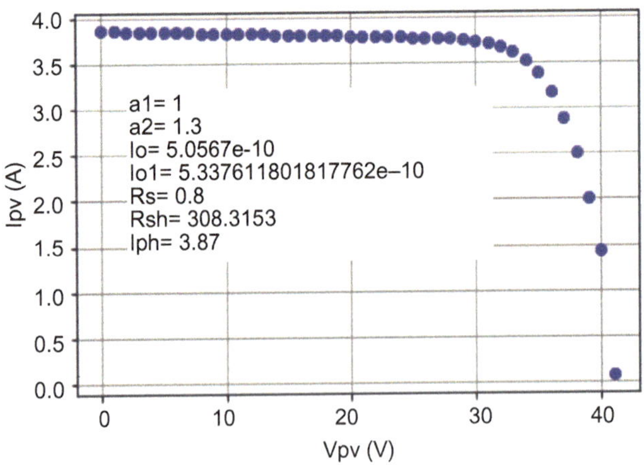

Fig. 1.17 *I-V* curve at STC conditions with the extracted parameters.

Listing 1.6: I-V curve modelling using an explicit model

```
clear all, clc, close all, format short
n=1; step=0.01; G=1, Tc=25,
base=220; tech=0; module=220;
[FF,m, Voc_base, Isc_base]=ff_m_bases_new(base,G,Tc,tech);
[Voc,Isc,base,W,Z]=pv_module_spec(module,tech);
w=(W*Voc)/(100*Voc_base);  z=(Z*Isc)/(100*Isc_base);
V=0:step:1;
[In]=In_model_new(G,Tc,m,V,z,w);
plot(V,In);xlabel('Module voltage [p.u]'); ylabel('module current
  [p.u]');
function [In]=In_model_new(Gj,Tj,m,V,z,w)
    In=Gj+z.*(-Tj+25)-(exp(m.*(V+w.*(-Tj+25)))-1)/(exp(m)-1);
function  [FF,m,  Voc_base,  Isc_base]=ff_m_bases_new(base1,G,Tc,
  tech);
if tech==0;
    if base1==210,
 Vmp_base=28.3; Imp_base=7.45;Voc_base=35.9; Isc_base=8.1;
  Z_base=0.05; W_base=-0.37;Pm_base=Vmp_base*Imp_base;
  base=Voc_base*Isc_base;
[FF]=fill_factor_new(T,G,Pm_base,Isc_base,Voc_base,W_base);
[m]=m_factor(FF);
end
end
function [FF]=fill_factor_new(Tc,G,Pn,Isc,Voc,w);
% constants
format long
G=G*1000;
c13=60;h10=20;c10=44;c12=25;c5=1000;k10=800;
W=-0.37;w=(W/100)*Voc;
d18=T+(c10-h10)/k10*G;d19=Voc+w*(d18-25);
d16=G*Isc/c5;d22=(25+273)/300*c12;
d24=Voc/(d22/1000*c13);d26=(d24-log(d24+0.72))/(d24+1);
d27=Pn/(Voc*Isc);d28=1-d27/d26;
d29=Voc/Isc*d28;d30=d16/d19*d29;
d21=(d18+273)/300*c12;d23=d19/((d21/1000)*c13);
d25=(d23-(log(d23+0.72)))/(d23+1);
FF=d25*(1-d30);
if FF>=1,FF=1,
else FF=FF;
end
function [m]=m_factor(FF)
m=1;
[mpp]=clacul_m(m);
while abs(mpp-FF)>0.0001,
m=m+0.01;
[mpp]=clacul_m(m);
End
```

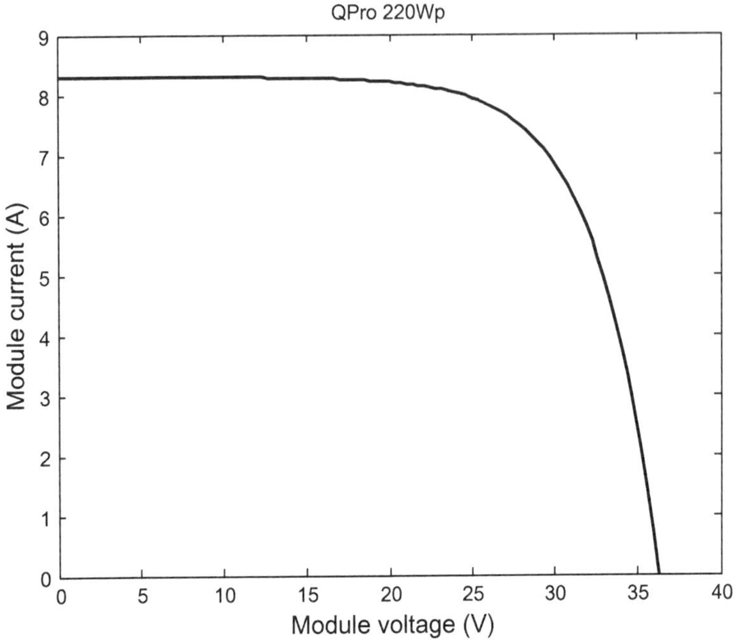

Fig. 1.18 *I-V* curve based on an explicit model at STC conditions for a multi-crystalline Q-Pro220 (220 W_p).

Example 1.7: *I-V* curve based on an explicit model [37,40] using Python

In this example, we implement the same model shown in the last example 1.5, but here the code is written in Python language. The code with functions that are written is shown in the following **Listing 1.7**). Fig. 1.19 shows the modeled *I-V* curve at STC conditions under the Google Colab environment [41].

Listing 1.7: I-V curve modeling based on explicit model (Python language)

```
import pandas as pd
import numpy as np
import seaborn as sns
import matplotlib.pyplot as plt
import math
%matplotlib inline
Tc=25;G=1;tech=6;base1=130;module=base1; Ns=60;
Vmp=29.2; Imp=7.85; Voc=36.4;Isc=8.45; Z=0.05; W=-0.37;
def fill_factor_new(T,G,Pm_base,Isc_base,Voc_base,W_base,Ns):
   Pn=Pm_base;G=G*1000;T25=25; G0=1000;
   Voc_T=Voc+W*(T-T25);           #the module'sopen-circuit voltage
atemperatureT;
   Isc_G=(G/G0)*Isc;               #the module'sshort-circuit current
at irradianceG;
   d22=(Tc+273)/300*T25;          #Normalized temp
```

```
    voc_25=Voc_T/((d22/G0)*Ns);   #the normalised open-
 circuit voltage at a cell temperature of 25
    FF0_25=(voc_25-
 math.log10(voc_25+0.72))/(voc_25+1);#The normalized fill factor at
   temperature 25%
    FF_m=Pn/(Voc*Isc);                  #themodule's fill factor at STC;
    rs=1-FF_m/FF0_25;   # the normalised solar cell's
 series resistance
    Rs=(Voc/Isc)*rs;                    #the solar cell's series resistance
    rs_GT=(Isc_G/Voc_T)*Rs;             #Normalized series resistance
    d21=(Tc+273)/300*T25;               #Normalized temp
    voc_T=Voc_T/((d21/G0)*Ns);          #the normalised open-
 circuit voltage at a cell temperature of Tc
    FF0_T=(voc_T-
 (math.log10(voc_T+0.72)))/(voc_T+1);     #The normalised fill factor
 at the solar cell temperature;
    FF=FF0_T*(1-rs_GT);
    if FF>=1:
      FF=1
    else:
      FF=FF
    return FF

 def m_factor(FF,z,w,Tc,G):
    w=0;z=0;m=1;mpp=1;eps=0.001;step=0.01;
    while abs(mpp-FF)>eps:
      m=m+0.1;V=0.01;
      while (G+z*(Tc-25)-((np.exp(m*(V+w*(Tc-25)))-1)/(np.exp(m)-
 1)))*V>(G+z*(-Tc+25)-((np.exp(m*(V-step+w*(-Tc+25)))-1)/(np.exp
 (m)-1)))*(V-step):
           V=V+step;Volt=V-step;
           Ampere=G+z*(+T-25)-((np.exp(m*(V-step+w*(-Tc+25)))-1)/
         (np.exp(m)-1));
           Watt=Ampere*Volt;
      mpp=Watt;
      if m>10:
         break
    return m

 def ff_m_bases_new(base1,G,Tc,tech,Ns):
   base1=230;Vmp_base=29.56;Imp_base=7.84;Voc_base=36.61;
Isc_base=8.30;Z_base=0.04; W_base=-0.3;
   #base1=120;Vmp_base=17; Imp_base=7.06; Voc_base=21.2;
Isc_base=7.92; Z_base=0.003; W_base=-0.08;
   Pm_base=Vmp_base*Imp_base;base=Voc_base*Isc_base;
   FF=fill_factor_new(Tc,G,Pm_base,Isc_base,Voc_base,W_base,Ns)
   m=m_factor(FF,Z_base,W_base,Tc,G)
   return Vmp_base,Isc_base,Voc_base,Pm_base,FF,m
```

```
Vmp_base,Isc_base,Voc_base,Pm_base,FF,m=ff_m_bases_new(base1,G,T,
  tech,Ns)
#module=130
Vmp=29.2; Imp=7.85; Voc=36.4;Isc=8.45; Z=0.05; W=-0.37;
#module120
#Vmp=17; Imp=7.06; Voc=21.2;Isc=7.92; Z=0.003; W=-0.08;
Pm=Vmp*Imp; base=Voc*Isc;
#Voc_base=36.61;Isc_base=8.30;
w=(W*Voc)/(100*Voc_base);  z=(Z*Isc)/(100*Isc_base);

def In_model_new(G,Tc,m,V,z,w):
    In=G+z*(-25+Tc)-(np.exp(m*(V+w*(25-Tc)))-1)/(np.exp(m)-1)
    return In
def mppt_f_new(G,T,m,z,w):
   V=0;step=0.01;
   while In_model_new(G,T,m,V,z,w)*V>In_model_new(G,T,m,V-step,z,w)
   *(V-step):
     V=V+step; Volt=V-step;
     Ampere=In_model_new(G,Tc,m,V-step,z,w);
     Watt=In_model_new(G,Tc,m,V-step,z,w)*(V-step);
   Vmpp=Volt;
   Impp=Ampere;
   mppt=Watt;
   return Vmpp,Impp,mppt
Vmpp,Impp,mppt=mppt_f_new(G,Tc,m,z,w)

V1=np.array([i for i in range(101)]); V=V1/100
In=In_model_new(G,Tc,m,V,z,w);
plt.plot(V,In,'bo')
plt.xlabel('Vpv (P.U)',fontsize=14)
plt.ylabel('Ipv (P.U)',fontsize=14)
plt.title('I-Vcurve',fontsize=14)
plt.grid(True)
)
```

Example 1.8: *I-V* curve based on Matlab-SimeScape (one-diode model) [42]

In this example, a physical model based on Matlab-Simscape toolbox used to obtain the *I-V* curve is presented. Simscape components (solar cells, bypass diodes, etc.) can be seamlessly used to design any PV module (i.e., built by interconnecting individual solar cells), PV string (i.e., built by connecting PV modules in series to get a sufficient voltage), or array (i.e., built by connecting PV strings in parallel to get a sufficient current) [42]. Fig. 1.20 shows an example of using a Simscape model to obtain the *I-V* curve of a PV module (ShellSP150). It consists of:

i. A Q.Pro 230 W_p PV module composed of 72 solar cells (see Fig. 1.20A) connected together by physical networks (see Fig. 1.20C).
ii. A variable resistor and ramp block (see Fig. 1.20A) are used to sweep the *I-V* and *P-V* characteristics.

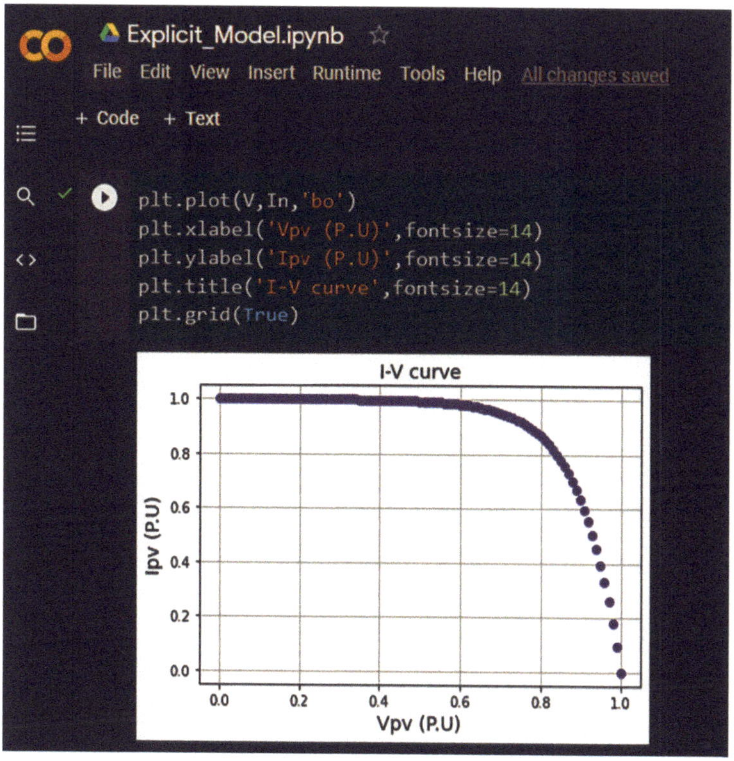

Fig. 1.19 *I-V* curve based on an explicit model at STC conditions for a multi-crystalline Q-Pro230 (230W_p) under Google Colab.

iii. Current and voltage sensors are employed to measure and plot the *I-V* curve, respectively (see Fig. 1.20A).

iv. A solver-configuration-block is used to determine the unknown variables for the entire system modeled; it can be connected anywhere on the physical network circuit by creating a branching point. It is imperative to fine-tune the solver parameters before running the simulation in order to avoid warning or error messages. The solver parameters are chosen as follows: the step-type is a fixed step, the solver-type is Runge-Kutta and the fixed step size is 0.001 (see Fig. 1.20A).

The *I-V* curve obtained is shown in Fig. 1.20E.

Example 1.9: *I-V* curve based on two-diode model

In this example, a two-diode model is developed using the following Matlab code (**Listing 1.8**) [31]. Fig. 1.21 shows the *I-V* obtained at STC conditions.

Listing 1.8: I-V curve modelling using Matlab (two diode model)

```
clear all, close all, clc
%% Information from the KD135SX UPU solar array datasheet
Isc_ref = 8.37; Voc_ref = 22.1; Imp = 7.63;Vmp = 17.7;Pmax_e = 135;
  kv = -8e-2;ki = 5.02e-3;Ns = 36;
G0=1000; T0=25+273.15;     T=T0; k = 1.3806503e-23;
```

```
q = 1.60217646e-19;          Vt0 = Ns*k*T0/ q;
Vt=k*T*Ns/q; a1=1;a2=1.3;p=2.3;
Rs=0; Rs_max=(Voc_ref-Vmp)/Imp;Rp_min=Vmp/(Isc_ref-Imp)-Rs_max;
Rp = Rp_min;tol = 0.001; error1 = inf;
I01=Isc_ref/(exp(Voc_ref/(((a1+a2)/p)*Vt))-1); I0=I01;tol=0.001;
Isc=Isc_ref;Iph=Isc_ref;
while (error1>tol)
Rp=(Vmp+(Imp*Rs))/(Iph-I0*(exp((Vmp+(Imp*Rs))/Vt)+exp((Vmp
  +(Imp*Rs))/((p-1)*Vt))+2)-(Pmax_e/Vmp));
I(1)=Isc_ref;V=0:1:Voc_ref+1;II=[];
 for n=1:length(V);
      for j=1:10;
         dg(j)=-1-I0*(((Rs/(Vt))*exp((V(n)+(I(j)*Rs))/(Vt)))+((Rs/
  ((p-1)*Vt))*exp((V(n)+(I(j)*Rs))/((p-1)*Vt))))-(Rs/Rp);
         g(j)=Iph-I(j)-I0*((exp((V(n)+(I(j)*Rs))/(Vt)))+(exp((V(n)
  +(I(j)*Rs))/((p-1)*Vt)))+2)-((V(n)+(I(j)*Rs))/Rp);
         I(j+1)=I(j)-(g(j)/dg(j));
  end
  II=[II I(j+1)];
 end
end
plot(V,II,'k','linewidth',2) %
xlabel('Module voltage (V)','fontsize',12); ylabel('Module current
  (A)','fontsize',12);
title('KD135SX-UPU solar module','fontsize',10)
xlim([0 Voc_ref+1]); ylim([0 Isc_ref+1]);
```

1.3.3.3 PV array modeling under Matlab-Simscape

A PV string consists of a number of PV modules connected in series. The current across the string is the same, while the current is the sum of the voltage produced by the PV modules. A PV array consists of a number of strings connected in series; the produced current is the sum of the PV current produced by each string. Bypass diodes are used to protect the PV modules from partially shaded PV cells from the fully operating cells in full sun within the same solar panel when used in high-voltage-series arrays. A blocking diode could be also used to avoid inverse current from the load.

Example 1.10: PV string simulation using Matlab-Simscape

Fig. 1.22A shows an example of a PV array consisting of three PV modules connected in series (PV string) using Matlab/Simulink for three values of solar irradiance: 1000, 500, and 100 W/m^2, respectively, for a constant temperature of 25°C.

As can be observed from Fig. 1.22B, three MPPs are obtained (one global MPP and two local MPPs), due to different values of solar irradiance on the three PV modules.

Example 1.11: PV array simulation using Matlab-Simscape

Fig. 1.23A depicts a Matlab-Simscape model for a PV array of 3×2 PV modules at STC conditions.

As shown in Fig. 1.23B, only one MPP is observed, as all PV arrays receive the same value of irradiance ($G = 1000\,\text{W/m}^2$).

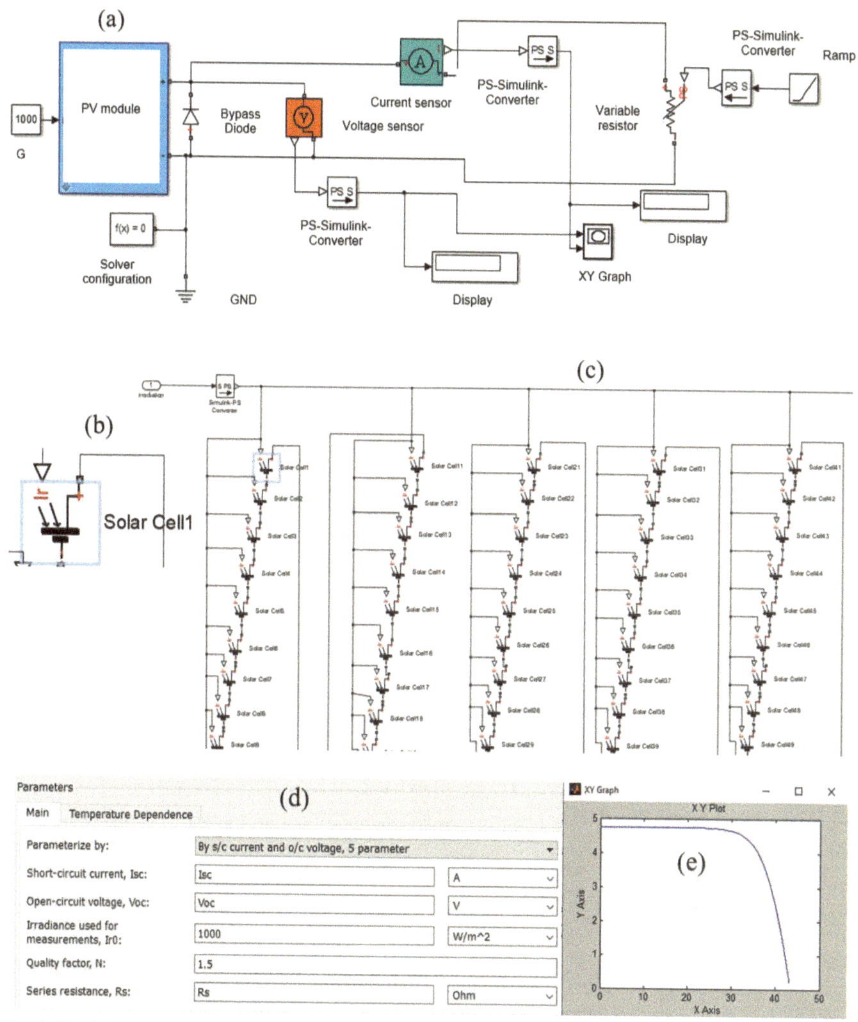

Fig. 1.20 (A) Whole system, (B) the structure of the PV module, (C) the basic solar cell, (D) box for parameters, and (E) the *I-V* curve.

1.3.4 Photovoltaic systems

A PV array or field consists of several PV modules connected in series and parallel. A PV system consists mainly of a PV array, batteries, converters, and inverters. From the capacity point of view, PV systems can be classified as [43]

(1) small-scale, include residential systems, rooftop, up to a few kW
(2) medium-scale, includes commercial systems, up to hundreds of kW
(3) large-scale utility grid, with tens of MW capacity

Globally, these can be classified into three types: (1) stand-alone alone PV systems, (2) grid-connected PV systems, and (3) hybrid PV systems.

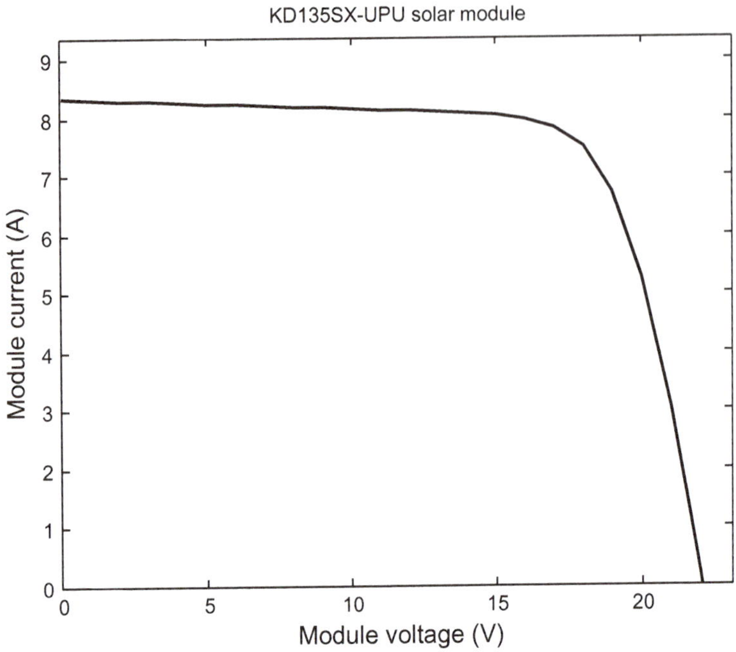

Fig. 1.21 *I-V* curve of a PV module (KD135SX-UPU) at STC conditions using a two-diode model at STC conditions.

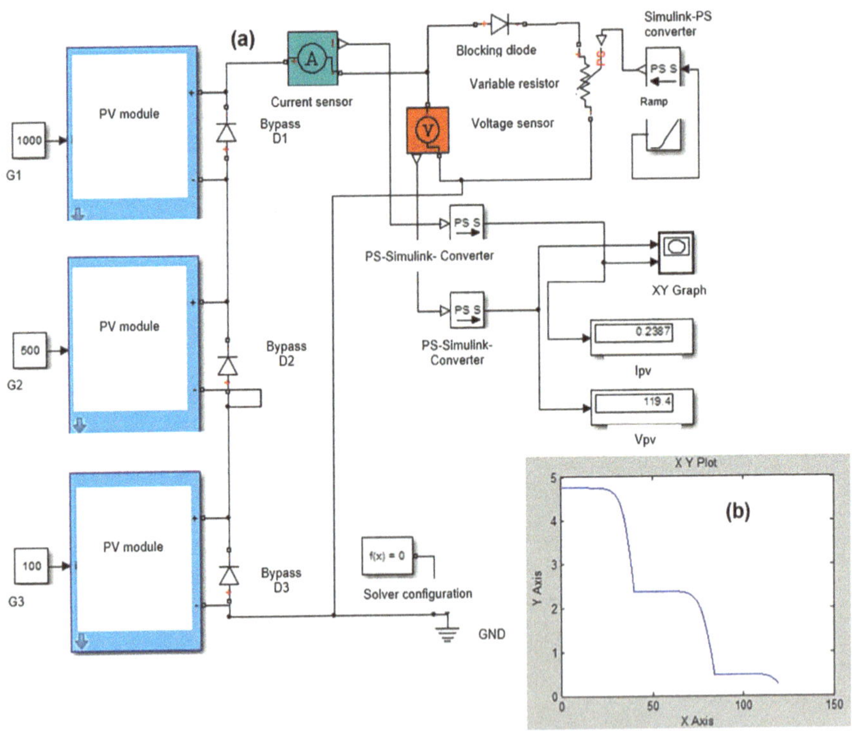

Fig. 1.22 (A) Matlab-Simscape-based model for a PV string of three PV modules, and (B) *I-V* curve under different values of G (1000, 500, and 100 W/m^2) and a constant value of T=25°C.

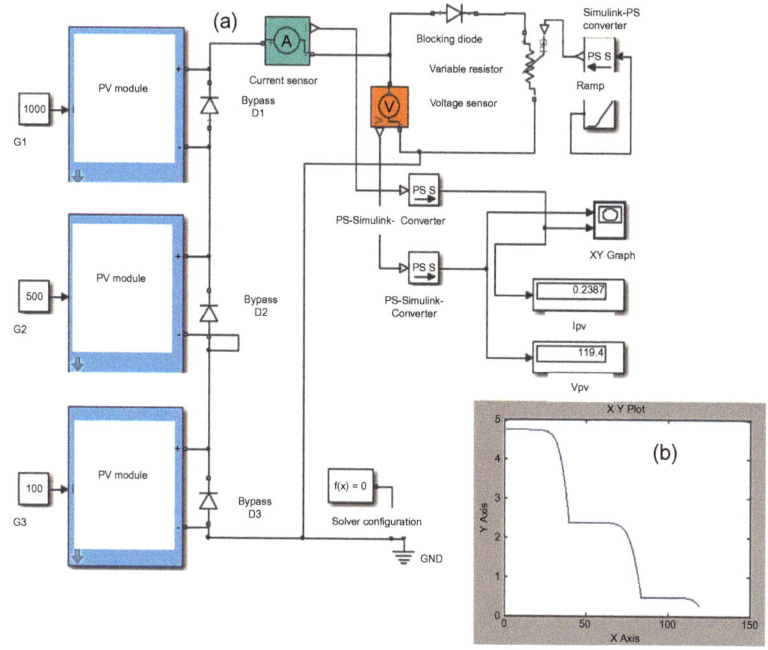

Fig. 1.23 (A) Matlab-Simscape-based model for a PV array of 3 × 2 modules, and (B) *I-V* curve at STC conditions.

1.3.4.1 Stand-alone (off-grid) PV systems

In this type of installation, batteries are always used, and the system can be employed particularly in remote areas, with no access to the main grid. Fig. 1.24 shows a simplified schematic of the stand-alone PV system [44].

Example 1.12: Simulation of a stand-alone PV system with resistive load

Fig. 1.25 shows a Matlab/Simulink model of a stand-alone PV system; it consists of DC-DC boost converter with a resistive load. An InCond algorithm is used to track the maximum power point.

The obtained simulation results are shown in Fig. 1.26. It shows the variation of I_{pv} (Fig. 1.26A) and V_{pv} (Fig. 1.26B) under various values of G (see Fig. 1.26C) and T is kept constant at 25°C.

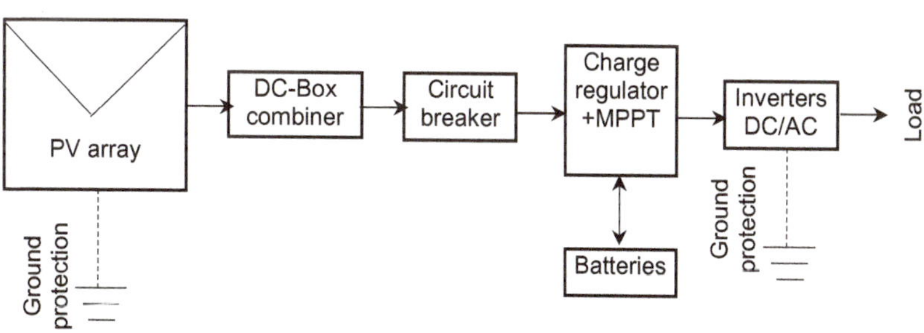

Fig. 1.24 Stand-alone PV system.

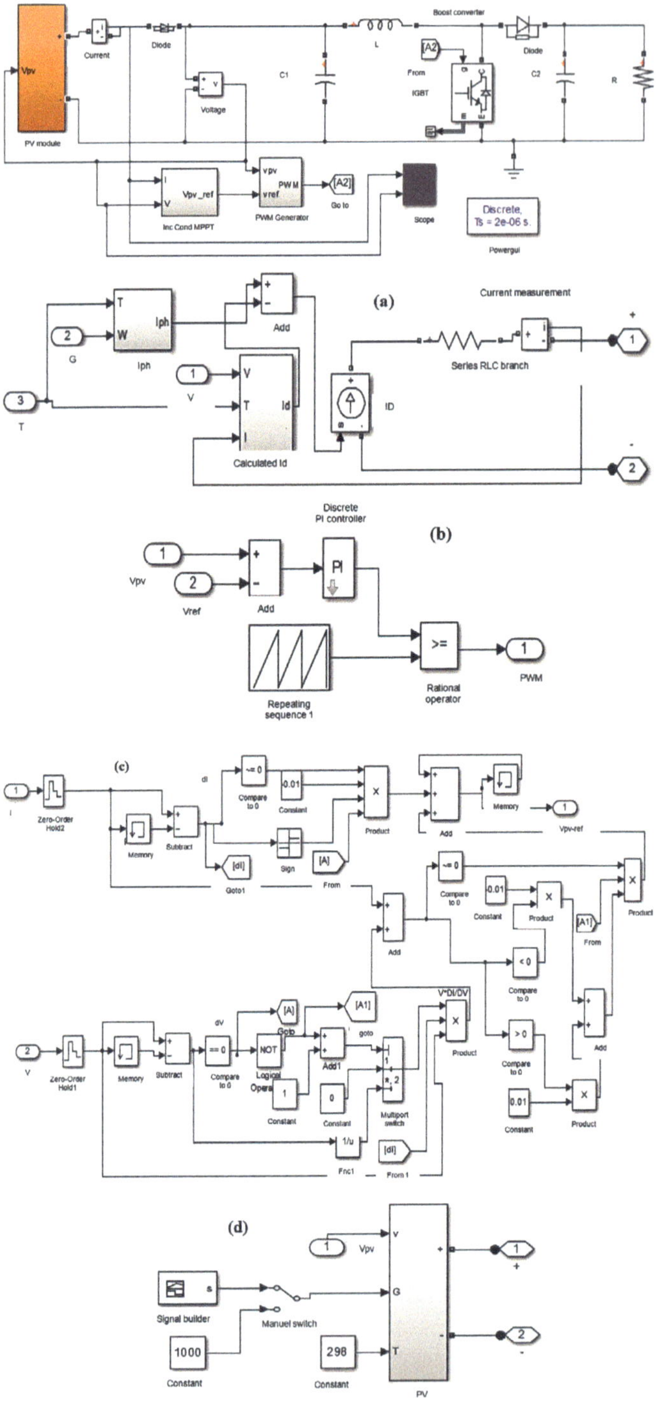

Fig. 1.25 Matlab/Simulink model of a stand-alone PV system with a restive load (100 Ω). (A) PV generator model, (B) PWM generator, (C) IncCond algorithm, and (D) block diagram for switching of *G* (constant or variable).

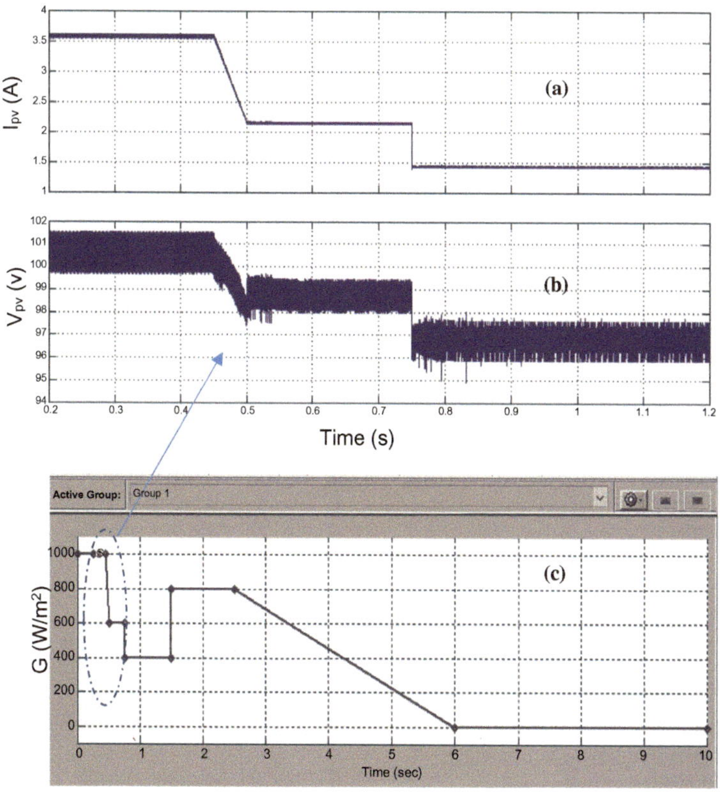

Fig. 1.26 Simulation results for variable values of G: (A) I_{pv} and (B) V_{pv}, (C) variation of G (1000, 600, 400, 600, and $0\,W/m^2$).

Example 1.13: Simulation of a stand-alone PV system with a battery
In this case, a battery is used to simulate the PV system under variable irradiance levels. The Simulink model is shown in Fig. 1.27.

The results obtained are depicted in Fig. 1.28.

Example 1.14: Simulation of a stand-alone PV system with batteries and a DC-AC converter
In this example, a two-stage PV system is simulated. It includes a DC-DC, a DC-AC, and a battery. The Simulink model is shown in Fig. 1.29.

As an example, Fig. 1.30 shows the simulated I_{pv}, V_{pv} and P_{load} (load power) at STC conditions.

1.3.4.2 Grid-connected PV systems

A block diagram of a grid-connected PV system is shown in Fig. 1.31. This is the most used type of system in medium- and large-scale PV plants, such as residences, commercial facilities, and utility companies. Over the past years, utility-scale PV systems have dominated the PV market; however, distributed PV systems, on commercial and

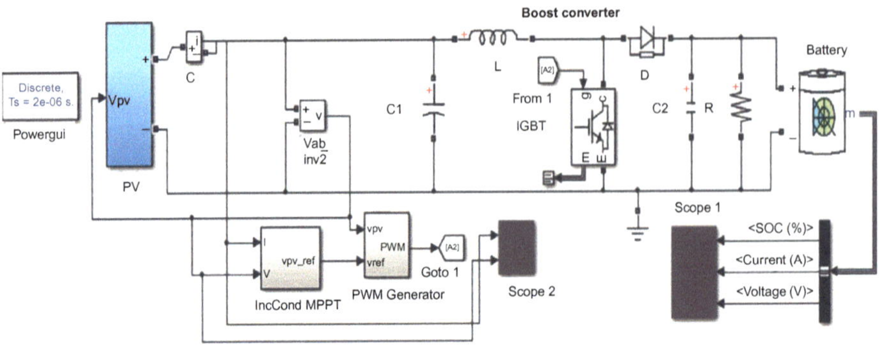

Fig. 1.27 Simulink model of a stand-alone PV system with batteries

industrial premises, are becoming more important in many countries, due to their favorable economics, in particular, when combined with increased self-consumption [44].

In point-of-view configuration, grid-connected PV systems can be mainly classified into four structures (see Fig. 1.31):

(1) Centralized structure, which is mainly used for very large-scale PV plants: In this configuration, all PV string, are connected to a three-phase DC-AC inverter (Fig. 1.32A). This structure is economic but does not guarantee the maximum power.
(2) Micro-inverter: As DC-DC with MPPT is integrated at the module level, which permits to extract the maximum power at each PV module (Fig. 1.32B): This structure can guarantee the maximum power but increases the cost of the system considerably.
(3) String structure: In this structure, each PV string has a proper MPPT and DC-AC inverter (Fig. 1.32C), mainly used in small- and medium-scale PV plants. This structure can guarantee the maximum power but increases the cost of the system.
(4) Multi-string structure: This one has only one centralized DC-AC inverter and multiple DC-DC converter for each string (Fig. 1.32D); it could be used for large-scale PV plants. This configuration is better than central and string structures as it combines the advantages of both configurations.

Example 1.15: Simulation of a grid-connected PV system

Fig. 1.33 depicts a Matlab/Simulink model of a grid-connected PV system (multi-string inverter). It consists of two PV generators with two DC-DC boost converters and a DC-AC inverter. Voltage-oriented control (VOC) is used to control the inverter, and the perturb and observe (P&O) algorithm is used to track the maximum power point of the PV generator by controlling the DC-DC converter.

The obtained simulation results at STC conditions are shown in Fig. 1.34.

1.3.4.3 Hybrid photovoltaic systems

The third type is named hybrid PV systems, because more than one renewable source can be integrated, e.g., wind generator, or fossil source such as diesel, as shown in the schematic diagram in Fig. 1.35.

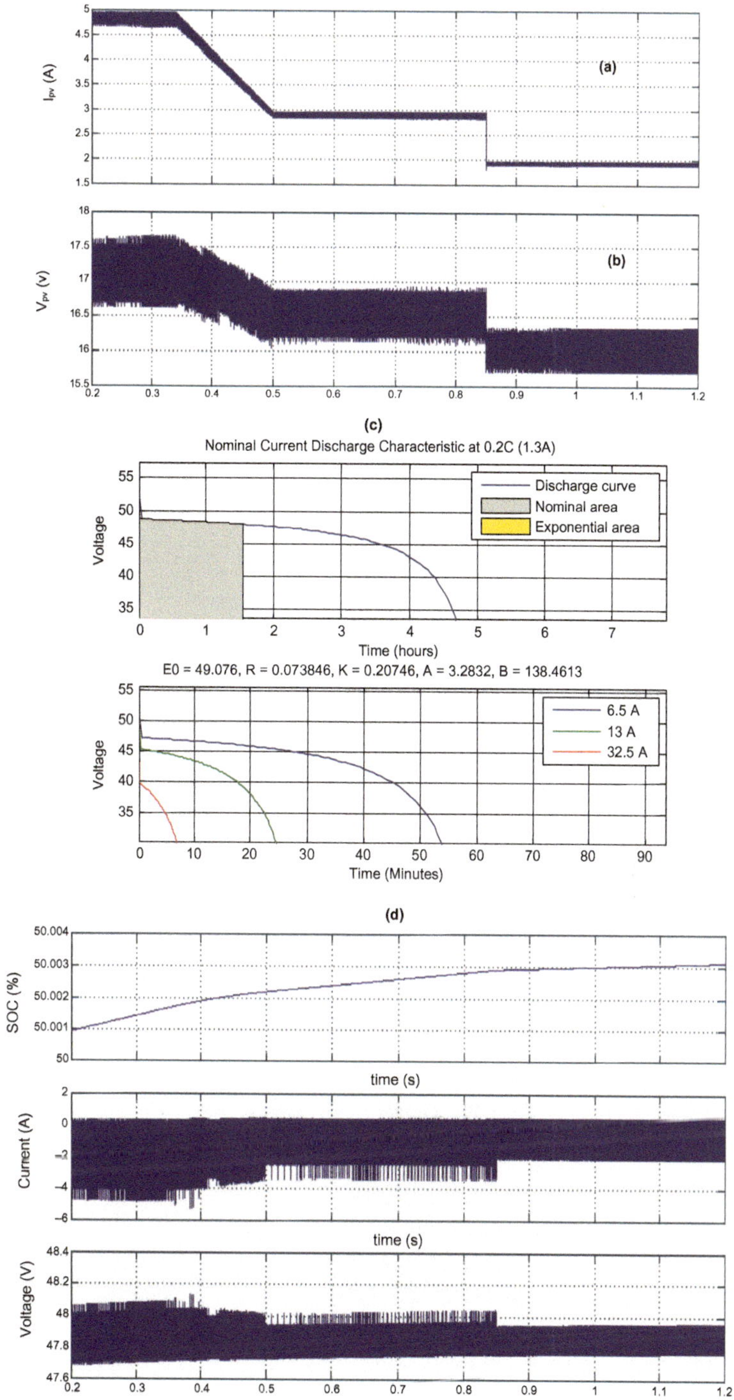

Fig. 1.28 Simulation results for a variable values of G: (A) I_{pv}, (B) V_{pv}, (C) battery discharge characteristic, and (D) SOC, current, and voltage of the battery during the simulation step.

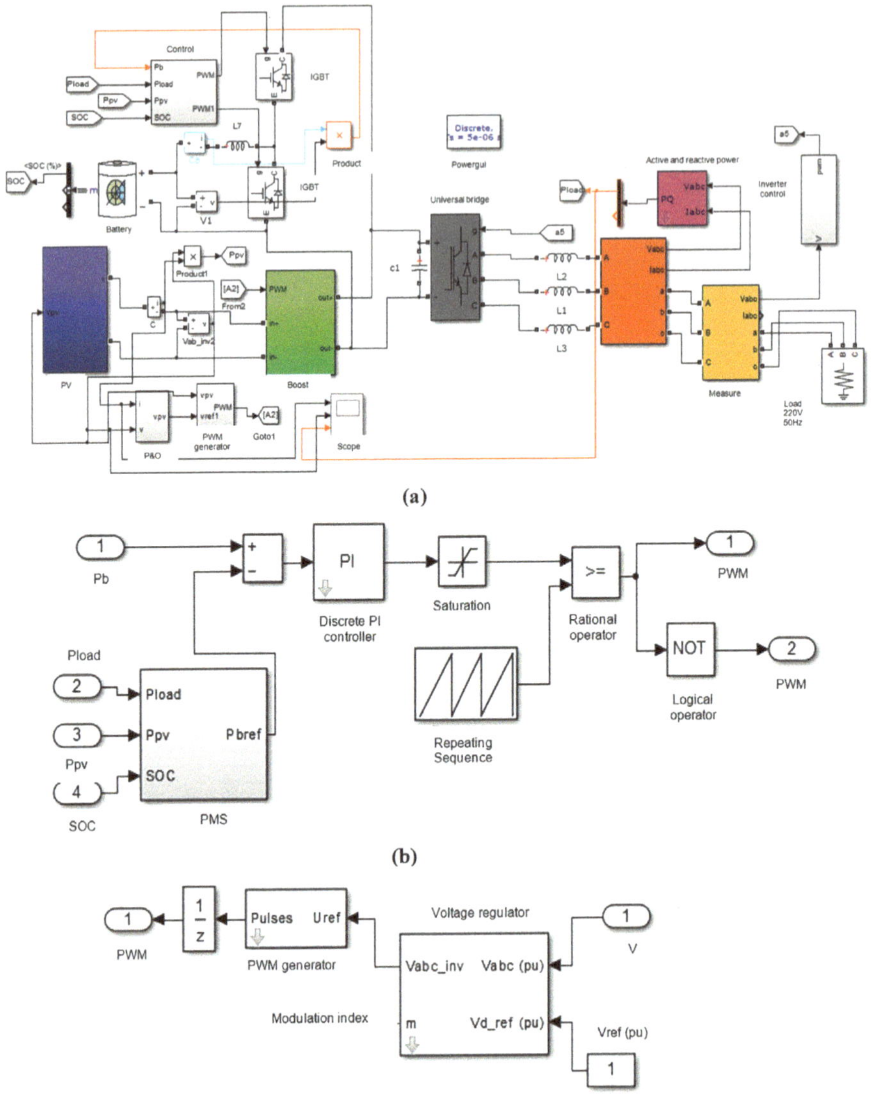

Fig. 1.29 Simulink model: (A) control block and (B) inverter control.

Example 1.16: Simulation of hybrid-PV system

Fig. 1.36 shows an example of a Matlab/Simulink model for a hybrid-system (PV and wind turbine). It consists of two-PV array two DC-DC boosters, and two wind turbines.

The simulation results obtained are shown in Fig. 1.37.

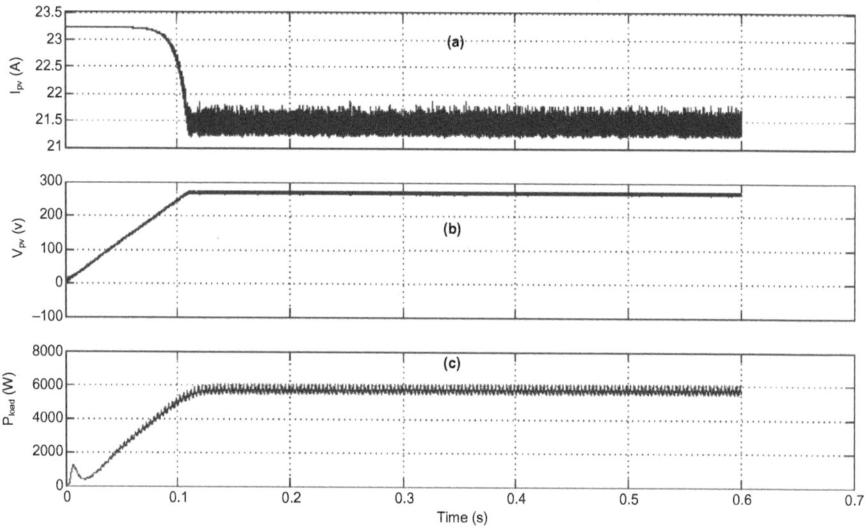

Fig. 1.30 Simulation results at STC conditions: (A) I_{pv}, (B) V_{pv}, and (C) P_{load}.

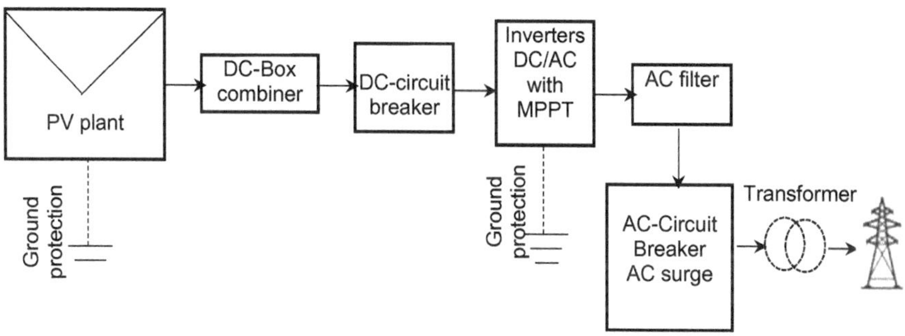

Fig. 1.31 Schematic representation of a grid-connected PV system.

1.4 Main issues of photovoltaic systems

Photovoltaic systems are mainly affected by two factors: (a) weather and environment, and (b) cell degradation (aging). Due to these factors, photovoltaic systems cannot work efficiently. Therefore, such systems need to be protected from faults, such as ground, arc, line-to-line faults, and other faults on PV modules [44] by integrating protection devices and automatic fault diagnosis techniques. PV systems need to be optimized, by extracting the maximum power [45], avoiding shading effects, dust accumulation (dirty PV modules), etc. [46]. They need to be managed by the development of the optimal energy management systems [47]. They need to

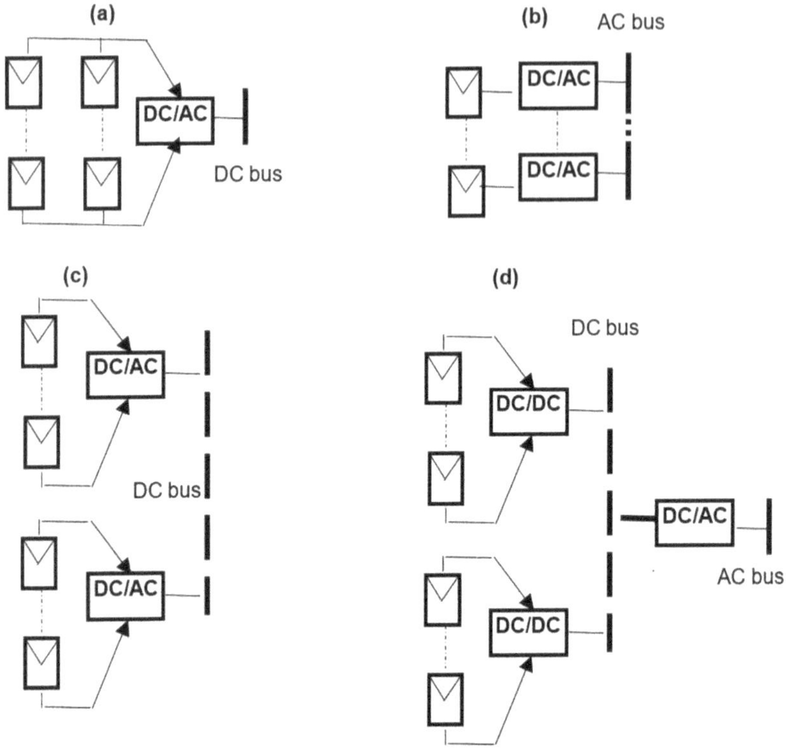

Fig. 1.32 PV array structures: (A) central inverter, (B) micro-inverter, (C) string inverter, and (D) multi-string inverter.

be monitored for yield performance analysis [48,49], by design remote monitoring systems, and also need to be controlled for grid protection and stability [50], by designing and integrating control techniques. In addition, the power produced by this kind of systems needs also to be predicted for energy management, planning, and distribution [51].

To address these issues, various papers were published in the literature based on conventional methods, and recently many attempts have been done to solve these problems using artificial intelligence techniques and metaheuristic methods. Real-time implementation of such advanced techniques remains an open challenge.

In the following chapters, these issues will be addressed and discussed by showing how to use artificial intelligence (AI) techniques and metaheuristic approaches to solve them. We will focus mainly on the application of AI techniques by providing various examples with explanations.

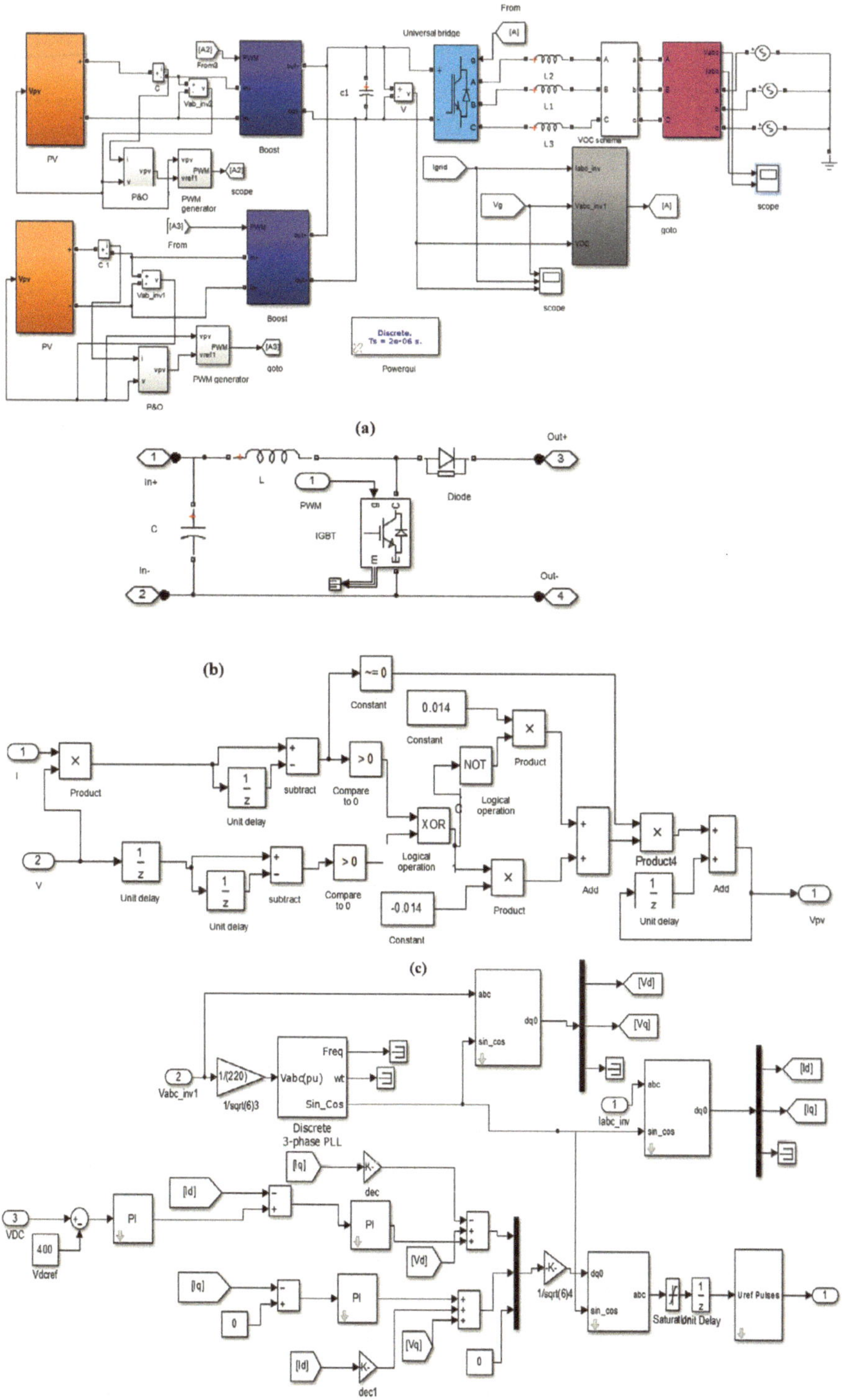

Fig. 1.33 Matlab/Simulink model of a grid-connected PV system: (A) DC-DC boost converter, (B) P&O algorithm, and (C) VOC scheme.

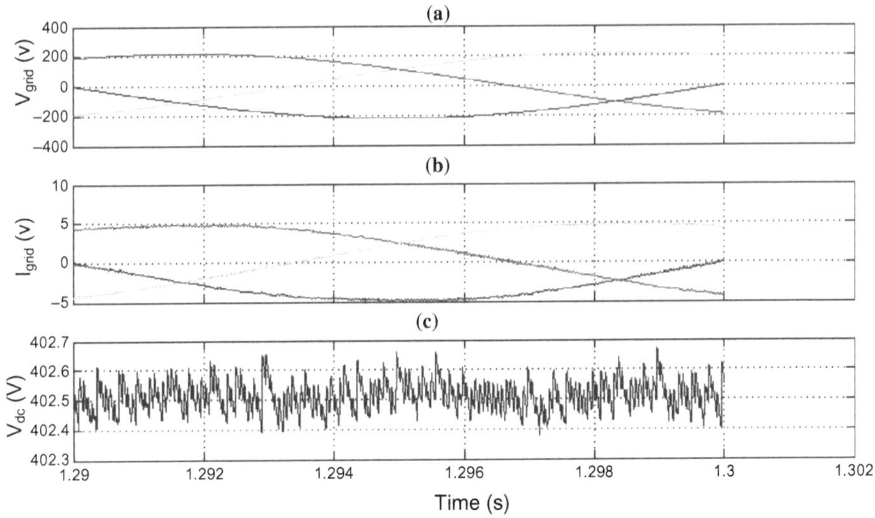

Fig. 1.34 Simulation results of the grid-connected PV system under STC conditions (A) V_{grid}, (B) I_{grid}, and (C) V_{dc}.

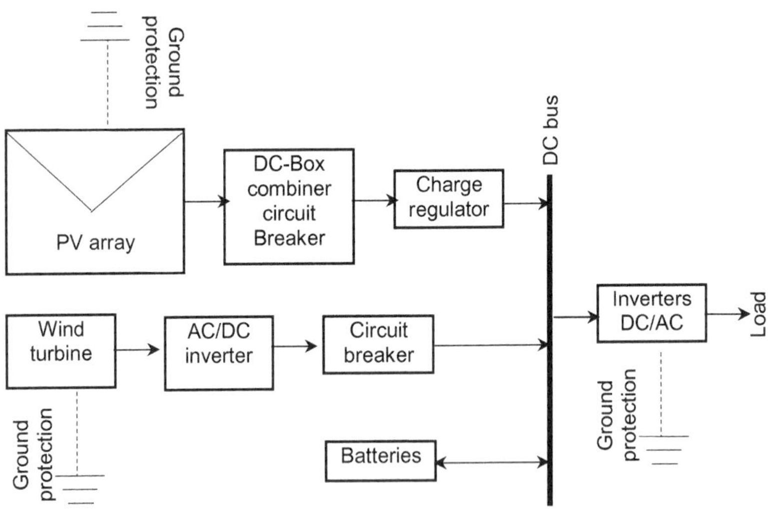

Fig. 1.35 Hybrid photovoltaic-wind turbine system configuration.

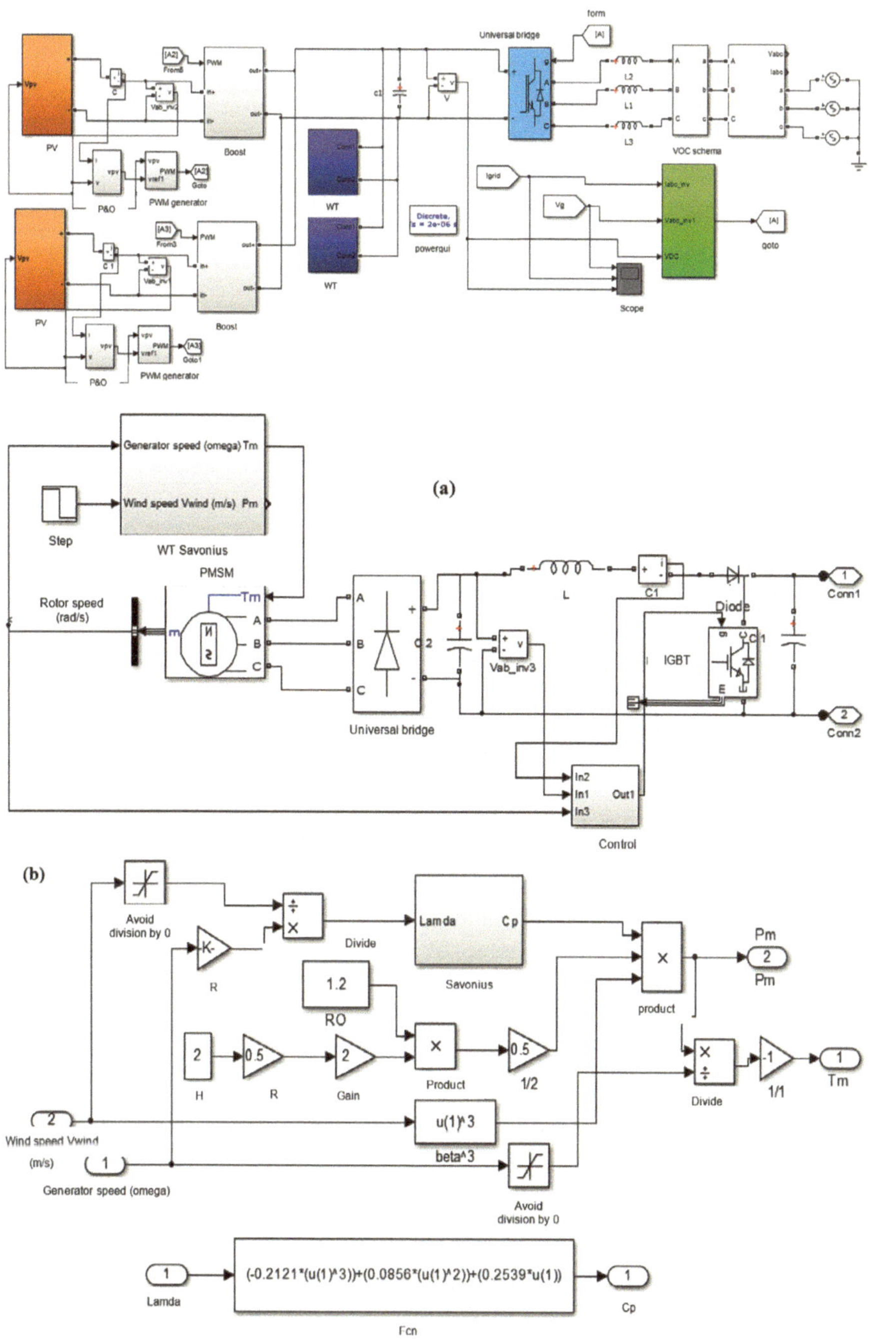

Fig. 1.36 Matlab/Simulink model of a hybrid-system (PV + wind turbine): (A) wind turbine and (B) wind turbine model (Savonius).

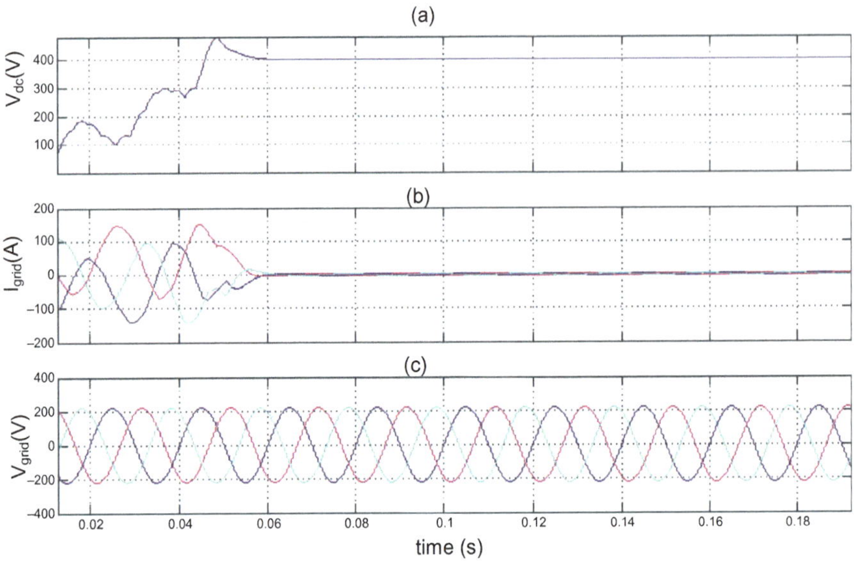

Fig. 1.37 Simulation results: (A) V_{dc}, (b) I_{grid}, and (C) V_{grid}.

1.5 Summary

This chapter comprises two parts: in the first part, an introduction to solar radiation is presented followed by some examples of codes in Python and Matlab, on how to model the solar spectrum, extraterrestrial solar irradiance, insolation, and clearness index. In the second part of the book, a general introduction to photovoltaics is presented, including PV devices, PV technologies, and PV system configurations. Then, 16 examples of how to simulate a PV cell, PV module, PV string, PV array, and different PV systems based on Matlab/Simulink environment and Python language have been described in detail. The main issues of photovoltaic systems have been reported in Section 1.4.

References

[1] IEA, Snapshot of global photovoltaic markets, Report IEA-PVPS T1-39, 2021 (Accessed April 2021).
[2] International Renewable Energy Agency (IRENA), Renewable Energy Statistics 2020, 2020, Available at: *https://www.irena.org/-/media/Files/IRENA/Agency/Publication/2020/Jul/IRENA_Renewable_Energy*.
[3] IEA, Sunspot of global PBV markets, 2020. Report IEA-PVPS T1-37:2020 (Accessed April 2020).
[4] IEA, Trend in photovoltaic applications, 2019. IEA PVPS T1-36: 2019.
[5] C.A. Gueymard, D.R. Myers, Solar radiation measurement: progress in radiometry for improved modeling, in: Modeling Solar Radiation at the Earth's Surface, Springer, Berlin, Heidelberg, 2008, pp. 1–27.

[6] R. Bird, C. Riordan, Simple solar spectral model for direct and diffuse irradiance on horizontal and tilted planes at the earth's surface for cloudless atmospheres, NREL Technical Report TR-215-2436, 1984, https://doi.org/10.2172/5986936.
[7] J.S. Stein, W.F. Holmgren, J. Forbess, C.W. Hansen, PVLIB: open source photovoltaic performance modeling functions for Matlab and Python, in: 43rd Photovoltaic Specialists Conference, 2016.
[8] J. Meydbray, K. Emery, S. Kurtz, Pyranometers and reference cells, what's the difference? (No. NREL/JA-5200-54498), National Renewable Energy Lab. (NREL), Golden, CO, United States, 2012.
[9] P. Gevorkian, Large-Scale Solar Power System Design: An Engineering Guide for Grid-Connected Solar Power Generation, McGraw-Hill Education, 2011.
[10] A.A.M. Sayigh, Solar Energy Engineering, Academic Press, New York, 1977 (Chapter 4).
[11] Z. Sen, Solar Energy Fundamentals and Modeling Techniques: Atmosphere, Environment, Climate Change and Renewable Energy, Springer Science & Business Media, 2008.
[12] J. Fan, L. Wu, F. Zhang, H. Cai, X. Ma, H. Bai, Evaluation and development of empirical models for estimating daily and monthly mean daily diffuse horizontal solar radiation for different climatic regions of China, Renew. Sust. Energ. Rev. 105 (2019) 168–186.
[13] T. Muneer, Solar Radiation and Daylight Models, second ed., Elsevier, Butterworth, Heineman, Oxford, UK, 2004.
[14] C.A. Gueymard, The sun's total and spectral irradiance for solar energy applications and solar radiation models, Sol. Energy 76 (2003) 423–453.
[15] V. Badescu, Modeling Solar Radiation at the Earth's Surface, vol. 1, Springer, Berlin, Heidelberg, 2014.
[16] A. Angström, Solar terrestrial radiation, Q. J. R. Meteorol. Soc. 50 (1924) 121–126.
[17] A. Angström, On the computation of global radiation from records of sunshine, Arkiv. Geof. 2 (1956) 471–479.
[18] C. Gueymard, P. Jidra, V. Eatrada-Cajigai, A critical look at recent interpretations of the Angström approach and its future in global solar irradiation prediction, Sol. Energy 54 (1995) 357–363.
[19] M.S. Gul, T. Muneer, H.D. Kambezidis, Models for obtaining solar radiation data from other meteorological data, Sol. Energy 64 (1–3) (1998) 98–108.
[20] J.S.G. Ehnberg, M.H.J. Bollen, Simulation of global solar radiation based on cloud observations, Sol. Energy 78 (2005) 157–162.
[21] R. Perez, P. Ineichen, K. Moore, M. Kmiecik, C. Chain, R. George, F. Vignola, A new operational model for satellite derived irradiances: description and validation, Sol. Energy 73 (5) (2002) 307–317.
[22] M.E. Becquerel, Mémoire sur les effets électriques produits sous l'influence des rayons solaires, in: Comptes rendus hebdomadaires des séances de l'Académie des Sciences, vol. 9, 1839, pp. 561–567.
[23] M.H. Shubbak, Advances in solar photovoltaics: technology review and patent trends, Renew. Sust. Energ. Rev. 115 (2019) 109383.
[24] S. Philipps, W. Warmuth, Photovoltaics report fraunhofer institute for solar energy systems, in: ISE With Support of PSE GmbH November 14th, 2019.
[25] V. Benda, L. Černá, PV cells and modules–state of the art, limits and trends, Heliyon 6 (12) (2020) e05666.
[26] J. Ramanujam, et al., Flexible CIGS, CdTe and a-Si: H based thin film solar cells: a review, Prog. Mater. Sci. 110 (2020) 100619.
[27] T.D. Lee, A.U. Ebong, A review of thin film solar cell technologies and challenges, Renew. Sust. Energ. Rev. 70 (2017) 1286–1297.

[28] S.A. Kalogirou, Solar Energy Engineering: Processes and Systems, second ed., Academic Press, 2014.
[29] M.A. Green, E.D. Dunlop, J. Hohl-Ebinger, M. Yoshita, N. Kopidakis, X. Hao, Solar cell efficiency tables (version 56), Prog. Photovolt. Res. Appl. 28 (2020).
[30] From: https://www.nrel.gov/pv/cell-efficiency.html.
[31] K. Ishaque, Z. Salam, H. Taheri, Simple, fast and accurate two-diode model for photovoltaic modules, Sol. Energy Mater. Sol. Cells 95 (2) (2011) 586–594.
[32] D. Sera, R. Teodorescu, P. Rodriguez, PV panel model based on datasheet values, in: Industrial Electronics, 2007. ISIE 2007. IEEE International Symposium on, IEEE, 2007, pp. 2392–2396.
[33] M.G. Villalva, J.R. Gazoli, Comprehensive approach to modeling and simulation of photovoltaic arrays, IEEE Trans. Power Electron. 24 (2009) 1198–1208.
[34] S. Karmalkar, S. Haneefa, A physically based explicit–model of a solar cell for simple design calculations, IEEE Electron Device Lett. 29 (2008) 449–451.
[35] A.K. Das, Analytical expression of the physical parameters of an illuminated solar cell using explicit J–V model, Renew. Energy 52 (2013) 95–98.
[36] S. Karmalkar, H. Saleem, The power law J–V model of an illuminated solar cell, Sol. Energy Mater. Sol. Cells 95 (2011) 1076–1084.
[37] A.M. Pavan, A. Mellit, V. Lughi, Explicit empirical model for general photovoltaic devices: experimental validation at maximum power point, Sol. Energy 101 (2014) 105–116.
[38] W. De Soto, S.A. Klein, W.A. Beckman, Improvement and validation of a model for photovoltaic array performance, Sol. Energy 80 (1) (2006) 78–88.
[39] W.F. Holmgren, C.W. Hansen, M.A. Mikofski, pvlib python: a python package for modeling solar energy systems, J. Open Source Softw. 3 (29) (2018) 884.
[40] A. Massi Pavan, A. Mellit, D. De Pieri, V. Lughi, A study on the mismatch effect due to the use of different photovoltaic modules classes in large-scale solar parks, Prog. Photovolt. Res. Appl. 22 (3) (2014) 332–345.
[41] https://colab.research.google.com/.
[42] R. Boukenoui, A. Mellit, A.M. Pavan, Photovoltaic model based on manufacture's datasheet: experimental validation for different photovoltaic technologies, in: A Practical Guide for Advanced Methods in Solar Photovoltaic Systems, Springer, Cham, 2020, pp. 239–256.
[43] A. Mellit, S. Kalogirou, Artificial intelligence and internet of things to improve efficacy of diagnosis and remote sensing of solar photovoltaic systems: challenges, recommendations and future directions, Renew. Sust. Energ. Rev. 143 (2021) 110889.
[44] A. Mellit, G.M. Tina, S.A. Kalogirou, Fault detection and diagnosis methods for photovoltaic systems: a review, Renew. Sust. Energ. Rev. 91 (2018) 1–17.
[45] A. Mellit, S.A. Kalogirou, MPPT-based artificial intelligence techniques for photovoltaic systems and its implementation into field programmable gate array chips: review of current status and future perspectives, Energy 70 (2014) 1–21.
[46] A.M. Pavan, A. Mellit, D. De Pieri, S.A. Kalogirou, A comparison between BNN and regression polynomial methods for the evaluation of the effect of soiling in large scale photovoltaic plants, Appl. Energy 108 (2013) 392–401.
[47] T. Poompavai, M. Kowsalya, Control and energy management strategies applied for solar photovoltaic and wind energy fed water pumping system: a review, Renew. Sust. Energ. Rev. 107 (2019) 108–122.
[48] S. Daliento, A. Chouder, P. Guerriero, A.M. Pavan, A. Mellit, R. Moeini, P. Tricoli, Monitoring, diagnosis, and power forecasting for photovoltaic fields: a review, Int. J. Photoenergy 2017 (2017) 1–13.

[49] M. Kumar, A. Kumar, Performance assessment and degradation analysis of solar photovoltaic technologies: a review, Renew. Sust. Energ. Rev. 78 (2017) 554–587.
[50] M.K. Singh, M.D. Equbal, B. Kumar, Stability of grid with the penetration of solar PV based generation in power systems, in: 2019 International Conference on Power Electronics, Control and Automation (ICPECA), IEEE, 2019, November, pp. 1–6.
[51] A. Nespoli, E. Ogliari, S. Leva, A. Massi Pavan, A. Mellit, V. Lughi, A. Dolara, Day-ahead photovoltaic forecasting: a comparison of the most effective techniques, Energies 12 (9) (2019) 1621.

Artificial intelligence techniques: Machine learning and deep learning algorithms

2

Chapter outline

2.1 Introduction

Artificial intelligence (AI) techniques have evolved significantly over the past few years and applications can be found in many different areas such as medicine, modern agriculture, health care, marketing, robotics, economics, energy, and other sectors. Additionally, AI techniques have also found applications in solar photovoltaic (PV) systems [1].

Currently, AI techniques including machine learning (ML) and deep learning (DL) algorithms could be easily implemented using Python programming language [2], due to the availability of some open source ML and DL frameworks (e.g., Scikit-Learn [3],

Handbook of Artificial Intelligence Techniques in Photovoltaic Systems. https://doi.org/10.1016/B978-0-12-820641-6.00002-8
Copyright © 2022 Elsevier Ltd. All rights reserved.

Keras [4], and Tensorflow [5]). Scikit-Learn and Keras libraries are chosen in this book for two reasons:

(1) They are free open source libraries with Python programming language.
(2) Both are dedicated to accelerate the implementation of ML and DL models.

Additional packages such as Scikit-fuzzy [6], Scikit-image [7], and OpenCV [8] are used in this book. Most of the functions used in this book are summarized in Appendix B.

The aim of this chapter is to present an introduction of AI techniques, especially machine learning, ensemble learning (EL), and deep learning algorithms. This chapter is not intended to provide the theoretical development of these algorithms, but it is focused on the application of these algorithms to solve certain PV system problems. Major supervised and unsupervised algorithms are used in a variety of PV applications, such as modeling, control, optimization, prediction, energy management, fault diagnosis, fault detection and localization, and forecasting.

The algorithms considered include naïve Bayes (NB), support vector machine (SVM), K-nearest neighbors (K-NN), linear regression (LR), logistic regression (LoR), decision tree (DT), random forest (RF), K-means (KM), fuzzy c-means (KCM), principal component analysis (PCA), hierarchical clustering (HC), neural networks (NNs), deep conventional neural networks (DCNNs), and advanced recurrent neural networks. The examples presented in this chapter have been written and developed using the Python programming language. The examples presented in the following sections will help the readers understand how to apply ML and DL algorithms to alleviate the drawbacks of classical methods usually applied in PV systems.

Matlab® in the MathWorks environment could also be used for solving problems in PV systems, as many ML, EL, and DL algorithms are currently available in different toolboxes, such as the deep learning toolbox and statistics and machine learning toolbox [9,10]. To help the readers to understand how to apply ML and DL algorithms to alleviate the drawbacks of classical methods usually applied in PV systems, some examples will be presented in the next sections.

2.2 Artificial intelligence

There are many definitions of artificial intelligence; basically, artificial intelligence refers to the simulation of human intelligence in machines that are programmed to think like humans and mimic their actions. In other words, AI is a branch of computer science, which makes a computer system mimic human intelligence [11]. As shown in Fig. 2.1, machine learning is a subset of AI, and deep learning is a subset of machine learning, so it can be noted that both machine learning and deep learning are used to achieve artificial intelligence [12].

Currently, deep learning is gaining much popularity due to its accuracy in classification and prediction when trained with huge amount of data and solving complex problems that machine learning is not able to solve. It should be noted that when a small amount of data is available, machine learning algorithms are more suitable than deep learning algorithms. However, with large amount of data deep learning

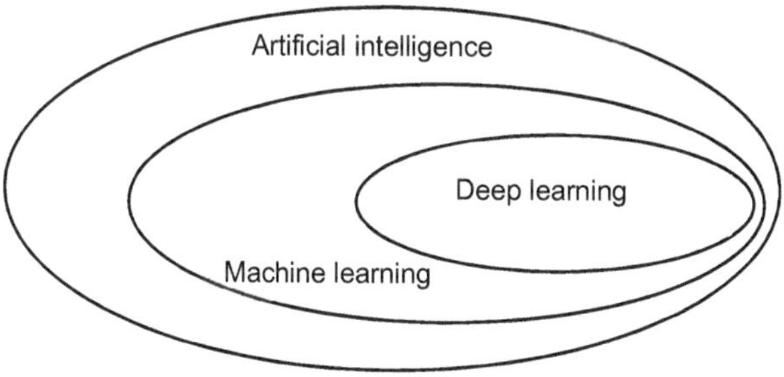

Fig. 2.1 A Venn-diagram of artificial intelligence: link between artificial intelligence, machine learning, and deep learning.

algorithms are the most appropriate. In addition, deep learning algorithms use directly raw data. Thus, relevant features are extracted automatically, unlike in ML algorithms.

2.3 Machine learning

Machine learning was originally introduced by Arthur [13]. It enables a machine to automatically learn from data, improve performance from experience, and predict things without being explicitly programmed. Fig. 2.2 shows the different steps of machine learning implementation, starting from the raw data, features extraction and preprocessing, training and evaluation, and model deployment.

Machine learning is used in many applications such as image and speech recognition, traffic prediction, weather forecasting, products recommendations, medical diagnosis, fault detection, stock market trading, air control, self-driving, robotic surgery, email spam filtering, search engine results, adverting and businesses intelligent, surveillance systems, automatic translation, and other applications. Machine learning is broadly classified into three major class of algorithms [14]: (a) supervised learning, (b) unsupervised learning, and (c) reinforcement learning. These are introduced briefly in the following sections.

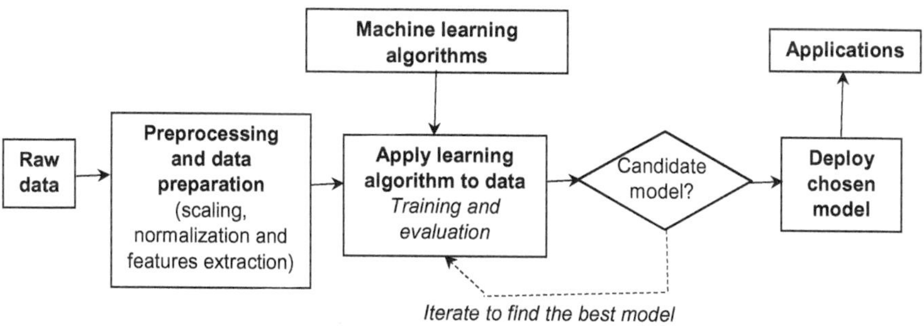

Fig. 2.2 Basic configuration of machine learning workflow.

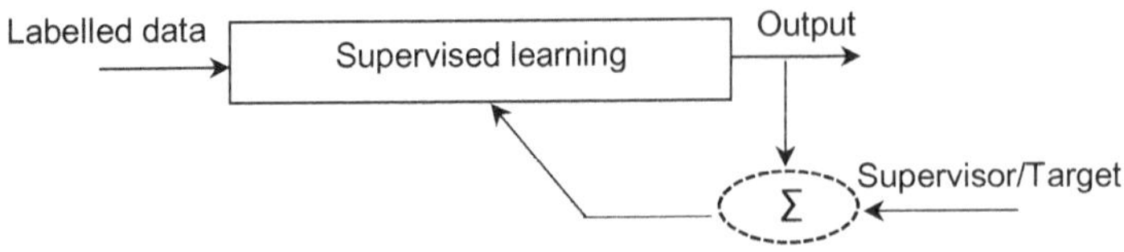

Fig. 2.3 Supervised learning.

2.3.1 Supervised learning

In this type of learning, shown schematically in Fig. 2.3, an algorithm attempts to create some relationships and dependencies between the input and output features, such that to be able to estimate the output values for new data [15]. The data used are labeled and they can be used to solve both regression and classification problems. This kind of machine learning algorithm is the one most applied in photovoltaic applications.

Major supervised learning algorithms used to address photovoltaic systems issues are as follows.

2.3.1.1 Linear regression

Linear regression (LR) is the most classic and simplest method based on the least-squares algorithm, which performs a regression task. Effectively, it predicts a dependent variable y based on a given independent variable x [16]. The LR can be divided into two algorithms: (1) simple LR and (2) multiple LR. Fig. 2.4 shows an example of a linear regression between dependent and independent variables.

Example 2.1: Regression problem

This example presents the application of the linear regression algorithm to find regression between the ambient temperature (T_a) and the PV cell temperature (T_C). Fig. 2.5 shows the dataset used, which consists of a PV module and air temperatures and comprises 8900 samples.

To train the model, 80% of the samples are been used and the rest 20% are used for testing the model. The results are shown in Fig. 2.6. To check the accuracy of prediction, three error metrics are used, the root mean squared error (RMSE), mean absolute error (MAE), and correlation coefficient (r). More details on how to calculate the error metrics are shown in Appendix C. The corresponding basic code using the Python programming language and *sklearn* library is provided in **Listing 2.1**:

```
#Listing 2.1: Linear regression problem
#import libraries
import numpy as np
import pandas as pd
from sklearn.model_selection import train_test_split
from sklearn.preprocessing import StandardScaler
```

```
from sklearn.linear_model import LinearRegression
from sklearn import metrics
import matplotlib.pyplot as plt
import seaborn as seabornInstance
#loading data
xy = pd.read_csv('E:\\data_result\\climate.csv',delimiter=';')
x  = xy['Tc'].values.reshape(-1,1)
y  = xy['Ta'].values.reshape(-1,1)
#database split
x_train, x_test, y_train, y_test = train_test_split(x, y,
  test_size=0.20,random_state=0)
#call linear regression function
lr1 = LinearRegression()
# fitting
lr1.fit(x_train, y_train)
# prediction
y_pred = lr1.predict(x_test)
# printing results MSE, RMSE and r
print('mae =', metrics.mean_absolute_error(y_test, y_pred))
mse = metrics.mean_squared_error(y_test, y_pred)
print('rmse =', np.sqrt(mse))
r2 = metrics.r2_score(y_test, y_pred)
print('r=',np.sqrt(r2))
```

The correlation between the module temperature and air temperature is shown in Fig. 2.6. As can be seen, good correlation is obtained with $r=98.2\%$. The relationship between T_C and T_a is given as $T_C = 0.698\ T_a$.

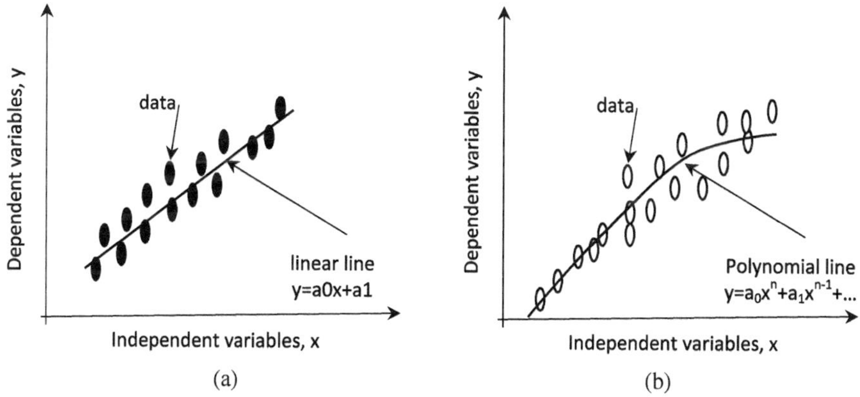

Fig. 2.4 Relationship between dependent and independent variables: (A) linear regression and (B) multiple regression.

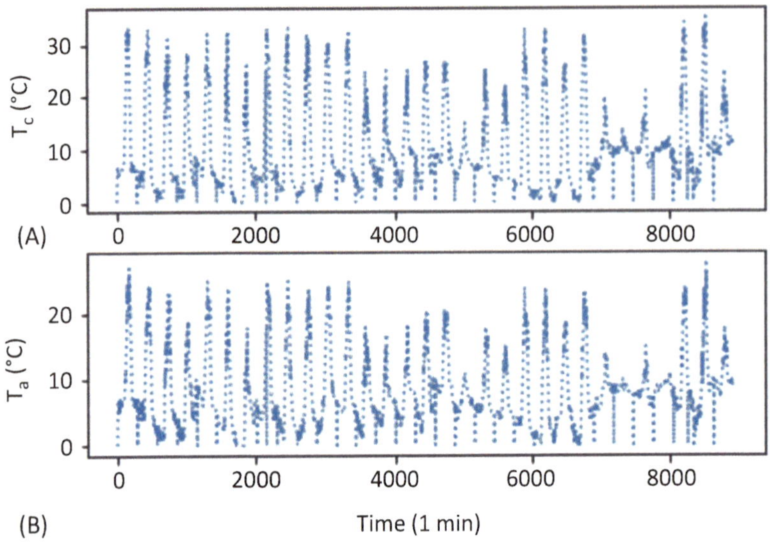

Fig. 2.5 (A) Module temperature and (B) air temperature variation (time step is 1 min).

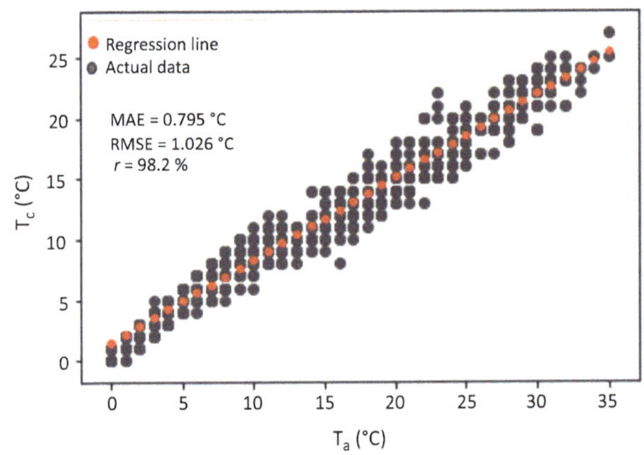

Fig. 2.6 Scatter curve between T_a and T_C using a linear function (*LinearRegression*)

2.3.1.2 *Logistic regression*

Logistic regressions (LoRs) are discrete values. They are used to predict the categorical dependent variable with the help of independent variables and it is based on the concept of maximum likelihood estimation [16]. The logistic regression problem output can be only between the 0 and 1, which is very suitable for classification problems. Fig. 2.7 shows an example of a logistic function.

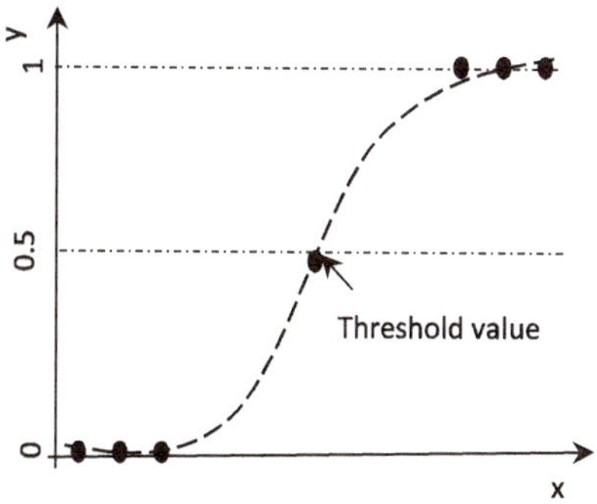

Fig. 2.7 Logistic function.

Example 2.2: Regression problem

In this example, the correlation between T_a and T_C is found using logistic regression function (*LogisticRegression*). A similar code as the one shown in **Listing 2.1** is used to find the regression. Thus, the logistic function *LogisticRegression* is used instead of the linear function. The following command line should be added to import the *LogisticRegression* from sklearn library as

```
from sklearn.linear_model import LogisticRegression
```

As can be seen from the results presented in Fig. 2.8, the logistic function does not perform well compared with linear regression (MAE=0.822 °C, RMSE=1.214 °C, and r=97.5%).

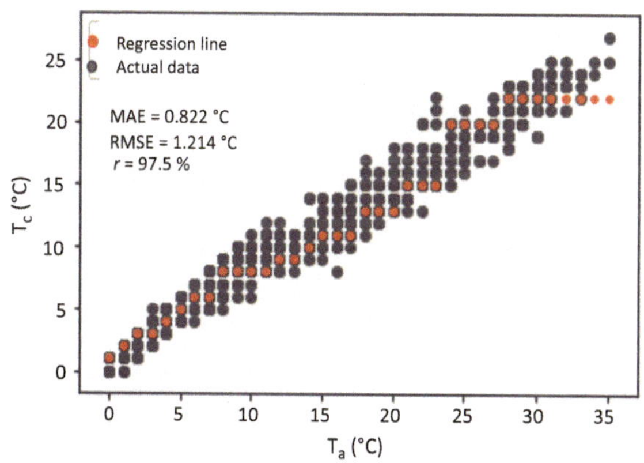

Fig. 2.8 Scatter curve between T_a and T_C using the logistic function (*LogisticRegression*).

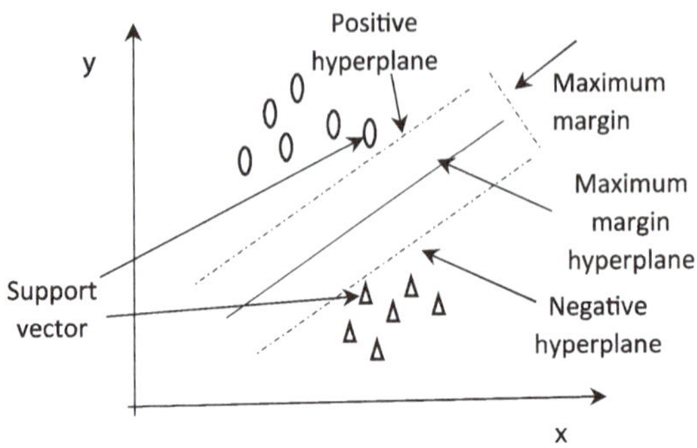

Fig. 2.9 Support vector machine workflow.

2.3.1.3 Support vector machine

The support vector machine (SVM) is the most widely used machine learning technique in PV systems. It is initially introduced by Drucker et al. [17] and is mainly used for classification and regression problems. As shown in Fig. 2.9, the SVM process attempts to plot a hyperplane that separates two classes optimally such that the margin between the hyperplane and the observations is maximum. Thus, the final objective of SVM is to find the one that gives a high margin. There are two types of SVM, linear and non-linear.

Example 2.3: Regression problem

In this example, the SVM algorithm is used to estimate global solar radiation (G) from ambient temperature $\{G = f(T_a)\}$. The dataset used for this example is plotted in Fig. 2.10. To train the predictor, 80% of the data are used and 20% are used for the testing. The corresponding code is presented in **Listing 2.2**.

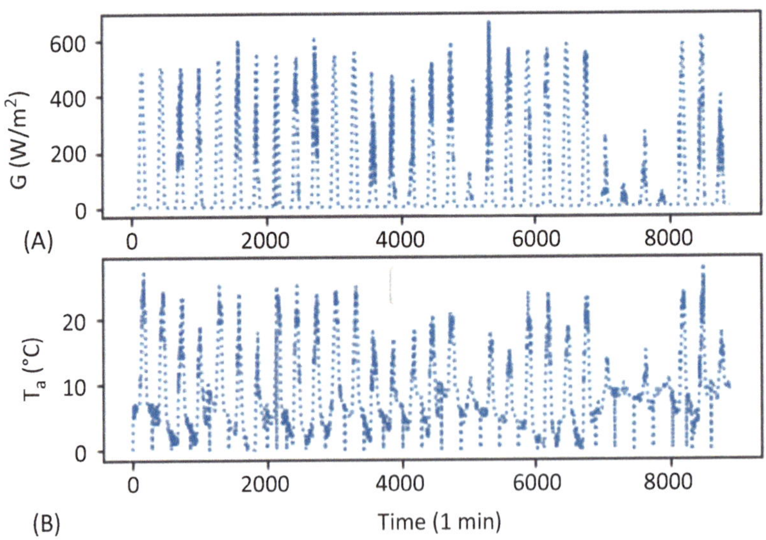

Fig. 2.10 (A) Global solar irradiance (G) and (B) air temperature (T_a).

```
#Listing 2.2: SVM for regression problem
#import libraries
from sklearn import svm
from sklearn.model_selection import train_test_split
#load data
xy= pd.read_csv('E:\\data_result\\climate.csv',delimiter=';')
x = xy['Ta'].values.reshape(-1,1)
y = xy['G'].values.reshape(-1,1)
#data split
x_train, x_test, y_train, y_test = train_test_split(x, y,
  test_size=0.20,random_state=0)
#call the svm model with rbf kernal
svm_rbf = svm.SVR(kernel ='rbf', C=100)
#fit=
svm_rbf.fit(x_train, y_train)
#prediction
y_pred = svm_rbf.predict(x_test)
```

As it can be observed from Fig. 2.11, there is an acceptable correlation between G and T_a with $r=93\%$.

2.3.1.4 K-Nearest neighbors

The K-nearest neighbor (K-NN) is generally the simplest ML algorithm, which can be used also to solve both classification and regression problems. It is initially introduced by Cover et al. [18] for classification problems, and then for regression problems by Altman [19]. In both, the input consists of the K closest training examples in the

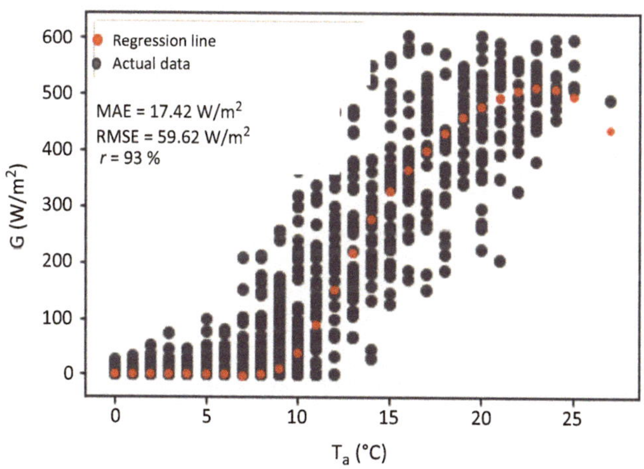

Fig. 2.11 Scatter curve between G and T_a using the SVR model (training an SVR model with an RBF kernel).

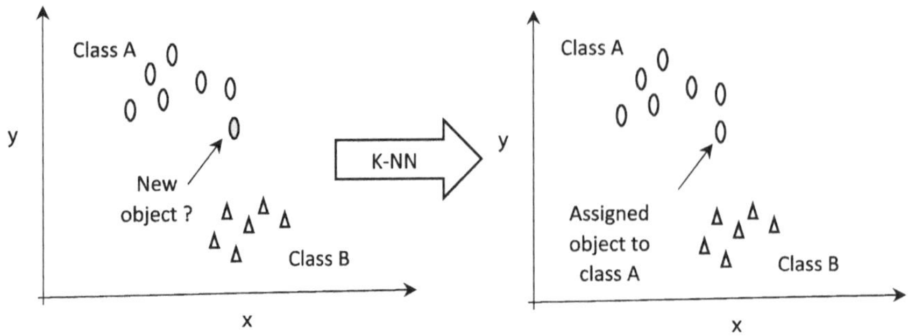

Fig. 2.12 K-NN work principle.

featured space. The output depends on whether the K-NN is used for classification or regression [17]. The K-NN algorithm is simple, can be easily implemented, versatile as it can be used for classification and regressions, and there is no need to build models [19]. Fig. 2.12 shows a simple example on how the algorithm works. The distance function can be Euclidean, Minkowski, or Hamming distance.

Example 2.4: Classification problem

In this example, the K-NN is employed to classify the type of days, i.e., sunny, overcast, cloudy, and rainy, based on the value of daily solar radiation and ambient temperature. A dataset of G, T_a, and class of day is used. To train the classifier, 80% of the data is used and the rest 20% are used for testing the classifier. The main function of the code is illustrated in **Listing 2.3**.

Listing 2.3: K-NN for classification problem

```
#import libraries
import numpy as np
import pandas as pd
import matplotlib.pyplot as plt
from sklearn.neighbors import KNeighborsClassifier
from sklearn.metrics import classification_report, confusion_matrix
from sklearn.model_selection import train_test_split
from sklearn import metrics
from sklearn.metrics import plot_confusion_matrix
dataset=pd.read_csv('E:\\data_result\\dayclass.csv' ,delimiter=';')
#load data
x= dataset.iloc[:, 0:2]
y=dataset.iloc[:, 2]
#data split
X_train, X_test, y_train, y_test=train_test_split(X, y,
  test_size=0.20,random_state=0)
#call for k-NN classifier
day_class=KNeighborsClassifier(n_neighbors=10)
#fit
```

```
day_class.fit(X_train, y_train)
#prediction
y_pred=day_class.predict(X_test)
#printingresults
print(classification_report(y_test, y_pred))
plot_confusion_matrix(day_class, X_test, y_test)
plt.show()
```

To evaluate this classifier the confusion matrix is used, which allows calculating the precision, recall, F1-score, and accuracy. More detail about these metrics is given in Appendix D. Fig. 2.13 shows the confusion matrix and results are listed in Table 2.1, showing clearly the accuracy of the K-NN classifier, being equal to 99%.

The recall of class 0 (sunny) is 310/311 (99.7%), one class {0} is incorrectly classified into class {1}, the precision is 310/317 (97.8%). The recall of class 1 (cloudy) is 360/369 (97.56%), 9 days were classified incorrectly 7 in class {0} and 2 in class {2},

Table 2.1 Error metrics: precession, recall, F1-score, and accuracy.

Class of day	Precision (%)	Recall (%)	F1-score (%)	Accuracy (%)
Sunny {class 0}	97.80	99.70	98.70	99
Cloudy {class 1}	98.90	97.56	98.22	
Overcast {class 2}	98.70	98.70	98.70	
Rainy {class 3}	99.33	99.00	99.13	

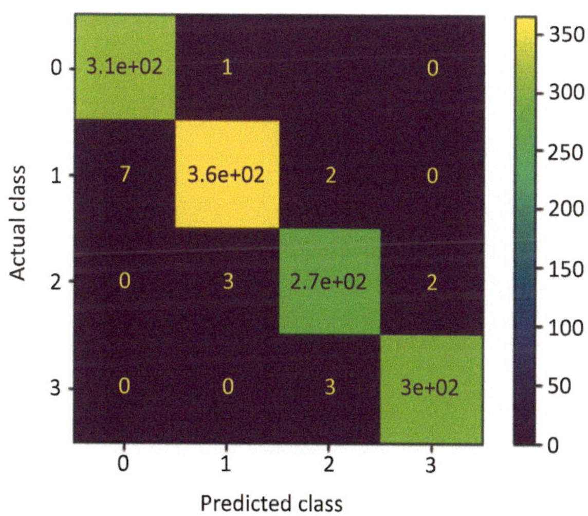

Fig. 2.13 Confusion matrix (sunny =0, cloudy=1, overcast=2, and rainy=3).

the precision is 360/364 (98.9%). The recall and the precision of class 2 (overcast) is 370/375 (98.7%), 5 days were classified incorrectly, 3 into class 1, and 2 into class 3. The recall of class 3 (sunny) is 300/303 (99%) 300/302 (99.33%), 3 days were classified incorrectly into class 2, the precision is 300/302 (99.33%).

2.3.1.5 Decision trees

Decision tree (DT) is a powerful tool, which can be used for solving both classification and regression problems. It is originally introduced by Quinlan [20]. Essentially, they learn the hierarchy of if/else questions, which lead to a decision. It consists mainly of three types of nodes [21]: (1) decision nodes, (2) chance nodes, and (3) end nodes. Decision trees are constructed via an algorithmic approach that identifies ways to split a dataset based on different conditions. Fig. 2.14 shows an example of weather forecasting using a simple three *if-then-else* decisions algorithm. Decision tree can handle non-linear datasets in an effective manner.

Based on the target variables, the decision tree algorithm is mainly classified into two groups, categorical variable decision trees and continuous variable decision trees.

Example 2.5: Classification problem

In this example, a decision tree algorithm is applied for days' classification. A similar code as the one of **Listing 2.3** is used, by replacing only the K-NN function by *DecisionTreeClassifier* and importing it from the library *sklearn.tree* using the following instruction.

```
from sklearn.tree import DecisionTreeClassifier
```

Fig. 2.15 shows the confusion matrix; it can be seen that all values are in diagonal elements of the matrix, which means there is no error and all cases have been classified

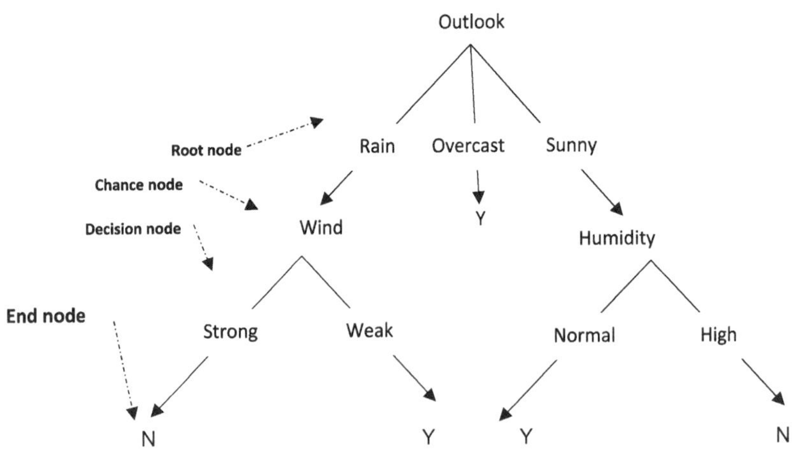

Fig. 2.14 Decision tree example for rain forecasting.

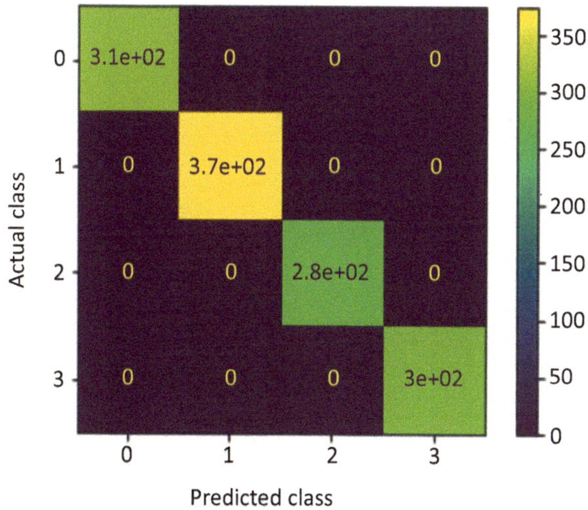

Fig. 2.15 Confusion matrix (sunny =0, cloudy=1, overcast=2, and rainy=3).

Table 2.2 Errors metrics: precision, recall, F1-score and accuracy.

Class of day	Precision (%)	Recall (%)	F1-score (%)	Accuracy (%)
Sunny {class 0}	100	100	100	100
Cloudy {class 1}	100	100	100	
Overcast{class 2}	100	100	100	
Rainy{class 3}	100	100	100	

correctly. Results are listed in Table 2.2 and reports clearly that the decision tree classifier performs better than the K-NN classifier with an accuracy of 100%.

2.3.1.6 Random forest

The literature shows that decision-tree algorithms have some problems, and their main issue is that they tend to overfit the training data [22]. Therefore, random forest (RF) or random decision trees can solve this issue, based on a collection of decision trees, where each tree is slightly different from the others. The RF algorithms were initially introduced by Ho in 1995 [23]. They are often accurate, do not require feature scaling, categorical feature encoding, and need little parameter tuning [24]. A simplified example of an RF algorithm for classification is shown in Fig. 2.16.

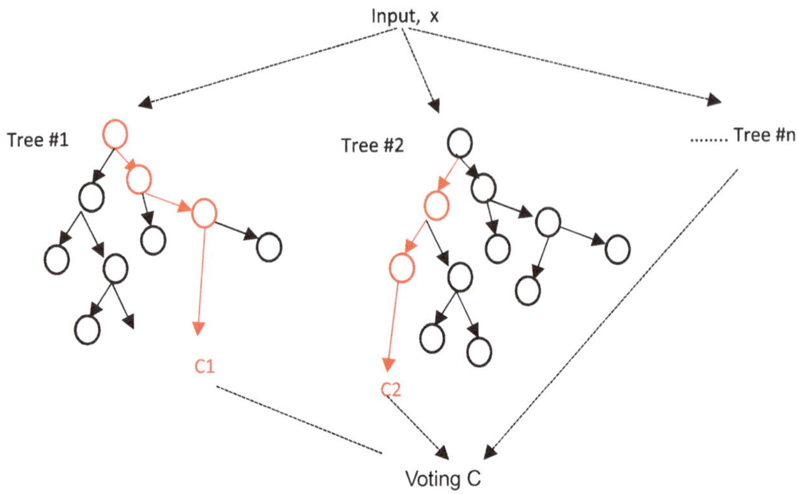

Fig. 2.16 A simplified example of an RF algorithm: classification example.

The random forest algorithm is based on the concept of ensemble learning, which is a process of combining multiple classifiers to solve a complex problem and to improve the performance of the model.

Example 2.6: Classification problem

In this example, a random forest algorithm is applied for the same problem as above. Simply in **Listing 2.3**, the decision tree function (*DecisionTreeClassifier*) is replaced by *RandomForestClassifier* and this function is imported from the library *sklearn.ensemble* by adding the following command line.

```
from sklearn.ensemble import RandomForestClassifier
```

After training and simulation of the model, a similar precision and accuracy are obtained as the one provided by the decision tree algorithm. Thus, the decision tree algorithm is adequate to solve this problem.

2.3.1.7 Naïve Bayes

Naïve Bayes (NB) algorithms are based on applying Bayes' theorem with the "naïve." It assumes that the presence of a particular feature in a class is unrelated to the presence of any other feature [25]. It is initially created for classification, but it can be used for solving regression problems as well [26]. The NB model is easy to build and is particularly useful for large datasets.

Example 2.7: Classification problem

In this example, the NB algorithm is applied for the day classification problem. In **Listing 2.3** only the function *GaussianNB* is replaced, imported from the package *sklearn.naive_bayes*. Thus the following command line should be added in the code:

Table 2.3 Errors metrics: precision, recall, F1-score and accuracy.

Class of day	Precision (%)	Recall (%)	F1-score (%)	Accuracy (%)
Sunny {class 0}	97	97	97	95
Cloudy {class 1}	95	96	92	
Overcast{class 2}	92	92	92	
Rainy{class 3}	96	96	96	

```
from sklearn.naive_bayes import GaussianNB
```

As listed in Table 2.3 this algorithm has low accuracy equal to 95%, compared with K-NN, DT, and RF.

From Table 2.3 with reference to F1-score values cloudy class and overcast are the worst classified classes (92%), while the classification rate of sunny class {0}is the best at 97%.

2.3.1.8 Neural networks

Neural networks (NNs) can be considered as simplified mathematical models of brain-like systems and they function as parallel-distributed computing networks. However, in contrast to conventional computers, which are programmed to perform specific task, most neural networks must be taught, or trained. They can learn new associations, new functional dependencies, and new patterns. The first computational model for NNs was introduced in Ref. [27] by McCulloch. There are many kinds of NNs including, Kohonen, self-organizing feed-forward NNs, recurrent NNs, radial basis function, modular NNs, conventional NNs, and others [28]. Fig. 2.17 shows an example of a feed-forward NN. This can be used for classification and regression problems. Where X_1, X_2, ..., X_N are the inputs and y_1, y_2, ..., y_M are the outputs.

Example 2.8: Function approximation problem

This example displays how to use a simple feed-forward NNs to estimate the power produced by a PV module (P_{PV}) based on solar irradiance {$P_{PV}=f(G)$}. Fig. 2.18 shows the dataset used for this function approximation. It consists of the solar irradiance and the power produced by a PV string of 500 W.

To train and validate the NN model, 85% of the samples were used and the rest 15% of the samples were used to test the *MLPRegressor*. The corresponding code is listed in **Listing 2.4**.

```
#Listing 2.4: MLP approximation function
#import necessary packages
import numpy as np
import pandas as pd
from sklearn.model_selection import train_test_split
```

```
from sklearn.neural_network import MLPRegressor
from sklearn import metrics
#loadingdata
xy=pd.read_csv('E:\\data_result\\PVarray.csv',delimiter=';')
x=xy.iloc[:, 3]
y=xy.iloc[:, 4]
x=x.values.reshape(-1,1)
y=y.values.reshape(-1,1)
#splitingdataset
x_train, x_test, y_train, y_test=train_test_split(x, y, test_size=
  0.15,random_state=1)
# modelestimator
Estimator=MLPRegressor(hidden_layer_sizes=(20,10),
  activation="logistic" ,random_state=1, max_iter=2000).fit
  (x_train, y_train)
#modeltest
y_pred=Estimator.predict(x_test)
```

The developed NN estimator consists of four layers, one input layer with one neuron, two hidden layers with 20, 10 neurons, respectively, and one neuron in the output layer. Results are shown in Fig. 2.19, the cost function is close to 0.2 (see Fig. 2.19A), the correlation coefficient r is equal to 98.5%, the RMSE = 4.74 W and MAE is 1.7 W, which are satisfactory (see Fig. 2.19B).

2.3.2 Unsupervised learning

In this type of learning, shown schematically in Fig. 2.20, the output is not labeled and the algorithm searches for rules and patterns in the available dataset in order to better describe the data. No idea about the expected results exists in this type of algorithms;

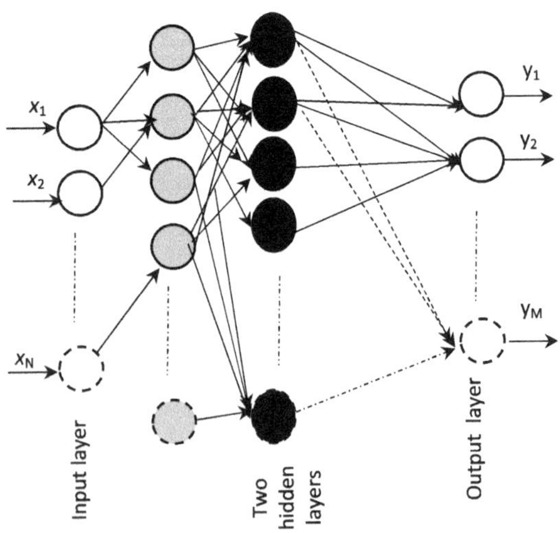

Fig. 2.17 Feed-forward NN with two hidden layers.

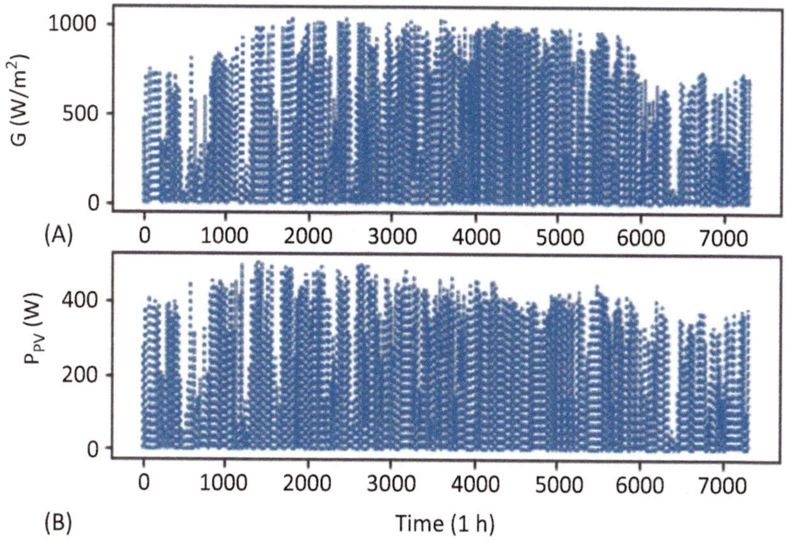

Fig. 2.18 Dataset: (A) solar irradiance and (B) photovoltaic power.

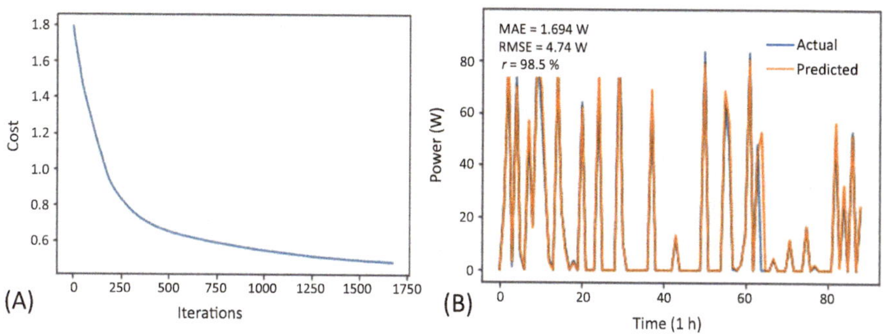

Fig. 2.19 (A) Loss (cost) function versus epochs and (B) comparison between actual and estimated power using the *MLPRegressor* model.

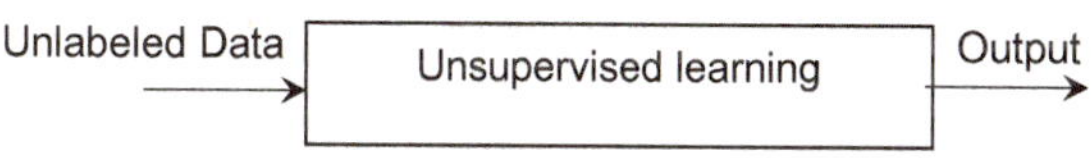

Fig. 2.20 Unsupervised learning.

so, the machine is trained with unlabeled data and it can mainly be used to solve problems related to clustering, association, and directionality reduction [15].

This type of machine learning is recently used in photovoltaic systems, and its main application concerns classification problems in photovoltaic arrays.

Unsupervised ML includes mainly the K-means, fuzzy c-means, principal component analysis, hierarchical clustering, and other algorithms.

2.3.2.1 K-means

The K-means (KM) is one of the simplest and popular clustering unsupervised ML [29] used for solving clustering problems. The KM algorithm tries to identify clusters of data objects in a dataset and is mainly used when unlabeled data are available [30]. In practice, clustering helps to identify two qualities of data: (1) meaningfulness and (2) usefulness. The results of the K-means clustering algorithm are the centroids of the K clusters, and labels for the training data.

Example 2.9: Clustering of weather days

The following code (**Listing 2.5**) permits to cluster the weather days based on G and T_a. The desired number of K subgroups is set to 4. A cluster refers to a collection of data points aggregated together because of certain similarities.

```
#Listing 2.5: K-means for clustering problem
#import libraries
import numpy as np
import pandas as pd
import matplotlib.pyplot as plt
from sklearn.cluster import KMeans
#K-means model
km = KMeans(n_clusters=4).fit(xy)
centroids = km.cluster_centers_
print(centroids)
#Plotting results
plt.scatter(xy['G'], xy['Ta'], c= kmeans.labels_.astype(float), s=50,
  alpha=0.5)
plt.scatter(centroids[:, 0], centroids[:, 1], c='red', s=50)
plt.ylabel('Ta (Degree C)')
plt.xlabel('G (W/m2)')
plt.show()
```

The results are shown in Fig. 2.21. The center of each cluster (shown in red) represents the mean of all the observations that belong to that cluster. The observations that belong to a given cluster are closer to the center of that cluster, in comparison to the centers of other clusters. The result of the value of the centroids is: {[3.35 4.51] [343.88 22.20] [494.11 28.18] [165.38 14.60]}.

2.3.2.2 Fuzzy c-means

The fuzzy-c-means (FCM) is a clustering algorithm, which allows one piece of data to belong to two or more clusters. The FCM algorithm is very similar to the KM, and it was developed originally by Dunn [31] and then improved by Bezdek [32]. The flowing code (**Listing 2.6**) is a general code of fuzzy c-means using the FCM function provided from *fcmeans*.

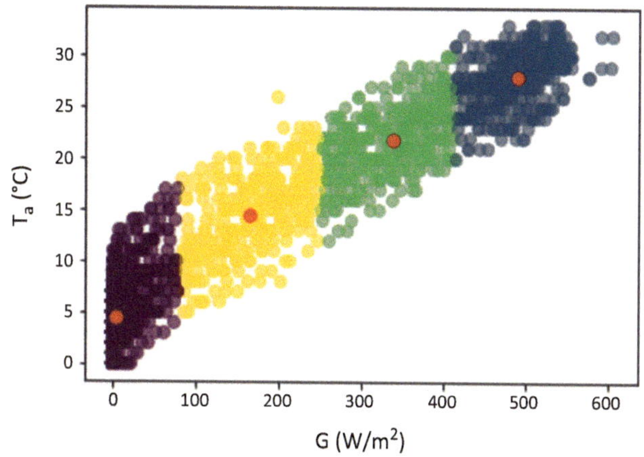

Fig. 2.21 Obtained clusters by K-means algorithm: *red* (centroids), *violet* (rainy), *yellow* (cloudy), *green* (overcast), and *blue* (sunny).

```
Listing 2.6: Fuzzy C-Means for clustering problem
# import libraries
import numpy as np
from fcmeans import FCM
from matplotlib import pyplot as plt
n_samples = 1000
X = np.concatenate((
    np.random.normal((-2, -2), size=(n_samples, 2)),
    np.random.normal((2, 2), size=(n_samples, 2))
))
# model fcm
fcm = FCM(n_clusters=2)
fcm.fit(X)
fcm_centers = fcm.centers
fcm_labels = fcm.predict(X)
#plotting results
f, axes = plt.subplots(1, 2, figsize=(10,5))
axes[0].scatter(X[:,0], X[:,1], alpha=.1)
axes[1].scatter(X[:,0], X[:,1], c=fcm_labels, alpha=.1)
axes[1].scatter(fcm_centers[:,0], fcm_centers[:,1], marker="+",
  s=500, c='w')
plt.show()
```

2.3.2.3 *Hierarchical clustering*

Hierarchical clustering (HC) is a class of clustering algorithms. It groups similar objects into groups called clusters [33], which are similar to the *k*-means but the number of clusters changes with iteration. It works well with smaller datasets [34]. There are mainly two types of HC: agglomerative and divisive.

Example 2.10: Clustering of weather days

To understand how to use the HC algorithm, in this example, an agglomerative hierarchical clustering algorithm is applied for weather clustering. The same dataset as above has been used and the code is displayed in **Listing 2.7**.

```
#Listing 2.7: Hierarchical clustering problem
#import libraries
import numpy as np
import pandas as pd
import seaborn as seabornInstance
import matplotlib.pyplot as plt
from sklearn.cluster import AgglomerativeClustering
from scipy.cluster.hierarchy import dendrogram, linkage
import scipy.cluster.hierarchy as sch
# dendrogram
dendrogram = sch.dendrogram(sch.linkage(xy, method = "ward"))
plt.title('Dendrogram')
plt.xlabel('Days')
plt.ylabel('Euclidean distances')
plt.show()
hc = AgglomerativeClustering(n_clusters = 4, affinity = 'euclidean',
  linkage ='ward')
y_hc=hc.fit_predict(xy)
#plotting results
plt.scatter(xy[y_hc==0, 0], xy[y_hc==0,1], s=20,c='red',
  label ='Cluster 1')
plt.scatter(xy[y_hc==1, 0], xy[y_hc==1,1], s=20,c='blue',
  label ='Cluster 2')
plt.scatter(xy[y_hc==2, 0], xy[y_hc==2,1], s=20,c='green',
  label ='Cluster 3')
plt.scatter(xy[y_hc==3, 0], xy[y_hc==3,1], s=20,c='black',
  label ='Cluster 4')
plt.xlabel('G (W/$mo2$)')
plt.ylabel('Ta (Degrees C)')
plt.show()
```

As shown in Fig. 2.22A (dendrogram), four clusters are available corresponding to rainy, cloudy, overcast, and sunny days (see Fig. 2.22B).

2.3.2.4 Principal component analysis

Principal component analysis (PCA) is used to make data easy to explore and visualize by reducing the number of variables. It is used to decompose a multivariate dataset in a set of successive orthogonal components that explain the maximum amount of the variance. It is more intuitive, and most frequently it uses methods of dimensionality reduction [35].

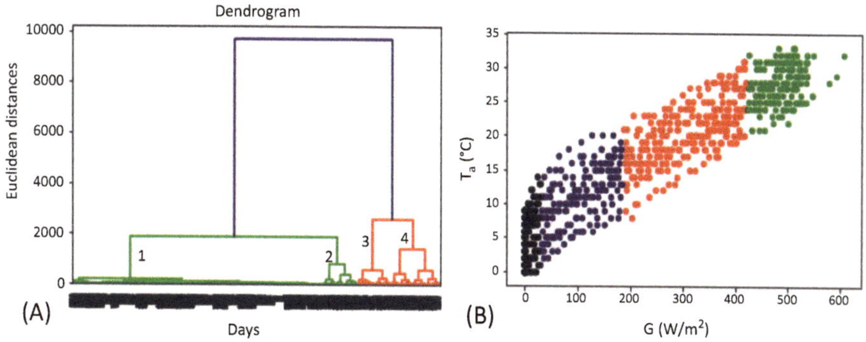

Fig. 2.22 (A) Dendrogram variable linkage, (B) obtained clusters by hierarchical clustering method; *black* (rainy), *blue* (cloudy), *red* (overcast), and *green* (sunny).

Example 2.11: Dimensionality reduction

In this example, the PCA algorithm and logistic regression are used to predict the power produced by PV plants. As input the G, T_a, T_C, wind speed (W_s), and relative humidity (R_h) are used. In this case, the PCA algorithm is used to reduce the number of the inputs up to the two most relevant inputs. The corresponding code is provided in the following code (**Listing 2.8**).

```
#Listing 2.8: Principal component analysis for dimensionality reduction
#import libraries
import numpy as np
import pandas as pd
import matplotlib.pyplot as plt
from sklearn.preprocessing import StandardScaler
from sklearn.metrics import accuracy_score
from sklearn.decomposition import PCA
from sklearn.model_selection import train_test_split
from sklearn.linear_model import LogisticRegression
#dataset
database= pd.read_csv('E:\\data_result\\PVdata1.csv',delimiter=';')
x= database.iloc[:, [0,1,2,3,4,5,6]].values
y= database.iloc[:,[6]].values
#splitting data
x_train, x_test, y_train, y_test = train_test_split(x, y,
  test_size=0.2, random_state = 0)
sc = StandardScaler()
x_train = sc.fit_transform(x_train)
x_test = sc.transform(x_test)
pca = PCA(n_components=1)
x_train = pca.fit_transform(x_train)
x_test = pca.transform(x_test)
explained_variane = pca.explained_variance_ratio_
```

```
pred = LogisticRegression(random_state = 0)
pred.fit(x_train, y_train)
#model evaluation
y_pred = pred.predict(x_test)
```

The correlation coefficient obtained is 79%, and Fig. 2.23 shows a comparison between the measured and predicted PV power based on the PCA and logistic regression.

2.3.3 Reinforcement learning

Reinforcement learning is shown schematically in Fig. 2.24. Here also, there are no training datasets, but an agent continuously learns from the environment in an iterative fashion. Reinforcement learning problems are formulated as Markov decision processes (MDPs) [36]. This type of learning can also be divided into two kinds of problems, classification, and control.

Reinforcement learning includes Q-learning (QL) [37] and other deep QL algorithms. This kind of ML algorithms is not widely investigated and applied in photovoltaic areas.

2.4 Ensemble learning

The ensemble learning (EL) method is a concept of ML; in simple words, it helps improve ML results by combining several models [38]. EL is originally developed to improve classification accuracy by reducing the variance in classifier outputs [39]. There are mainly three methods that dominate the field of EL methods: bagging, boosting, and stacking methods.

2.4.1 Bagging

A simplified block diagram of bagging ensemble (or bootstrap aggregation) is shown in Fig. 2.25. The bootstrapped samples are drawn from the training set, and the submodels are tested using the testing set. The final output prediction is combined across

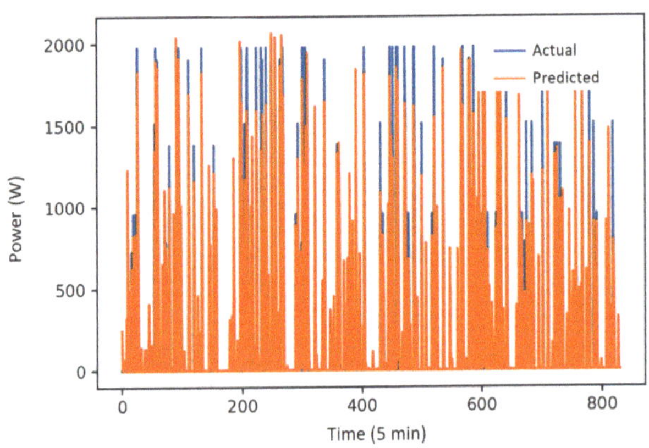

Fig. 2.23 Comparison between actual and predicted power using PCA and logistic regression .

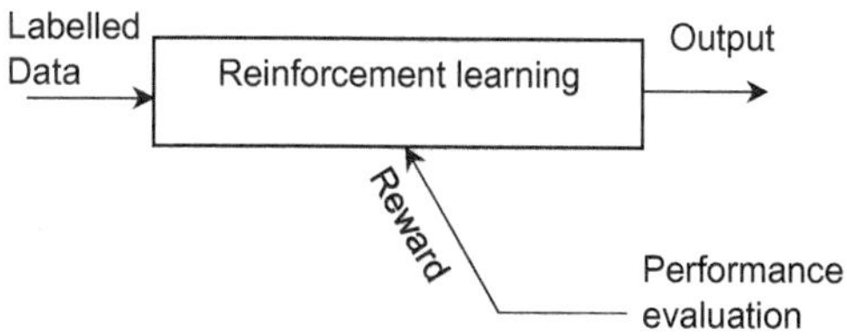

Fig. 2.24 Reinforcement learning.

the projections of all the sub-models. Algorithms based on this concept are: Random forest and Extra tree algorithms.

A basic procedure for a bagging algorithm is given below:

Step #1: A sample from observation is selected randomly with replacement (Bootstrapping), assuming that N observations and M features exist.
Step #2: A subset of features is selected to create a model with sample of observations and subset of features.
Step #3: A feature from the subset is selected, which gives the best split on the training data. This step is repeated to create many models and every model is trained in parallel.
Step #4: Prediction is given based on the aggregation of predictions from all the models.

2.4.2 Boosting

Boosting was introduced by Schapire [40]. A simplified block diagram of the boosting ensemble method is shown in Fig. 2.26. The algorithm works by training a model with

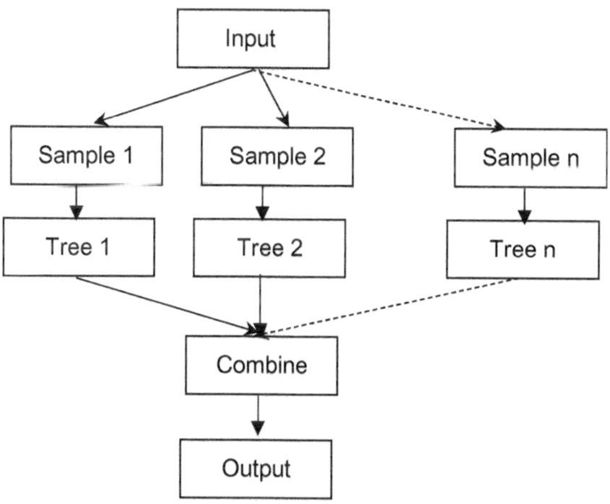

Fig. 2.25 Bagging ensemble diagram.

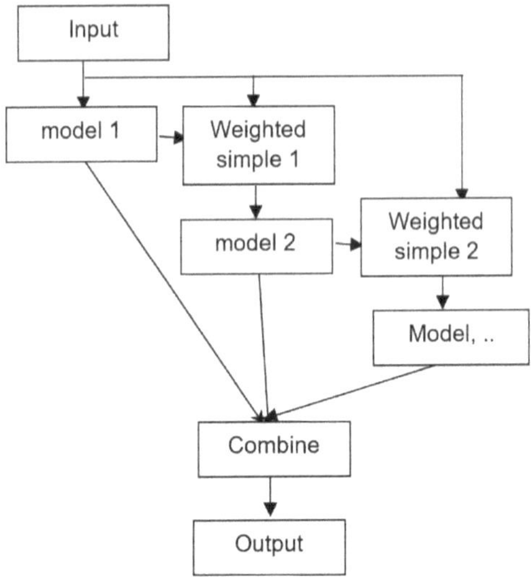

Fig. 2.26 Boosting ensemble diagram.

the entire training set, and subsequent models are constructed by fitting the residual error values of the initial model.

A basic procedure for a boosting algorithm is given below:

Step #1: The base learner takes all the distributions and assigns equal weight or attention to each observation.
Step #2: If there is any prediction error caused by the first base learning algorithm, then greater attention is paid to observations having prediction error. Then, the next base learning algorithm is applied.
Step #3: Iterate **Step #2** till the limit of the base learning algorithm is reached or higher accuracy is achieved.
Step #4: Finally, the algorithm combines the outputs from weak learners and creates a strong learner, which eventually improves the prediction power of the model. Boosting pays greater focus on examples, which are either misclassified or have higher errors by preceding weak rules.

Algorithms based on this concept are: stochastic gradient boosting, gradient boosting, and AdaBoost.

2.4.3 Stacking

Stacking-a Meta Modeling Technique was introduced by Wolpert [41]. A basic block diagram of the stacking-ensemble method is shown in Fig. 2.27. Stacking is an EL technique that combines multiple models (predictions); it tries to learn how to

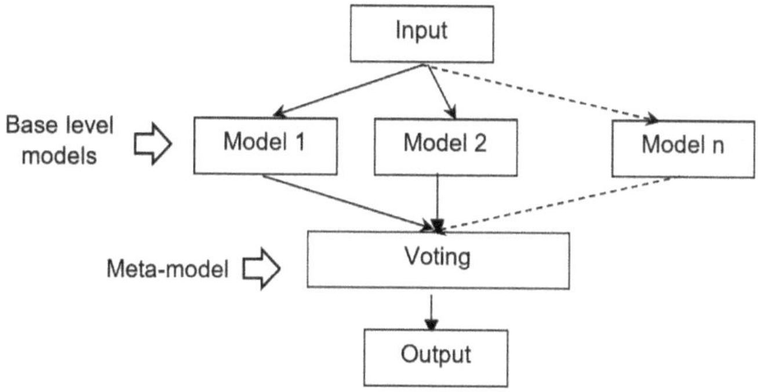

Fig. 2.27 Stacking ensemble diagram.

consider these predictions in the best possible way. The base-level models are trained based on a complete training set; then, the meta-model is trained on the outputs of the base level model as features. It consists mainly of three steps: (1) learn base level, (2) construct a new dataset for predictions, and (3) learn a meta-model.

The steps of a stacking method are:

Step #1: Split the training data into two disjoint sets.
Step #2: Train several base learners on the first part.
Step #3: Test the base learners on the second part and make predictions.
Step #4: Using the predictions from step #3 as inputs, the correct responses from the output, train the higher level learner or meta level learner.

Example 2.12: Classification problem

In this example, a stacking ensemble method is applied to classify days. The models used are K-NN, DT, and SVM. To do this, the available *votingClassifier* function in the Scikit-Learn library is used. The main code is shown in **Listing 2.9**.

```
#Listing 2.9 EL voting method, classification problem
from sklearn.tree import DecisionTreeClassifier
from sklearn.svm import SVC
from sklearn.neighbors import KNeighborsClassifier
from sklearn.ensemble import VotingClassifier
# Training classifiers

clf1 = DecisionTreeClassifier(max_depth=3)
clf2 = KNeighborsClassifier(n_neighbors=3)
clf3 = SVC(gamma=.1, kernel='rbf', probability=True)
eclf = VotingClassifier(estimators=[('dt', clf1),
                                    ('knn', clf2),
                                    ('svc', clf3)],
                        voting='soft', weights= [2, 1, 2])
```

```
clf1.fit(X_train, y_train)
clf2.fit(X_train, y_train)
clf3.fit(X_train, y_train)
eclf.fit(X_train, y_train)
y_pred = eclf.predict(X_test)
print("Accuracy:",metrics.accuracy_score(y_test, y_pred))
```

As shown in Fig. 2.28 the accuracy is 100%; the four classes of days are correctly classified.

2.5 Deep learning

Fig. 2.29 shows a simplified process diagram of deep learning. As can be seen, the feature extraction step is removed as it is not needed in this kind of learning; this is because deep learning algorithms are able to learn and automatically extract features from raw (original) input data.

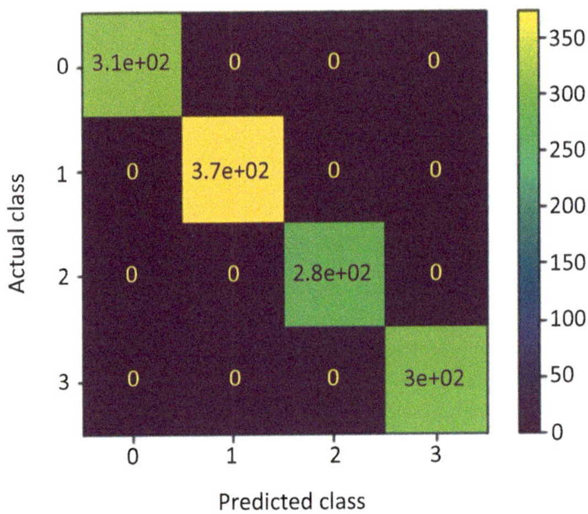

Fig. 2.28 Confusion matrix: days' classification using EL voting methods.

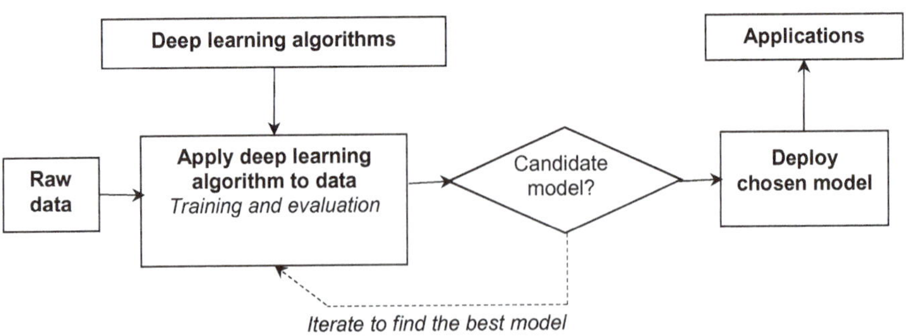

Fig. 2.29 Deep learning workflow.

Deep learning is a relatively new advancement in NN and represents a way to train deep neural networks (DNNs), as traditional NN-based methods might be affected by problems such as overfitting and diminishing gradients [42]. Essentially, any NN with more than two layers is called deep. On the other hand, deep learning is a semi-supervised training approach, suitable for DNN training [43]. The main deep learning methods are deep convolutional neural network (DCNN), long short-term memory (LSTM), gated recurrent unit (GRU), deep belief networks (DBN), generative adversarial networks (GANs), deep conventional GAN (DCGAN), and other hybrid combinations [42].

2.5.1 Convolutional neural networks

Convolutional neural networks are a specialized kind of NN for processing data that have a known, grid-like topology. These are simply NNs that use convolution in place of general matrix multiplication in at least one of their layers [43]. Like other NNs, a CNN is composed of an input layer, an output layer, and many hidden layers; mainly, convolution and pooling layers can be found. Fig. 2.30 shows a schematic representation of the basic structure of a DCNN.

The rectified linear unit (ReLU) is a popular activation function, which helps accelerating the convergence speed, and Softmax function is used before the output layer.

The following code (**Listing 2.10**) is an example for a CNN used for image classification, which can be used, for example, to diagnose and identify hot spot faults in photovoltaic arrays, shown in Fig. 2.31.

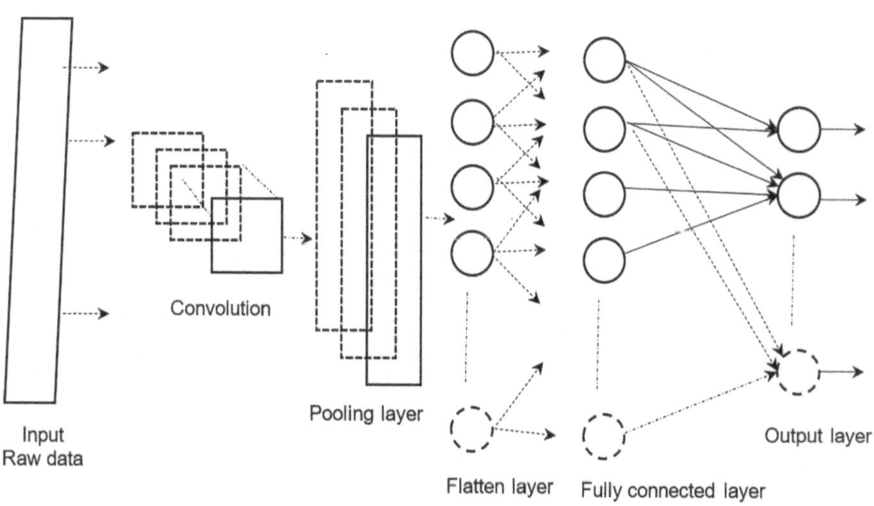

Fig. 2.30 Deep learning convolutional neural network structure.

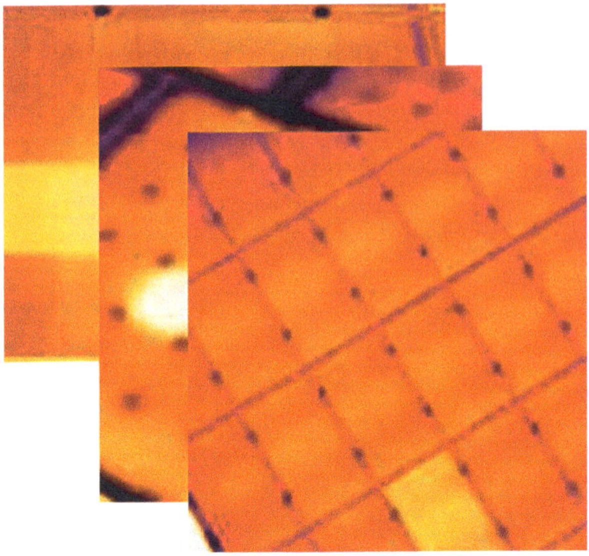

Fig. 2.31 PV module with hotspot problem identification (images taken by a thermal camera).

Listing 2.10 Deep CNN model, classification problem

```
#libraries
import os
import numpy as np
import pandas as pd
import tensorflow as tf
from tensorflow import keras
from tensorflow.keras.layers import Dropout, Activation, Flatten,
  Dense
from tensorflow.keras.models import Sequential
from tensorflow.keras.layers import Conv2D, MaxPooling2D
#model
cnn = Sequential()
cnn.add(Conv2D(64,3, activation="relu", input_shape=(155,155,3)),
  padding='same')
cnn.add(MaxPool2D())
cnn.add(Conv2D(64, 3, activation="relu"))
cnn.add(MaxPool2D())
cnn.add(Dropout(0.2))
cnn.add(BatchNormalization())
cnn.add(Flatten())
cnn.add(Dense(128,activation="relu"))
cnn.add(Dense(1, activation="softmax"))
print(model.summary())
# compiling model
```

```
cnn.compile(loss = 'binary_crossentropy',optimizer = 'rmsprop',
  metrics = ['accuracy'])
cnn.fit(X_train, y_train, validation_data=(X_test, y_test),
  epochs=epochs, batch_size=64)
# model evaluation
scores = cnn.evaluate(X_test, y_test, verbose=0)
```

2.5.2 Long short-term memory

The main drawback of recurrent neural networks is that they practically fail to handle long-term dependencies and vanishing [44]. The introduced long short-term memory (LSTM) network is capable of handling these long-term dependencies. An LSTM block typically has a memory cell, input gate, output gate, and a forget gate in addition to the hidden state used in traditional recurrent neural networks. A simple structure consists of some LSTM cells and a dense output layer, as shown in Fig. 2.32.

where X_{cs} is a matrix of the number of features, H_{cs} is a matrix of the number of hidden units, x_t is the input, h_t is the hidden state, c_t is the cell state, g is the memory cell, o is the output gate, and f is the forget gate.

Example 2.13: Forecasting problem

This example explains how to predict the next value of the ambient temperature based on the past three values. A dataset of 4000 samples has been used; 85% of the samples are used for training and 15% for testing the LSTM-model. A basic code is listed in the following code (**Listing 2.11**).

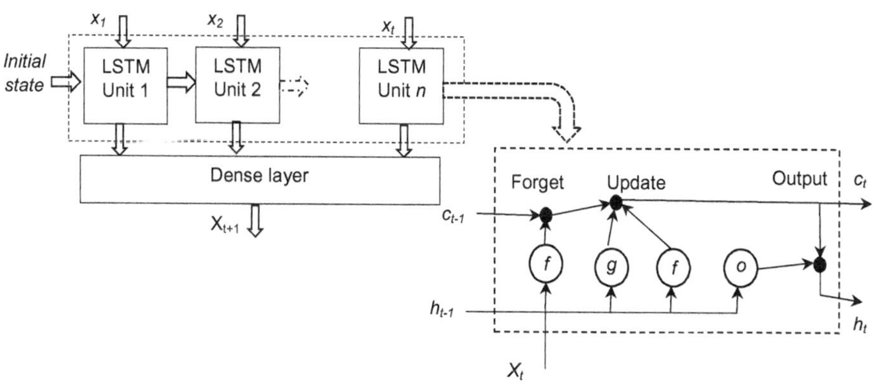

Fig. 2.32 LSTM architecture, one hidden layer with n units and one dense layer.

```
#Listing 2.11 LSTM model for forecasting problem
# import libraries
from sklearn.preprocessing import MinMaxScaler
from tensorflow.keras.layers import LSTM
from tensorflow.keras.models import Sequential
import sklearn.metrics as metrics
# function definition splitting data
def fun_sequences(seq, seq_size):
    vect = []
    for index in range(len(seq) - seq_size):
    vect.append(seq[index: index + seq_size])
    return np.array(vect)
def fun_preprocess(seq_raw, seq_size, train_split):
    seq = fun_sequences(seq_raw, seq_size)
    num_train = int(train_split * seq.shape[0])
    X_train = seq[:num_train, :-1, :]
    y_train = seq[:num_train, -1, :]
    X_test = seq[num_train:, :-1, :]
    y_test = seq[num_train:, -1, :]
    return X_train, y_train, X_test, y_test
X_train, y_train, X_test, y_test = fun_preprocess(scaled_Ta,
  SEQ_SIZE, train_split = 0.85)
# model definition
Temp_for = Sequential()
Temp_for.add(LSTM(units = 100, activation='relu',input_shape =
  (X_train.shape[1], 1)))
Temp_for.add(Dense(units = 1))
Temp_for.summary()
% compling model
Temp_for.compile(loss='mean_squared_error', optimizer='adam')
# results
results= Temp_for.fit( X_train, y_train, epochs=100, batch_size=64)
y_prediction = Temp_for.predict(X_test)
```

As shown in Fig. 2.33A, the model converges after about 100 epochs to 0.001. The $r=99\%$, RMSE=0.24 °C, and MAE=0.01 °C, which show a very good accuracy. Good correlation is also shown in Fig. 2.33B.

2.5.3 Gated recurrent unit

Gated recurrent unit (GRU), introduced initially by Cho et al. [45], is similar to LSTM with fewer parameters than the LSTM. The parameters are learned through the gating mechanism of GRU. The GRU is computationally more efficient considering that fewer parameters are required and need less data to generalize. It can also learn long-terms dependencies. The same diagram with the one in Fig.2.32 is used but

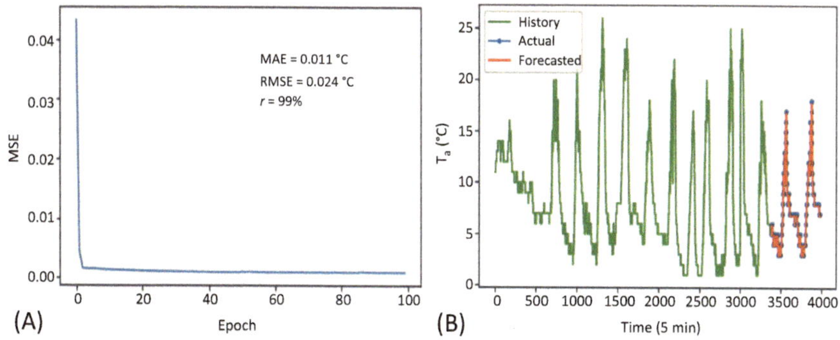

Fig. 2.33 (A) Loss evolution and (B) comparison between actual and forecasted temperature (*red*: forecasted, *blue*: measured).

the LSTM unit is replaced by a GRU. The code is also similar to the LSTM one, just the LSTM is replaced by the GRU as shown in **Listing 2.11** and importing GRU from Keras, as shown by the following command lines.

```
from tensorflow.keras.layers import GRU
Temp_for.add(LSTM(units=100, activation='relu',input_shape=
(X_train.shape[1], 1)))
```

2.5.4 Bidirectional LSTM and GRU

Bidirectional LSTM (BiLSTM) is initially introduced by Schuster and Paliwal [46] and is a modified version of LSTM. It consists of two separate hidden layers; firstly, it computes the forward hidden sequence, then it calculates the backward hidden sequence, and finally combines both to calculate the output, as shown in Fig. 2.34. The bidirectional GRU (BiGRU) algorithm is similar to the BiLSTM, in which the LSTM units are replaced by the GRU.

The following command lines provide an idea on how to use BiGRU or BiLSTM to predict the future value of T_a.

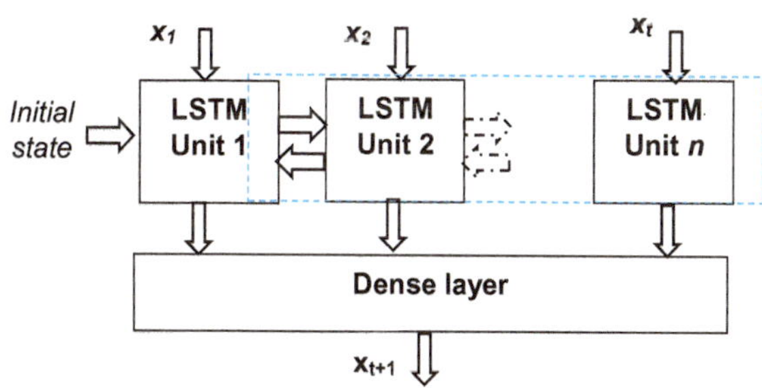

Fig. 2.34 BiLSTM structure of one hidden layer with *n* units and one dense layer.

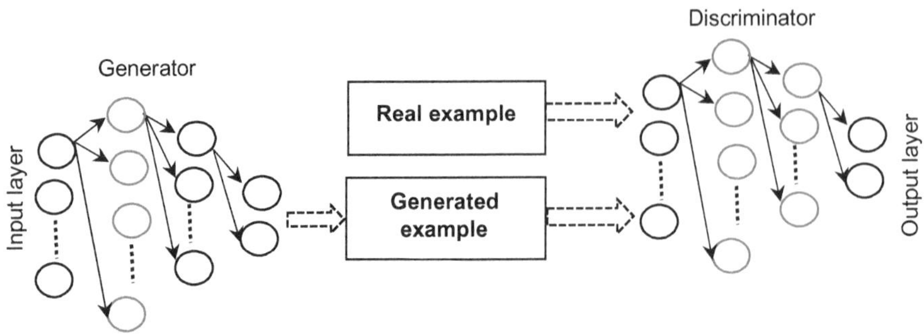

Fig. 2.35 GAN structure.

```
from tensorflow.keras.layers import LSTM, GRU, Bidirectional
Temp_for.add(Bidirectional(LSTM(200, activation='relu'),
input_shape=( X_train.shape[1], 1)))
Temp_for.add(Bidirectional(GRU(200, activation='relu'),
input_shape=( X_train.shape[1], 1)))
```

2.5.5 *Generative adversarial network*

The generative adversarial network (GAN) was introduced by Goodfellow et al. [47,48]. Fig. 2.35 shows a general schematic diagram of GAN. It consists of two networks, a generative and a discriminative; the first one is used to generate new data instances, whereas the second to evaluate the data for authenticity. Each should be trained against a static adversary. The GAN use supervised learning in order to reconstruct the input layer.

This configuration can be used for fault detection and diagnosis in photovoltaic systems. The following code (**Listing 2.12**) is a general code for defining the generator and discriminator.

```
#Listing 2.12GANmodel
# importlibraries
import keras
from keras.layers import Dense, Conv2DTranspose, Reshape,
  BatchNormalization, Conv2D
from keras.models import Sequential
from keras.layers import Flatten, Dropout,LeakyReLU
# definefunctiongenerator
def gen_ima():
    gen=Sequential()
    gen.add(Dense(256*3*3, input_shape=(50,)))
    gen.add(LeakyReLU())
    gen.add(Reshape((3,3,256)))
    gen.add(Conv2DTranspose(128,kernel_size=2, strides=1,
      padding="same"))
    gen.add(LeakyReLU(alpha=0.1))
```

```
    gen.add(Conv2D(3,kernel_size=3, padding="same",
      activation='tanh'))
    return(gen)
mg=gen_ima()
mg.summary()
#definefunctiondiscriminator
def discr_ima():
    discr=Sequential()
    discr.add(Conv2D(64, kernel_size=2, padding="same", input_shape=
     (16,16,2)))
    discr.add(LeakyReLU(alpha=0.2))
    discr.add(Conv2D(256, kernel_size=3, strides=(2,2),
     padding="same"))
    discr.add(LeakyReLU(alpha=0.2))
    discr.add(Flatten())
    discr.add(Dropout(0.2))
    discr.add(Dense(1, activation='sigmoid'))
    discr.compile(loss='binary_crossentropy', optimizer='adam',
     metrics=['accuracy'])
    return(discr)
mo_dic=discr_ima()
mo_dic.summary()
```

2.6 Advantages and disadvantages of ML, EL, and DL algorithms

This section aims to present a brief comparison between ML, EL, and DL algorithms. Table 2.4 lists the main advantages and limits based on the literature survey [49–51].

As can be seen from the methods explored, each algorithm has advantages and limits as well the problem that can solve. In this situation, readers can choose the appropriate method according to the type of the problem, as well the data available.

2.7 Summary

In this chapter, an introduction to artificial intelligence techniques, including machine learning and deep learning algorithms is presented. Most machine learning, ensemble learning, and deep learning algorithms that could be used for solving photovoltaic problems are provided in detail. The examples presented have been developed using the Python programming language. They offer a clear idea to readers on how to apply and use these algorithms for further applications. Furthermore, they help readers understand and select easily the suitable algorithm for solving their problems. It is worth nothing that low accuracy in some results depends mainly on many factors, such as the size of the dataset, uncertainty measurement, data quality, and tuned parameters

Table 2.4 Summary on the advantages and disadvantages of machine learning, ensemble and deep learning methods.

Methods	Advantages	Limits	Type of problem
Machine learning			
Naïve Bayes	– Small amount of training data – Simple and easy to implement	– Assumes that all predictors (or features) are independent	Regression and classification
K-Nearest neighbor	– Very easy to implement – No assumptions – No training step (does not explicitly build any model) – Variety of distance metrics (Euclidean, Minkowski, Manhattan distance, etc.)	– Need feature scaling (standardization and normalization) – Does not work well with large dataset – Does not work well with high dimensions – Sensitive to noisy data, missing values. and outliers – Optimal number of neighbors (the value of *k*) – Imbalanced data problems	Regression and classification
Decision tree	– Does not require normalization and scaling of data – Very intuitive and easy to explain to technical teams as well as stakeholders – Missed data do not affect the process – Comprehensive rules are generated in a decision tree – Simple to implement and easy to understand – Usually robust to outliers and can handle them automatically	– A small change in the data can cause a large change in the model stability – Involves relatively higher time to train the model – Consumes more memory – Overfitting problem – Infected by noise	Regression and classification
Random forest	– Works well with both categorical and continuous variables	– Not a state-of-the-art algorithm	Regression and classification

	– Automatically handle missing values – No feature scaling required (as it uses a rule-based approach) – Handles non-linear parameters efficiently – Robust to outliers and can handle them automatically – Less impacted by noise	– Longer training period and complexity – Not easily interpretable	
Support vector machine	– More effective in high-dimensional spaces – Effective in cases where the number of dimensions is greater – Regularization capabilities – Very effective even with high-dimensional data – Reduces overfitting	– Not suitable for large datasets – Difficult to choose the appropriate kernel solution – Long training time for large datasets – Extensive memory requirement – Difficult to understand and interpret by human-beings unlike decision trees – Requires feature scaling – Does not performs well with overlapping classes	Regression and classification
Logistic regression	– Simple and easy to implement – Very efficient when the dataset has features that are linearly separable – Logistic regression is less prone to overfitting – Interpret and very efficient to train	– Difficult to capture complex relationships using logistic regression – Only important and relevant features should be used – Requires relatively a large dataset – Overfit in high-dimensional datasets	Classification

Continued

Table 2.4 Continued

Methods	Advantages	Limits	Type of problem
Hierarchical clustering	– Easy to implement – Not necessary to specify the number of clusters a priori – Good results visualization	– Not suitable for large datasets – Computations are expensive – Initial seeds have a strong impact on the final results – Involves a lot of arbitrary decisions – The order of the data has an impact on the final results – Very sensitive to outliers	Clustering
K-means	– Easy to implement – An instance can change cluster (move to another cluster) when the centroids are recomputed – Scales to large datasets – Easily adapts to new examples	– Difficult to predict the number of clusters (K-value) – Initial seeds have a strong impact on the final results – The order of the data has an impact on the final results – Scaling with number of dimensions	Clustering
Principal component analysis	– Removes correlated features – Improve algorithm performance – Reduce overfitting and improve visualization	– Data standardization – Independent variables become less interpretable – Loss information	Dimensionality reduction

Neural networks	– Ability to work with incomplete knowledge – Parallel processing capability – Storing information on the entire network – Ability to learn in complex situation – Learning methods are quite robust to noise in the training data – Can bear long training times depending on factors such as the number of weights in the network, the number of training examples considered – Give a better result when they gather all the data and information – Dynamic and robust	– Unexplained behavior of the network – Difficulty of showing the problem to the network – Require a large amount of data – Computationally expensive	Regression, control and classification
EL methods			
Bagging	– Reduces model overfitting – Performs well on high-dimensional data – Not affected by missing data – Ensemble makes the model more robust and stable – Bagging takes advantage of ensemble learning wherein multiple weak learners outperform a single strong learner – It helps reduce variance and thus helps us avoid overfitting – Bagging uses Bootstrap sampling method	– Final prediction based on the mean predictions – Bagging is not helpful in case of bias or underfitting in the data – There is loss of interpretability of the model – There can possibly be a problem of high bias if not modeled properly – Computationally expensive – Ensemble reduces the model interpret-ability	Prediction and classification
Boosting	– Good at handling the missing data – Successful techniques in solving the two-class classification problems – Net error is evaluated in each learning steps. It works good with interactions	– High memory consumption during tree construction – Little performance gain from large training data time consuming	Prediction and classification

Continued

Table 2.4 Continued

Methods	Advantages	Limits	Type of problem
	– Boosting technique helps when we are dealing with bias or under-fitting in the dataset – Multiple boosting techniques are available. For example: AdaBoost, LPBoost, XGBoost, GradientBoost, BrownBoost	– Often ignores overfitting or variance issues in the dataset – Increase the complexity of the classification – Time and computation can be a bit expensive	
DL methods			
DCNN	– Features are automatically extracted unlike machine learning – Automatically learned and robust to natural variations in the data – Massive parallel computations can be performed using GPUs – Can manage a huge amount of dataset – The deep configuration ids flexible and scalable	– A very large database is needed, with small database cannot relatively get good results – Expensive to train the model and consuming time – Require a supercomputer with GPU – Not so easy to find the best deep configuration	Prediction, classification and control

of the algorithms. Most functions used in this area are reported in Appendix B, and more details will be presented in the subsequent chapters for each photovoltaic application.

Scikit-Learn, Keras, Scikit-fuzzy, and TensorFlow platforms have been employed to implement the codes. Typical models can be built in a few lines of code using these functions. Scikit-Learn provides a standard interface to many of the most commonly used machine learning algorithms, and is the most popular and frequently used library for machine learning for Python. Other deep learning algorithms such as LSTM-CNN, Seq2Seq, and the GAN can be implemented based on some functions of LSTM, GRU, and others. Matlab in the MathWorks environment could also be used for solving problems in photovoltaic systems, as many machine learning and deep learning algorithms are currently available in different toolboxes, such as the deep learning toolbox and statistics and the machine learning toolbox. The Python programming language and Matlab/Simulink will be used also in the examples presented in other chapters of this book.

References

[1] A. Mellit, S.A. Kalogirou, Artificial intelligence techniques for photovoltaic applications: a review, Prog. Energy Combust. Sci. 34 (5) (2008) 574–632.

[2] Python, Python.org, 2013, Available: https://www.python.org/.

[3] https://scikit-learn.org/stable/.

[4] https://keras.io/.

[5] https://www.tensorflow.org/.

[6] https://pythonhosted.org/scikit-fuzzy/overview.html.

[7] S. van der Walt, et al., Scikit-image: image processing in python, Peer J 2 (2014) e453. Available: https://peerj.com/articles/453.

[8] https://docs.opencv.org/4.5.0/d6/d00/tutorial_py_root.html.

[9] https://nl.mathworks.com/products/statistics.html.

[10] https://nl.mathworks.com/products/deep-learning.html.

[11] S. Russell, P. Norvig, Artificial Intelligence: A Modern Approach, third ed., Prentice-Hall, USA, 2010.

[12] F. Chollet, Deep Learning With Python, Simon and Schuster, 2017.

[13] S. Arthur, Some studies in machine learning using the game of checkers, IBM J. 3 (1959) 211–229.

[14] E. Alpaydin, Machine Learning: The New AI, MIT Press, USA, 2016.

[15] A. Géron, Hands-On Machine Learning With Scikit-Learn, Keras, and TensorFlow: Concepts, Tools, and Techniques to Build Intelligent Systems, O'Reilly Media, 2019.

[16] F.E. Harrell Jr., Regression Modeling Strategies: With Applications to Linear Models, Logistic and Ordinal Regression, and Survival Analysis, Springer, 2015.

[17] H. Drucker, C.J. Burges, L. Kaufman, A. Smola, V. Vapnik, Support vector regression machines, in: Advances in Neural Information Processing Systems, vol. 9, MIT Press, 1997, pp. 155–161.

[18] T. Cover, P. Hart, Nearest neighbor pattern classification, IEEE Trans. Inf. Theory 13 (1) (1967) 21–27.

[19] N.S. Altman, An introduction to kernel and nearest-neighbor nonparametric regression, Am. Stat. 46 (3) (1992) 175–185.

[20] J.R. Quinlan, Induction of decision trees, Mach. Learn. 1 (1) (1986) 81–106.
[21] C.J. Matheus, L.A. Rendell, Constructive induction on decision trees, in: IJCAI, vol. 89, 1989, August, pp. 645–650.
[22] A.C. Müller, S. Guido, Introduction to Machine Learning With Python: A Guide for Data Scientists, O'Reilly Media, 2016.
[23] T.K. Ho, Random decision forests, in: Proceedings of 3rd International Conference on Document Analysis and Recognition, vol. 1, IEEE, 1995, August, pp. 278–282.
[24] L. Breiman, Random forests, Mach. Learn. 45 (1) (2001) 5–32.
[25] A. McCallum, K. Nigam, A comparison of event models for naive bayes text classification, in: AAAI-98 Workshop on Learning for Text Categorization, vol. 752, No. 1, 1998, July, pp. 41–48.
[26] E. Frank, L. Trigg, G. Holmes, I.H. Witten, Naive Bayes for regression, Mach. Learn. 41 (1) (2000) 5–25.
[27] W. McCulloch, W. Pitts, A logical calculus of ideas immanent in nervous activity, Bull. Math. Biophys. 5 (4) (1943) 115–133.
[28] S. Hayking, Neural Networks and Learning Machines, third ed., Prentice Hall, USA, 2009.
[29] J.B. MacQueen, Some methods for classification and analysis of multivariate observations, in: Proceedings of 5-th Berkeley Symposium on Mathematical Statistics and Probability, vol. 1, University of California Press, Berkeley, 1967, pp. 281–297.
[30] J.A. Hartigan, M.A. Wong, Algorithm AS 136: a k-means clustering algorithm, J. R. Stat. Soc. Ser. C Appl. Stat. 28 (1) (1979) 100–108.
[31] J.C. Dunn, A fuzzy relative of the ISODATA process and its use in detecting compact well-separated clusters, J. Cybern. 3 (1973) 32–57.
[32] J.C. Bezdek, Pattern Recognition With Fuzzy Objective Function Algorithms, Plenum Press, New York, 1981.
[33] L. Rokach, O. Maimon, Clustering methods, in: Data Mining and Knowledge Discovery Handbook, Springer, USA, 2005, pp. 321–352.
[34] F.W. Wilmink, H.T. Uytterschaut, Cluster analysis, history, theory and applications, in: Multivariate Statistical Methods in Physical Anthropology, Springer, Dordrecht, 1984, pp. 135–175.
[35] J.I.T. Principal, Component Analysis, second ed., Series: Springer Series in Statistics, Springer, NY, 2002, ISBN: 978-0-387-95442-4. XXIX, 487 p. 28 illus.
[36] K. Arulkumaran, M.P. Deisenroth, M. Brundage, A.A. Bharath, Deep reinforcement learning: a brief survey, IEEE Signal Process. Mag. 34 (6) (2017) 26–38.
[37] C.J.C.H. Watkins, P. Dayan, Q-learning, Mach. Learn. 8 (3–4) (1992) 279–292.
[38] G. Kyriakides, K.G. Margaritis, Hands-On Ensemble Learning With Python: Build Highly Optimized Ensemble Machine Learning Models Using Scikit-Learn and Keras, Packt Publishing, 2019.
[39] C. Zhang, Y. Ma (Eds.), Ensemble Machine Learning: Methods and Applications, Springer Science & Business Media, 2012.
[40] R.E. Schapire, The strength of weak learnability, Mach. Learn. 5 (2) (1990) 197–227.
[41] D.H. Wolpert, Stacked generalization, Neural Netw. 5 (2) (1992) 241–259.
[42] I. Goodfellow, Y. Bengio, A. Courville, Deep Learning, MIT Press, New York, USA, 2016.
[43] Y. LeCun, Y. Bengio, G. Hinton, Deep learning, Nature 521 (2015) 436–444.
[44] S. Hochreiter, J. Schmidhuber, Long short-term memory, Neural Comput. 9 (1997) 1735–1780.

[45] K. Cho, B. van Merrienboer, C. Gulcehre, D. Bahdanau, F. Bougares, H. Schwenk, Y. Bengio, Learning phrase representations using RNN encoder-decoder for statistical machine translation, 2014. arXiv:1406.1078.
[46] M. Schuster, K.K. Paliwal, Bidirectional recurrent neural networks, IEEE Trans. Signal Process. 45 (11) (1997) 2673–2681.
[47] I. Goodfellow, J. Pouget-Abadie, M. Mirza, B. Xu, D. Warde-Farley, S. Ozair, Y. Bengio, Generative adversarial nets, in: Advances in Neural Information Processing Systems, MIT Press, 2014, pp. 2672–2680.
[48] R. Atienza, Advanced Deep Learning With Keras: Apply Deep Learning Techniques, Autoencoders, GANs, Variational Autoencoders, Deep Reinforcement Learning, Policy Gradients, and More, Packt Publishing, 2018, pp. 1–37.
[49] http://theprofessionalspoint.blogspot.com.
[50] https://towardsdatascience.com/.
[51] https://developers.google.com/machine-learning/.

Forecasting of solar radiation using machine learning and deep learning algorithms

3

Chapter outline

3.1 Introduction

Solar radiation is considered the most important parameter in solar energy applications, including thermal and photovoltaic (PV) systems. Normal solar radiation is the most appropriate component for thermal applications, especially in concentrating solar systems, while global solar radiation component is indispensable for PV applications [1].

Forecasting of global solar irradiance on a horizontal or inclined surface plays a very important role in

(1) Integrating of PV systems into the grid.
(2) PV system performance analysis.
(3) Sizing and optimization of PV systems.
(4) Helping designers to predict the produced PV power for the management of the electricity grids.

In the literature, numerous methods were developed to forecast solar radiation components such as beam and global radiation [2]. These can be classified mainly into three categories:

Handbook of Artificial Intelligence Techniques in Photovoltaic Systems. https://doi.org/10.1016/B978-0-12-820641-6.00003-X

Copyright © 2022 Elsevier Ltd. All rights reserved.

(1) Numerical weather prediction (NWP) models [3].
(2) Stochastic methods such as autoregressive (AR), autoregressive moving average (ARMA), autoregressive integrated moving average (ARIMA), seasonal autoregressive integrated moving average (SARIMA), Markov chains [4], and other hybrid combinations [5].
(3) Statistic and machine learning (ML) methods [6,7].

In the first category, forecasting is mainly based on a satellite image (cloud image) analysis and physical modeling, while in thc second and third approaches it is based on in situ measured data. In the literature hybrid methods that combine solar irradiance measurement, NWP models, and satellite-derived data can also be found [8].

Based on the time horizon, the forecasting methods can be grouped into short-term [9] and long-term [10]. Short-term forecasting, from 1 hour up to a few days, can be used for PV plant management and for estimating the power produced by the PV plants. However, long-term forecasting ranges from a few weeks to a month and can be used to estimate the degradation rate and to analyze the performances of PV plants.

Over the past two decades, forecasting of solar radiation using ML algorithms has been widely investigated and proven their capability in this topic [6,7,11]. Recently, researchers are more and more attracted by the application of deep learning algorithms, which is a subfield of artificial intelligence techniques, to improve the accuracy of the forecasters [12,13].

This chapter focuses on the third approach, as many researchers showed that methods based on artificial intelligence techniques including machine, ensemble, and deep learning methods performed better than stochastic methods [14–16]. The first approach requires the availability of satellite images, which are not always accessible and available in many regions.

The chapter covers four applications: the first application is about one-step ahead forecasting of the daily global horizontal irradiation (GHI), using ML algorithms such as multilinear regression (MLR), decision tree (DT), random forest (RF), support vector machine (SVM), K-nearest neighbors (K-NN), multilayer perceptron (MLP), and recurrent neural networks (RNNs). The second application is related to one-step ahead forecasting of in-plane solar irradiance using deep learning neural networks such as long-short term memory (LSTM), bidirectional LSTM (BiLSTM), gated recurrent unit (GRU), conventional neural network (CNN), and hybrid configurations. The third application concerns the investigation of deep learning neural networks (DCNN) for multistep ahead forecasting of in-plane solar irradiance. In the last application, emphasis is given on the use of two neural networks for one-step ahead of daily and hourly GHI from meteorological parameters (e.g., wind speed, air temperature, relative humidity, and pressure).

All models presented have been written and developed using Python programming language. In all, 16 examples are described in order to help readers understand the application of ML and deep learning in this topic. Four databases have been used to develop the models. These databases were collected from different regions and for different timescales (daily, hourly, and 5 min).

3.2 Solar radiation forecasting based on data-driven methods

Accurate forecasting of global solar irradiation (G) is a crucial task for designing optimal PV systems. Solar irradiation data are usually presented as a time series, and time series prediction is an important scientific task. Generally, forecasting of solar irradiation based on data-driven methods can be achieved in three ways.

The first approach uses only historical G records (e.g., ground measurements):

$$\{G(t+1),G(t+2),...,G(t+k)\}=f\{G(t-n),G(t-n+1),...,G(t)\} \tag{3.1}$$

where $G(t)$ is the measured value at time t, $G(t-n)$, $G(t-n+1)$, ..., $G(t)$ are the previous time values, $G(t+k)$ is the forecasted value at time step k, and f is a functional dependency between past and future samples. $t\in\{1,..., n\}$, n is the size of the sequence.

The second approach relies upon forecasted meteorological parameters by numerical weather prediction models (NWP) such as air temperature (T_a), relative humidity (R_h), wind speed (W_s), wind direction (W_d), pressure (P_r), and cloud cover (C_c).

$$\{G(t+1),G(t+2)...G(t+k)\}=f\{T_a(t),R_h(t),W_s(t),P_r(t),...\} \tag{3.2}$$

The third approach combines the use of historical G data records with meteorological parameters:

$$\{G(t+1),G(t+2),...,G(t+k)\}=f\{G(t-n),G(t-n+1),...,G(t),R_h(t),W_s(t),P_r(t),...\} \tag{3.3}$$

These approaches could be used for one-step ($k=1$) and multistep ($k>1$) ahead forecasting.

3.3 Datasets description and preparation

The datasets used in this chapter for the development of the various models are described as follows:

3.3.1 Dataset 3.1

The first dataset consists of daily global horizontal radiation (GHI), shown schematically in Fig. 3.1, which is the most common solar measurement recorded by weather stations worldwide. The data were collected from the National Aeronautics and Space Administration (NASA) [17] from January 1, 2000 to June 30, 2020, i.e., for 20 years and 6 months, for the Hail region in Saudi Arabia. The statistical properties that

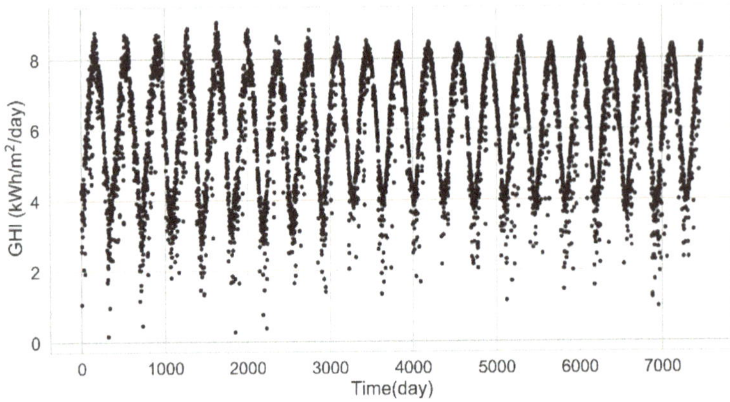

Fig. 3.1 Dataset 3.1: daily global horizontal irradiation (from January 2000 to June 2020).

Table 3.1 Mean, variance and deviation of the dataset 3.1.

Dataset 3.1	Variance	Mean	Standard deviation
Global horizontal irradiation (kWh/m^2/day)	2.861	5.988	1.691

include the mean, variance, and standard deviation of this dataset are shown in Table 3.1.

To understand the GHI distribution and seasonality of the dataset, the autocorrelation has been plotted, as shown in Fig. 3.2A. If the data are significantly autocorrelated, the previous values of the data named {lags: GHI($t-1$),...GHI($t-n$)} may be helpful in forecasting the current value{GHI(t)} (see Fig. 3.2A). Partial autocorrelation also conveys similar information as well as the pure correlation of the data and its lag as shown in Fig. 3.2B.

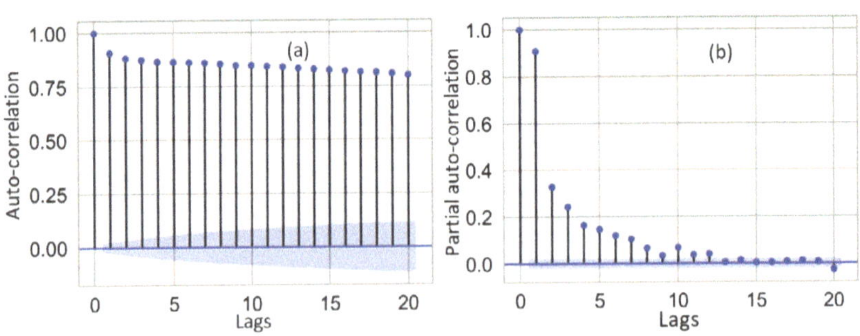

Fig. 3.2 Dataset 3.1: (A) autocorrelation and (B) partial autocorrelation.

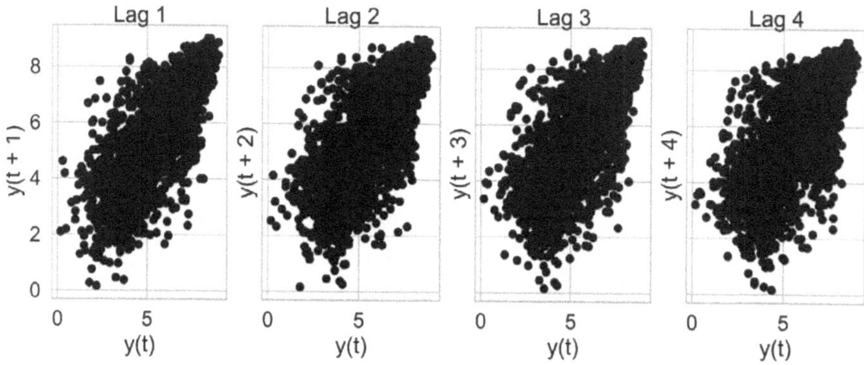

Fig. 3.3 Scatter curves of the GHI versus different lags ($y(t+1)$, $y(t+2)$, $y(t+3)$, and $y(t+4)$).

To check the autocorrelation of the data, the scatter curves between the present value of GHI at time t against a lag of itself are plotted as shown in Fig. 3.3, where $y(t)$ denotes the GHI(t). From Fig. 3.3, the autocorrelation between GHI(t), GHI($t+1$), GHI($t+2$), GHI($t+3$), and GHI($t+4$) is clearly observed.

3.3.2 Dataset 3.2

The second dataset consists of in-plane measured solar irradiance. These data were collected at Trieste University, Italy, from December 22, 2019 to February 18, 2021, with a timescale of 5 min. Fig. 3.4A shows the collected in-plane solar irradiance, and Fig. 3.4B depicts the autocorrelation between $G(t)$, $G(t+1)$, and $G(t+2)$, and as can be observed there is a good correlation between $G(t)$ and $G(t+1)$. The statistical properties including mean, variance, and standard deviation of this dataset are shown in Table 3.2.

3.3.3 Dataset 3.3

The third dataset regroups the daily global horizontal radiation with other meteorological parameters such as air temperature, wind speed, relative humidity, and pressure as shown in Fig. 3.5. The data were collected from January 1, 2000 to June 30, 2020 (20 years and 6 months) at the Hail region, Saudi Arabia. The statistical properties of the different parameters measured, i.e., W_s, T_a, R_h, and GHI, which include the mean, variance, and standard deviation of the dataset are listed in Table 3.3.

3.3.4 Dataset 3.4

The fourth dataset was collected at Medinah University, Saudi Arabia [17], during the period January 1 to December 31, 2011 with a timescale of 10 min. The data were then arranged in mean hourly values and consisted of hourly global

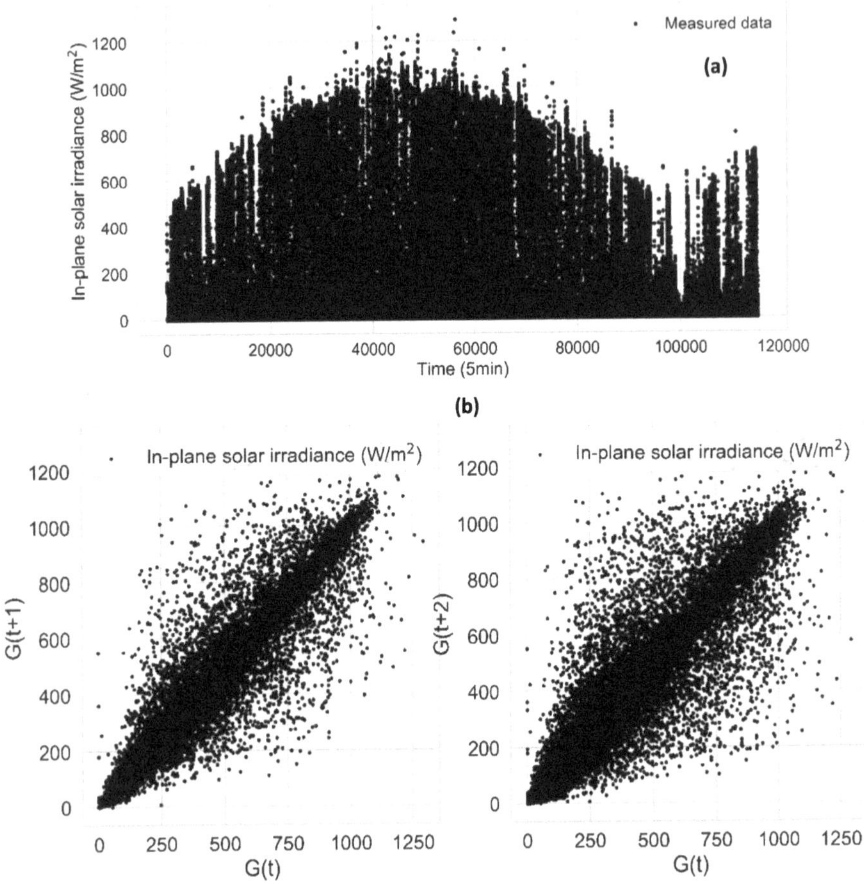

Fig. 3.4 Dataset 3.2: (A) measured in-plane solar irradiance at Trieste University, Italy and (B) autocorrelation between $G(t)$, $G(t+1)$, and $G(t+2)$.

Table 3.2 Mean, variance, and deviation for the dataset 3.2.

Dataset 3.2	Variance	Mean	Standard deviation
In plane solar irradiance (W/m^2)	73,231.460	270.61	164.736

horizontal irradiance with other meteorological parameters such as air temperature, wind speed, and relative humidity, as shown in Fig. 3.6. The statistical properties of the different measured parameters, i.e., W_s, T_a, R_h, and GHI, which include the mean, variance, and standard deviation of the dataset are reported in Table 3.4.

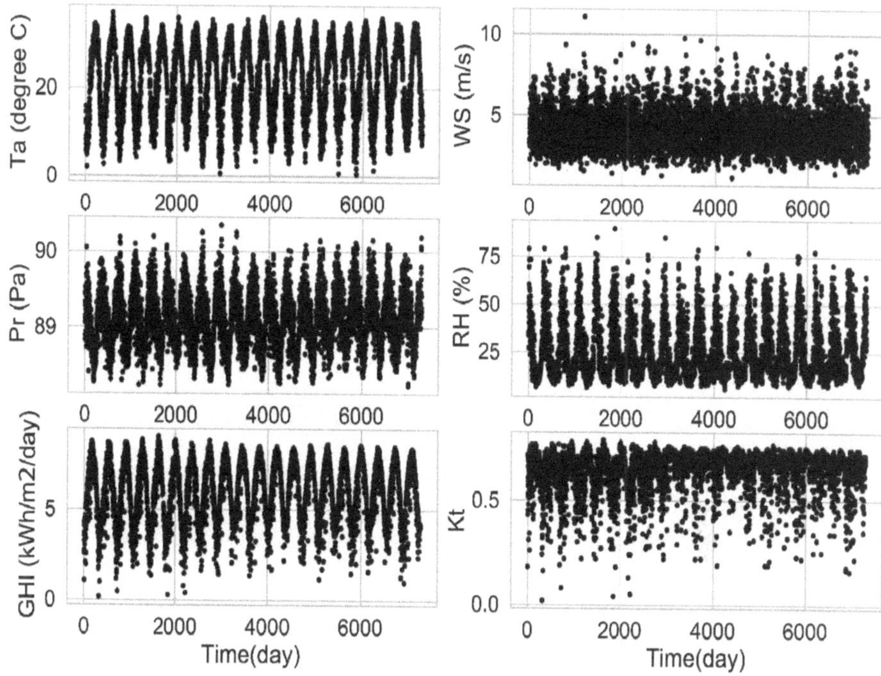

Fig. 3.5 Dataset 3.3: measured GHI, T_a, R_h, W_s, and calculated K_t at the Hail region, Saudi Arabia.

Table 3.3 Mean, variance, and deviation of the measured parameters (dataset 3.3).

Dataset 3.3	Variance	Mean	Standard deviation
GHI (kWh/m^2/day)	2.879	5.985	1.696
W_s (m/s)	1.341	4.173	1.158
T_a (°C)	71.374	22.068	8.448
R_h (%)	215.306	25.165	14.673
P_r (Pa)	0.131	89.102	0.363

3.4 Application of machine learning and deep learning algorithms for forecasting of solar radiation

The following code represents the main libraries and packages used in this chapter (**Listing 3.1**).

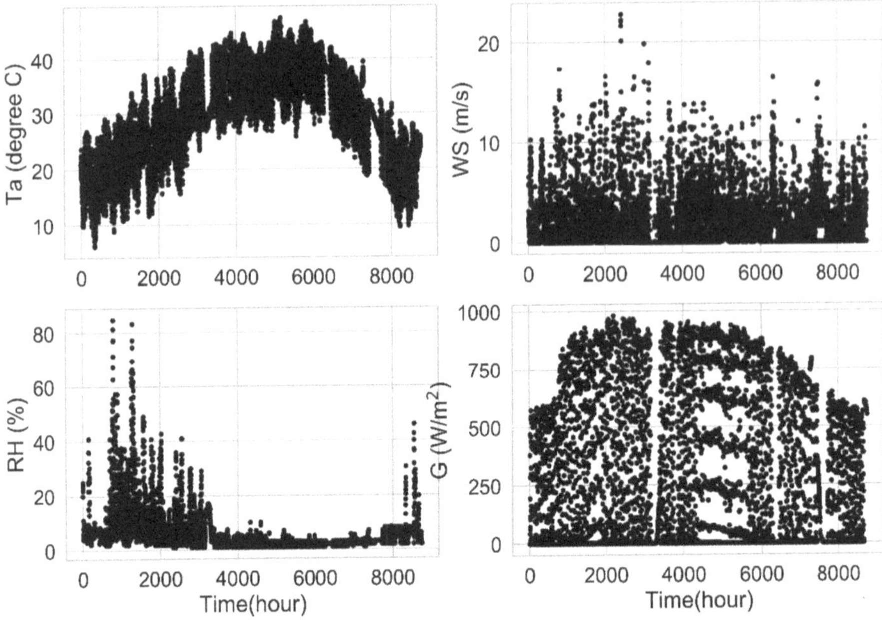

Fig. 3.6 Dataset 3.4: measured T_a, W_s, R_h, and G at Medinah, Saudi Arabia.

Table 3.4 Mean, variance, and deviation of the dataset 3.4.

Database 3.4	Variance	Mean	Standard deviation
T_a (°C)	65.280	29.132	8.079
W_s (m/s)	8.045	2.342	2.836
R_h (°C)	58.625	5.991	7.656
G (W/m^2)	90,397.21	219.969	300.661

```
# Listing 3.1 Libraries and packages
import numpy as np
import pandas as pd
from sklearn.model_selection import train_test_split
from sklearn.preprocessing import StandardScaler
from sklearn.neural_network import MLPRegressor
from sklearn.neighbors import KNeighborsClassifier
import matplotlib.pyplot as plt
import seaborn as seabornInstance
from sklearn import metrics
from pylab import rcParams
```

```
from keras.models import Sequential
from keras.layers.recurrent import SimpleRNN
from keras.layers.core import Dense, Activation
from keras.optimizers import SGD, adam
from scipy.stats import sem
from tensorflow.keras.layers import Bidirectional, Dropout,
  Activation, Dense, LSTM, Flatten
from tensorflow.python.keras.layers import CuDNNLSTM
from tensorflow.keras.models import Sequential
import xlsxwriter
%matplotlib inline
```

A function named {`static_test`} is created to evaluate the performance of the different algorithms investigated in this chapter, and includes the MSE, MAE, RMSE, *r*, wMAPE, MRPE, and SEM; the function is shown in the following code, **Listing 3.2**, (more details about these errors can be found in Appendix E).

```
#Listing 3.2 Performance evaluation function
def static_test(y_test,y_pred):
    mae=metrics.mean_absolute_error(y_test, y_pred)
    mse = metrics.mean_squared_error(y_test, y_pred)
    rmse=np.sqrt(mse)
    r2 = metrics.r2_score(y_test, y_pred)
    r=np.sqrt(r2)
    mape=np.mean(np.abs((y_test - y_pred) / y_test)) * 100
    y_pred=np.array(y_pred)
    m1=np.mean(y_pred)
    m2=np.mean(y_test)
    mrpe=((m2-m1)/m2)*100
    wmape = sum(np.abs(y_test - y_pred))/sum(y_test)
    semean=sem(y_pred)
    print("MRPE:",format(mrpe, ".3f"))
    print('MAE =', format(mae, ".3f"))
    print('RMSE =', format(rmse, ".3f")) #mse**(0.5)
    print('r=',format(r, ".3f"))
    print("MAPE:", format(mape, ".3f"))
    print("wMAPE:", format(wmape, ".3f"))
    print('Mean(Forecasted)=',format(m1, ".3f"))
    print('Mean(Measured)=',format(m2, ".3f"))
    print('SEM=',format(semean, ".3f"))
    return mae,rmse,r,mape,m1,m2,mrpe,wmape,semean

mrpe,mae,rmse,r,mape,wmape,m1,m2,semean=static_test(y_test,y_pred)
```

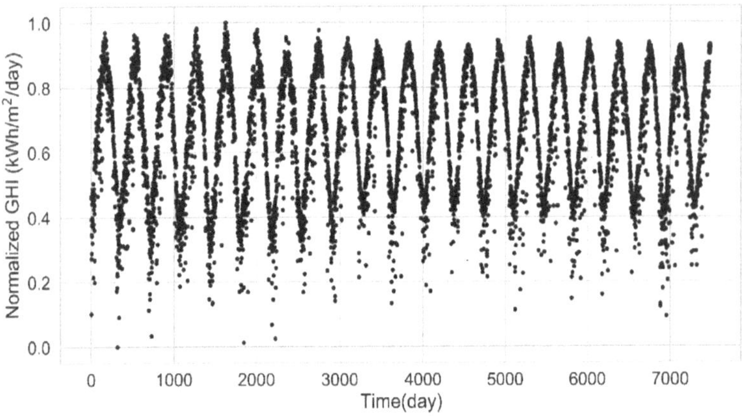

Fig. 3.7 Normalized GHI data (dataset 3.1).

3.4.1 Machine learning for forecasting of daily global horizontal irradiation (GHI)

In this section, it is shown how to forecast the next daily value of GHI (one-step ahead) based on the previous daily values of historical GHI. In the examples that follow, different ML algorithms (e.g., multiple linear regression, DT, SVM, K-NN, RF, MLP neural network, and RNNs) have been considered and described. The ML algorithms considered are described in Chapter 2. Dataset 3.1 is used in this section to develop the models. In these examples, 85% of data have been used to train and validate the models and the rest 15% are for testing. The dataset was normalized using the *MinMaxScaler* function. The following code permits normalizing and plotting the dataset shown in Fig. 3.7 (**Listing 3.3**).

```
#Listing 3.3 Data normalization, plotting curves and string figure.
scaler = MinMaxScaler()
GHIN = data.G.values.reshape(-1, 1)
Norm_G = scaler.fit_transform(GHIN)
plt.plot(Norm_G,'.',color='gray')
plt.xlabel('Time(day)')
plt.ylabel('Normalized GHI (kWh/m$o2$/day)')
plt.show()
plt.savefig('E:\\AI-Book\\Cha3_results\\Dataset1GHINorm.png',
dpi=600)
```

In the following examples, the number of input samples has been adopted during the experiment by evaluating the convergence of the mean squared error (MSE). In fact, there are no rules to set the optimal structure before conducting the experiments, for example, to determine the number of input samples. This number is generally obtained by a trial-and-error approach, varying the number of samples until obtaining good accuracy of MSE.

To split the dataset into training and testing sets, two functions have been implemented (`create_sequences and train_test_split_GHI`), as shown in **Listing 3.4**.

```
#Listing 3.4 dataset split
GHI=data.values
seq_len=15 #n-1
def create_sequences(data, seq_len):
    vect = []
    for index in range(len(data) - seq_len):
        vect.append(data[index: index + seq_len])
    return np.array(vect)
vect=create_sequences(GHI, seq_len)
def train_test_split_GHI(data,train_split):
    num_train = int(train_split * data.shape[0])
    x_train = data[:num_train, :-1]
    y_train = data[:num_train,-1]
    x_test = data[num_train:, :-1]
    y_test = data[num_train:,-1]
        return x_train, y_train, x_test, y_test
x_train, y_train, x_test, y_test = train_test_split_GHI(vect,
 train_split = 0.85)
```

The function (`train_test_split_GHI`) permits varying the number of inputs {GHI($t-1$), GHI($t-2$),...GHI(t)}, while the output is set to GHI($t+1$). For example, `seq_len=15` means that 15 samples are available {GHI(1), GHI(2), GHI(3),...,GHI (15)}, and the purpose is to forecast GHI(16) corresponding to day 16.

It is also possible to use the available function (`train_test_split`) in the *sklearn* library, by implementing the following command lines; however, in this case the data are selected randomly.

```
 from sklearn.model_selection import train_test_split
 x_train,x_test,y_train,y_test=train_test_split(x1,y1,
test_size =0.15, random_state=1)
```

Example 3.1: Multilinear regression (MLR)

In this example, a multiple linear regression algorithm is applied, named *{LinearRegression()}*. The main command lines are:

```
LR = LinearRegression()
LR.fit(x_train, y_train)
y_pred = LR.predict(x_test)
```

Example 3.2: Decision tree (DT)

Here, the DT algorithm is developed by using the *DecisionTreeRegressor()* function. The main command lines are:

```
DT = DecisionTreeRegressor(random_state = 0)
DT.fit(x_train, y_train)
y_pred = DT.predict(x_test)
```

Example 3.3: Random forest (RF)

In this example, we investigate the RF algorithm; the *RandomForestRegressor()* function is used. The main command lines are:

```
RF= RandomForestRegressor(n_estimators=100,max_features=1,
oob_score=True)
RF.fit(x_train, y_train)
y_pred = RF.predict(x_test)
```

Example 3.4: Support vector machine (SVM)

In this example, an SVM algorithm is developed by using *svm.SVR()* function. Various kernel functions like the *rbf, linear, ploy,* and *sigmoid* have been evaluated. The main command lines are:

```
SVM= svm.SVR(kernel ='rbf', C=100)
SVM.fit(x_train, y_train)
y_pred = SVM.predict(x_test)
```

Example 3.5: K-Nearest Neighbors (K-NN)

In this example, the K-NN algorithm, named {KNeighborsRegressor()} function, is used to develop the model. The important parameter here is the choice of K, and for this, the K is between 1 and 50 and it has been found that the optimal one is for K = 10. The code is given in the following box.

```
Knn = KNeighborsRegressor(n_neighbors=10,leaf_size=50,p=3)
Knn.fit(x_train, y_train)
y_pred = knn.predict(x_test)
```

Example 3.6: Multilayer perceptron (MLP) neural networks

MLP neural network function, named {*MLPRegressor()*}, is used to develop the forecaster. Various activation functions have been examined. The main command lines are:

```
MLP=MLPRegressor(hidden_layer_sizes=(7,5),activation="logistic" ,
random_state=1, max_iter=1000)
MLP.fit(x_train, y_train)
y_pred=MLP.predict(x_test)
```

Example 3.7: Recurrent neural networks (RNNs)

In this example, a simple recurrent neural newel (Elman) is used based on the implemented function {*SimpleRNN()*}. Various configurations have been tested by modifying the number of nodes in the hidden layer and the type of the activation functions. The main command lines are:

```
RNN=Sequential()
RNN.add(SimpleRNN ( units=12, activation="tanh", input_shape=
(seq_len-1 , 1)))
RNN.add( Dense ( units=1, activation='linear'))
sgd=SGD (lr=0.0001 , momentum=0.95 ,nesterov=True )
RNN.compile ( loss='mean_squared_error',optimizer=sgd )
RNN.fit( x_train , y_train , batch_size=10 ,epochs=100)
```

For example, Fig. 3.8A shows a comparison between the measured and forecasted GHI by a DT algorithm. Fig. 3.8B presents the DT regression (blue color) during the training process. Fig. 3.8C depicts the corresponding scatter curves. From these curves, it can be seen that there is an acceptable correlation between measured and forecasted GHI for 1 day ahead. Calculated error metrics for the different ML algorithms investigated are summarized in Table 3.5.

As can be seen from Table 3.5, the correlation coefficient r ranges between 91.1% and 94.7%, and the lower value of RMSE is obtained by the RF algorithm. For all modes investigated, the MAE does not exceed 0.5 kWh/m^2/day, and SEM is the same for all models. However, other error metrics such as the mean and MAPE are relatively similar.

It should be pointed that the ML algorithms investigated are simple from the structure point of view and can be easily implemented. However, the accuracy needs to be improved by tuning the hyperparameters carefully. Hyperparameters are parameters whose values control the learning process and determine the model parameter values that a learning algorithm eventually learns (e.g., n_estimator, max_dpth, kernel, activation function,batch size, etc.)

3.4.2 Application of deep learning algorithms for in-plane solar irradiance forecasting

In this section, deep learning algorithms such as LSTM, stacked LSTM, GRU, BiLSTM, and CNN, and a hybrid CNN-LSTM have been developed and evaluated. For these models, the second dataset is used; it consists of 115,045 samples measured at a 5-min timescale. A total of 85% (97,784 samples) of the data are used to train the models and the rest 15% (17,257 samples) were used for testing the models. The dataset was normalized using the *MinMaxScaler* function. Fig. 3.9 shows the normalized in-plane solar irradiance values.

The following functions permit splitting the database into training and testing portions (**Listing 3.5**).

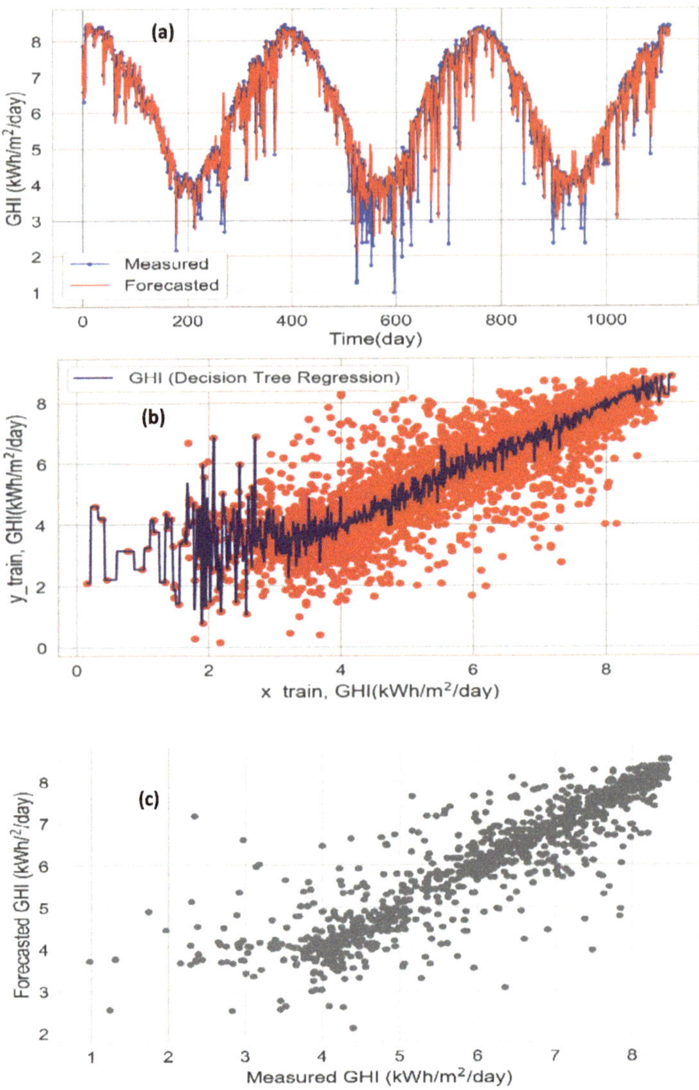

Fig. 3.8 (A) Measured versus forecasted GHI, (B) DT regression during the training process, and (C) scattered curve between measured and forecasted.

Table 3.5 Error metrics of the ML algorithms investigated for daily forecasting of GHI.

Model parameters	Mean (measured) ($kWh/m^2/day$)	Mean (forecasted) ($kWh/m^2/day$)	*r* (%)	RMSE ($kWh/m^2/day$)	MAPE (%)	wMAPE (%)	SEM ($kWh/m^2/day$)	MAE ($kWh/m^2/day$)
MLR								
copy_X=True, fit_intercept=True, n_jobs=None, normalize=False NI=3	6.092	6.088	94.1	0.609	8.33	0.066	0.045	0.384
DT								
criterion='mse' splitter='best' max_depth=10 NI=1	6.092	6.081	91.4	0.701	9.250	0.071	0.046	0.435
RF								
n_estimators=100, max_features=1 NI=11	6.092	6.065	94.4	0.589	8.415	0.062	0.045	0.381
SVM								
kernel='rbf', C=100 NI=6	6.092	6.179	94.6	0.612	8.139	0.058	0.045	0.354
K-NN								
K=10, leaf_size=50,p=3	6.092	6.046	94.7	0.602	8.521	0.065	0.046	0.395
MLP								
hidden_layer_sizes=(7,5), activation="logistic",	6.092	6.081	94.3	0.595	8.239	0.062	0.045	0.377
RNN								
units=7, activation="tanh", optimizer=sgd, los=mse	***6.092***	6.084	94.5	0.631	8.217	0.154	0.045	0.381

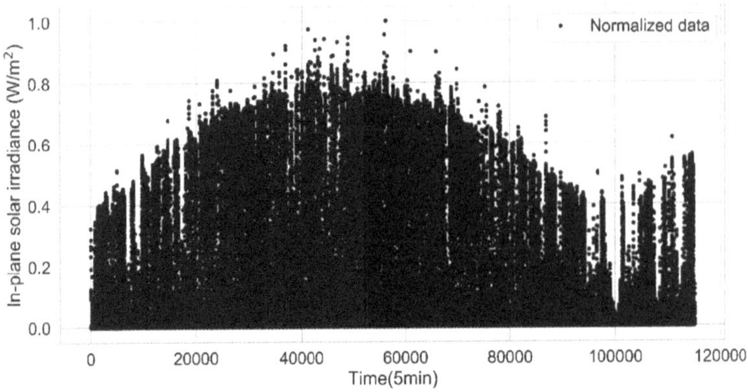

Fig. 3.9 Normalized in-plane solar irradiance (115,045 samples).

```
#Listing 3.5: Split function
Input_Samples =4
def creat_sequences(data, seq_len):
    vect = []
    for index in range(len(data) - seq_len):
        vect.append(data[index: index + seq_len])
    return np.array(vect)
def datasplit(data_raw, seq_len, train_split):
    data = creat_sequences (data_raw, seq_len)
    num_train = int(train_split * data.shape[0])
    X_train = data[:num_train, :-1, :]
    y_train = data[:num_train, -1, :]
    X_test = data[num_train:, :-1, :]
    y_test = data[num_train:, -1, :]
    return X_train, y_train, X_test, y_test
X_train, y_train, X_test, y_test = datasplit (Norm_data,
Input_Samples, train_split = 0.85)
```

In the next examples, a sequential API (application programming interface) is used as this is the easiest model to build and run in Keras. A sequential model allows the user to create the models, layer by layer, in a step-by-step manner. Initially, this *model* variable is just an empty neural network model until more layers are added sequentially starting from the beginning; thus, it allows someone to create a neural network architecture by adding layers into the model, as indicated by the codes shown in the box below. The Adam optimizer and the rectified linear unit (ReLU) are used.

Example 3.8: LSTM model

The main line commands for developing an LSTM model are summarized as follows (**Listing 3.6**).

```
#listing 3.6: LSTM model
model = Sequential ()
model.add(LSTM(100, activation='relu',input_shape=(X_train.shape
[1], 1)))
model.add(Dense(1))
model.summary()
model.compile(loss='mean_squared_error', optimizer='adam')
Performance= model.fit( X_train, y_train, epochs=50,
batch_size=128, shuffle=False,validation_split=0.15)
```

Example 3.9: GRU model

Similar to the LSTM code (**Listing 3.6**), the LSTM is replaced in the second line by GRU.

```
# GRU model
model.add(GRU(100, activation='relu',input_shape=(X_train.shape
[1], 1)))
```

Example 3.10: Stacked LSTM model

Similar to the LSTM code (**Listing 3.6**), more layers are added, as shown in the following code.

```
#StackedLSTMmodel
model=Sequential()
model.add(LSTM(100, activation='relu', return_sequences=True,
input_shape=(X_train.shape[1], 1)))
model.add(LSTM(50, activation='relu'))
model.add(Dense(1))
```

Example 3.11: Bidirectional LSTM model

Similar to LSTM code (**Listing 3.6**), the LSTM is replaced only by the BiLSTM, using the following command line.

```
model.add(Bidirectional(LSTM(units=100, activation='relu',
input_shape=(X_train.shape[1], X_train.shape[2]))))
```

Example 3.12: CNN model

The CNN model consists of a Conv1d layer, MaxPooling layer, Flatten layer, Dense (100) layer, and the last Dense layer with one node. The corresponding code (**Listing 3.7**) is:

```
#listing 3.7: CNN model
cnn = Sequential()
cnn.add(Conv1D(filters=64, kernel_size=2, activation='relu',
input_shape = (X_train.shape[1], 1)))
cnn.add(MaxPooling1D(pool_size=2))
cnn.add(Flatten())
cnn.add(Dense(100, activation='relu'))
cnn.add(Dropout(0.2))
cnn.add(Dense(1))
cnn.summary()
```

Example 3.13: Hybrid CNN-LSTM model

The hybrid CNN-LSTM comprises a CNN network with an LSTM of 50 cells:

```
#listing 3.8: CNN_LSTM model
model_cnn_lstm = Sequential()
model_cnn_lstm.add(TimeDistributed(Conv1D(filters=64, ker-
  nel_size=2, activation='relu'), input_shape=(None, X_train_sub.
  shape
[2], X_train_sub.shape[3])))
model_cnn_lstm.add(TimeDistributed(MaxPooling1D(pool_size=2)))
model_cnn_lstm.add(TimeDistributed(Flatten()))
model_cnn_lstm.add(LSTM(50, activation='relu'))
model_cnn_lstm.add(Dense(1))
model_cnn_lstm.compile(loss='mse', optimizer='adam')
model_cnn_lstm.summary()
```

As an example, Fig. 3.10A shows the loss function during the training and the testing process. As can be observed after 50 epochs, the cost function converges to 0.0012. To check the correlation between measured and forecasted data, the scatter curve is plotted as shown in Fig. 3.10B. Fig. 3.10C shows a superposition of measured and forecasted in-plane solar irradiance for a few samples. As can be seen from these figures, a very good correlation is obtained. To evaluate the accuracy of different algorithms investigated, various error metrics like the MAE, MRPE, RMSE, and r are calculated and reported in Table 3.6. The structure of each forecaster based on deep learning algorithm is reported in Table 3.6.

With reference to Table 3.6, a correlation coefficient equal to $r=98.7\%$ is practically the same for all the investigated algorithms (LSTM, GRU, BiLSTM, and CNN), except for the case of CNN-LSTM, which is equal to 90.8%. A lower value of MRPE is obtained for the stacked LSTM model, which is the most accurate model compared to the other models investigated. CNN and GRU converge very quickly to the fixed error. The CNN-LSTM model has the most complex configuration with a low accuracy based on the measured errors. Generally, the CNN and CNN-LSTM models are not recommended for this application (due to their relatively complex structures); however, their results could be improved by varying the different hyperparameters like the number of filters, number of layers, batch size, and kernel.

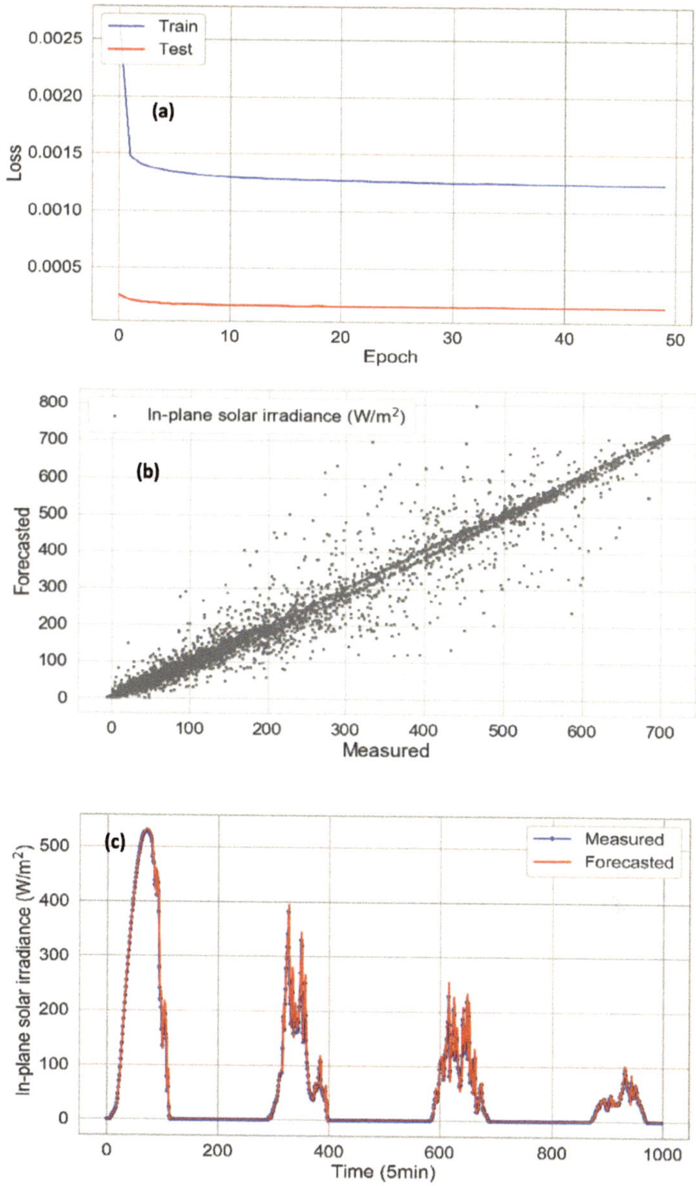

Fig. 3.10 (A) Loss function evolution, (B) scatter curve, and (C) superposition between measured and forecasted solar irradiance by a stacked LSTM model.

Table 3.6 Error metrics for the investigated deep leaning algorithms for one-step ahead forecasting of in-plane solar irradiance.

Models and structure Epoch = 50 Batch size = 128 Optimizer = Adam	**MRPE (%)**	**MAE (W/m^2)**	**RMSE (W/m^2)**	***r* (%)**	**Convergence time (min)**
LSTM: 3× 50×1, NI=3	−2.444	0.005	0.017	98.7	9
GRU: 3× 50× 1, NI=3	− 3.753	0.006	0.017	98.7	8
Stacked LSTM: 3× 100× 50× 1, NI=3	− 1.590	0.005	0.017	98.7	25
Bidirectional LSTM: 3× 100× 1, NI=3	− 6.024	0.007	0.017	98.7	16
CNN: NI=3, filters=64, kernel_size=2,	− 2.678	0.006	0.018	98.6	8
CNN-LSTM: NI=6, filters=64, kernel_size=2, LSTM=50	−43.98	0.033	0.044	90.8	35

3.4.3 Multistep ahead forecasting of 5 min in-plane solar irradiance using stacked LSTM

Multistep solar radiation forecasting is one of the most challenging issues in this topic. This section aims to apply a deep learning algorithm like the stacked LSTM, for multistep ahead forecasting (2, 3, 4, and 5 samples) of solar irradiance. To do this, the same dataset 3.2 is used.

Example 3.14: Stacked LSTM model

The following function allows to prepare the dataset (input/output) for training the network (**Listing 3.9**).

```
#listing 3.9: data preparation  function
n,m=6,4
def data_preparation(n,m,data_train):
    x_train, y_train=[], []
    for i in range(n,len(data_train)-n):
        x_train.append(data_train[i-n:i])
        y_train.append(data_train[i:i+n-m])
    return x_train, y_train

x_train,y_train=data_preparation(n,m,data_train)
```

where n is the number of input samples and $n\text{-}m$ is the number of forecasting steps. For example, $n=6$ and $m=4$ means we have six inputs and two outputs. The main code is given in **Listing 3.10**.

```
#listing 3.10: Stacked LSTM model
model= Sequential()
model.add(LSTM(units=50,activation='relu',return_sequences=True,
input_shape=(n,1)))
model.add(Dropout(rate=0.2))
model.add(LSTM(units=30, activation='relu',return_sequences=True))
model.add(Dropout(rate=0.2))
model.add(LSTM(10, activation='relu'))
model.add(Dense(n-m))
model.compile(loss='mean_squared_error', optimizer='adam')
modelfit = model.fit( x_train, y_train, epochs=5, batch_size=64,
verbose=1)
```

The correlation between the measured and forecasted solar irradiance using stacked LSTM is shown in Fig. 3.11A. Fig. 3.11B depicts comparison curves between the measured and forecasted data.

As reported in Table 3.7, when increasing the forecasting steps to 2, 3, and 4 samples, the accuracy is decreased, and the training process takes much more time to converge. Increasing the number of input and output increases significantly the training time. The developed model was executed only for 30 epochs, and by increasing the epoch could contribute to improve the forecasting accuracy. Dropout layers are added to prevent the overfitting problem.

3.4.4 Application of neural networks for solar radiation forecasting from meteorological parameters

In this section, a *MLPregressor* function and a MLP in a sequential model are applied to forecast the next daily and hourly global solar irradiance from meteorological parameters, using two datasets (dataset 3.3 and dataset 3.4).

Example 3.15: MLPregressor model

In this case, the third dataset is used to forecast the GHI from other meteorological parameters such as W_s, T_a, R_h, P_r, and the calculated clearness index K_t. The models have been created by combining different parameters; thus, the variation of the different inputs investigated in the various models are:

- Model 1: K_t, GHI, T_a, W_s, R_h, and P_r
- Model 2: GHI, T_a, W_s, R_h, and P_r
- Model 3: GHI, T_a, W_s, and R_h
- Model 4: GHI, T_a, and R_h
- Model 5 : K_t, GHI, and T_a

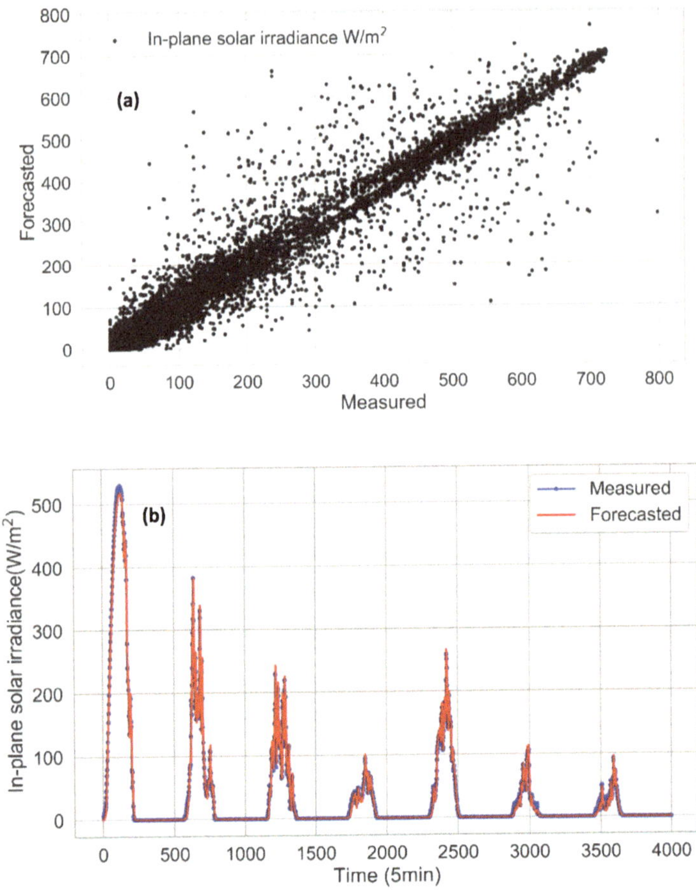

Fig. 3.11 (A) Scatter curve and (B) measured versus forecasted in-plane solar irradiance by the stacked LSTM model (two steps ahead).

Table 3.7 Error metrics of the investigated stacked LSTM algorithm for multi-steps ahead forecasting of in-plane solar irradiance.

Stacked LSTM structures	MRPE (%)	MAE (W/m^2)	RMSE (W/m^2)	*r* (%)	Time (min)
Structure LSTM: $6 \times 50 \times 30 \times 10 \times 2$, Epoch = 30 Batch size = 64, Optimizer = Adam	1.666	7.256	27.733	97.9	40
Structure LSTM: $9 \times 50 \times 30 \times 10 \times 3$, Epoch = 30 Batch size = 64, Optimizer = Adam	2.345	7.855	26.456	96.2	55
Structure LSTM: $8 \times 50 \times 30 \times 10 \times 4$, Epoch = 30 Batch size = 64, Optimizer = Adam	4.187	8.987	25.733	95.1	115

The data have been divided using the following code (**Listing 3.11**).

```
#listing 3.11: Dataset splitting function
def train_test_split1(x1,y1,train_split):
    y1=np.array(y), x1=np.array(X)
    num_train = int(train_split * data.shape[0])
    X_train = x1[:num_train, :]
    y_train = y1[:num_train]
    X_test = x1[num_train:, :]
    y_test = y1[num_train:]
    return X_train, y_train, X_test, y_test
X_train, y_train, X_test, y_test = train_test_split1(X,y,
train_split = 0.85)
```

The *MLPregressor* function is used to develop the forecaster as given by the command lines below:

```
 model=MLPRegressor(hidden_layer_sizes=(17,15,10),
activation="logistic" ,random_state=1, max_iter=3000)
 model.fit(X_train, y_train)
 y_pred=model.predict(X_test)
```

The error metrics calculated are listed in Table 3.8, and Fig. 3.12 shows a plot between measured and one-step ahead forecasted GHI.

As can be seen, the best accuracy is obtained by model 5 (r = 94.2% and MAE = 0.39 kWh/m^2 per day). It has been observed that the most relevant input parameters of the model are the clearness index, the previous value of G, and air temperature. Other models investigated do not perform as well, as the r ranges between 91.1% and 94.2%. It can be also observed that on cloudy days all models do not provide good results, which remain a challenging issue in this area as shown by the right side of

Table 3.8 Error metrics for the investigated MLP *regressor* for one step ahead forecasting of daily GHI.

Model parameters Epochs = 3000, Activation function: logistic, structure: input × 17 × 15 × 10	***r* (%)**	**RMSE (KWh/ m^2/day)**	**MAPE (%)**	**wMAPE (KWh/ m^2/day)**	**SEM (KWh/ m^2/ day)**	**MAE (KWh/ m^2/ day)**
MLP 1 (input = 6)	92.1	0.667	8.823	0.067	0.045	0.403
MLP 2 (input = 5)	91.9	0.674	8.956	0.068	0.045	0.412
MLP 3 (input = 4)	91.5	0.686	9.102	0.077	0.044	0.428
MLP 4 (input = 3)	91.1	0.699	9.194	0.077	0.046	0.423
MLP 5 (input = 3)	94.2	0.602	8.234	0.045	0.042	0.390

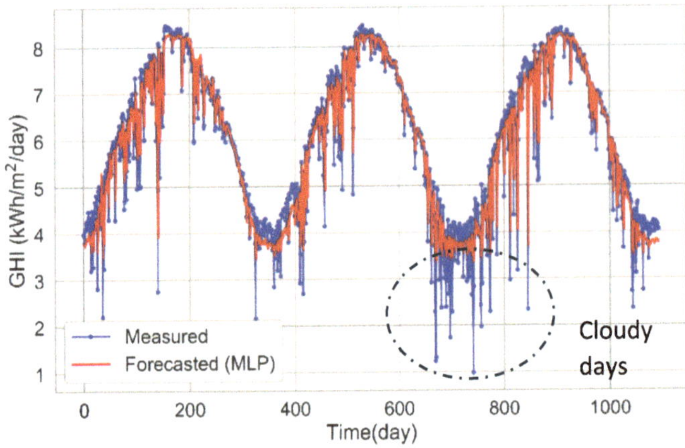

Fig. 3.12 Plot of the measured versus forecasted 1 day ahead of GHI using model 5.

Fig. 3.12. However, a few attempts have been made to solve this shortcoming by clustering and classification days [18,19].

Example 3.16: MLP-based sequential model

A set of 7446 (85%) of the whole dataset 3.4 was used to train the MLP and a set of 1313 for testing the model. In this case, a sequential mode-based MLP is used, indicated by the code in the box below. The fourth database is used in this model, which represents hourly data. The models have been created by combining different input parameters; thus, the variations of the different inputs investigated in the various models are:

- Model 1: G, T_a, W_s, and R_h
- Model 2: G, T_a, and R_h
- Model 3: G and T_a
- Model 4: G, T_a, and W_s

The same code was used to divide the dataset. The corresponding code of the model consists of three hidden layers with a dropout layer between each layer as shown in **Listing 3.12**.

```
#Listing 3.12: MLP sequential model
MLP_model = Sequential()
MLP_model.add(Dense(50, activation = 'tanh',input_dim = 2))
MLP_model.add(Dropout(0.2))
MLP_model.add(Dense(30, activation = 'tanh'))
MLP_model.add(Dropout(0.2))
MLP_model.add(Dense(10, activation = 'tanh'))
MLP_model.add(Dropout(0.2))
MLP_model.add(Dense(1, activation='linear'))
MLP_model.summary()
MLP_model.compile(loss='mean_squared_error', optimizer='adam')
history = MLP_model.fit( X_train, y_train, epochs=100,
validation_split=0.15, batch_size=64, verbose=1)
y_pred = MLP_model.predict(X_test)
```

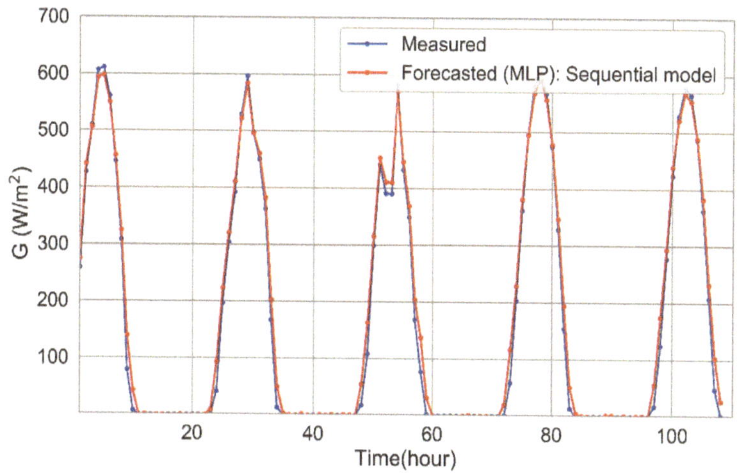

Fig. 3.13 Measured versus forecasted one-step ahead in-plane solar irradiance by model 3.

Table 3.9 Error metrics for the investigated MLP based sequential model for one-step ahead forecasting of in-plane solar irradiance from meteorological parameters.

Model parameters Epochs = 3000, Activation function Tanh, batch size = 64, structure: input × 50 × 30 × 10 × 1	***r* (%)**	**RMSE (W/m^2)**	**MAPE (%)**	**wMAPE (W/m^2)**	**SEM (W/m^2)**	**MAE (W/m^2)**
MLP 1 (input = 4)	93.1	0.085	6.643	0.334	0.0059	0.0654
MLP 2 (input = 3)	92.9	0.087	7.156	0.349	0.0065	0.0702
MLP 3 (input = 2)	95.1	0.079	4.102	0.281	0.0057	0.0521
MLP 4 (input = 3)	92.0	0.089	7.194	0.346	0.0065	0.0071

Fig. 3.13 shows an example of the measured versus forecasted G by the MLP model developed for a sequence of 110 h. As can be seen from Fig. 3.13, a good agreement is obtained, and from Table 3.9 it can be seen that the r ranges between 92.0% and 95.1%. With reference to Table 3.9, model 3 performs better than other models investigated (i.e., other input configurations). Thus, it can be concluded that the most relevant parameters are G and T_a.

3.5 Summary

This chapter is mainly focused on the application of ML and deep learning for solar radiation forecasting. A total of 16 examples, including codes with explanations, have been presented. Four databases with different timescales (daily, hourly, and 5 min) measured in three regions have also been used for this purpose. Among the various machine and deep learning algorithms, supervised neural network-based models are the most studied in this field.

Hybrid configurations such as CNN-LSTM and stacked LSTM require much time to train, particularly in the case of multistep forecasting, e.g., about 2 h in the case of three steps ahead. The algorithms developed have been implemented under the Intel Core i5-6300U CPU 2.40–3 GHz processor and 8 GB RAM. In this situation, a supercomputer is recommended to reduce the running time.

Supervised ML algorithms are the most used in this topic, while very limited works are related to the application of unsupervised ML algorithms. For example, the participial component analysis helps for finding the most relevant input parameters; *K*-means algorithm also could be used for day clustering (sunny, cloudy, etc.). Hybrid configurations (such as BiLSTM-CNN, BiGRU-CNN, etc.), encoder-decoder (Seq2Seq) are not investigated in this chapter, but they could be used as well.

The forecasting accuracy depends on many factors such as data quality, pretreatment of data (e.g., remove noises, missed data, normalization, etc.), size of the dataset, completeness of the dataset (i.e., important parameters such as cloud index are not always measured), efficiency of the algorithms, and uncertainty measurements (e.g., pyranometer accuracy).

Furthermore, combining satellite images with ML or deep learning and NWP models improve the forecasting accuracy. In this context, few attempts were made; however, so far multistep forecasting of solar irradiance, particularly for cloudy circumstances, remains a challenging issue.

References

[1] A. Mellit, S.A. Kalogirou, L. Hontoria, S. Shaari, Artificial intelligence techniques for sizing photovoltaic systems: a review, Renew. Sust. Energ. Rev. 13 (2) (2009) 406–419.

[2] M. Diagne, M. David, P. Lauret, J. Boland, N. Schmutz, Review of solar irradiance forecasting methods and a proposition for small-scale insular grids, Renew. Sust. Energ. Rev. 27 (2013) 65–76.

[3] R. Perez, E. Lorenz, S. Pelland, M. Beauharnois, G. Van Knowe, K. Hemker Jr., D. Heinemann, J. Remund, S.C. Müller, W. Traunmüller, G. Steinmauer, Comparison of numerical weather prediction solar irradiance forecasts in the US, Canada and Europe, Sol. Energy 94 (2013) 305–326.

[4] F.O. Hocaoglu, F. Serttas, A novel hybrid (Mycielski-Markov) model for hourly solar radiation forecasting, Renew. Energy 108 (2017) 635–643.

[5] W. Ji, K.C. Chee, Prediction of hourly solar radiation using a novel hybrid model of ARMA and TDNN, Sol. Energy 85 (5) (2011) 808–817.

[6] C. Voyant, G. Notton, S. Kalogirou, M.L. Nivet, C. Paoli, F. Motte, A. Fouilloy, Machine learning methods for solar radiation forecasting: a review, Renew. Energy 105 (2017) 569–582.

[7] A. Mellit, Artificial Intelligence technique for modelling and forecasting of solar radiation data: a review, Int. J. Artif. Intell. Soft Comput. 1 (1) (2008) 52–76.

[8] L.M. Aguiar, B. Pereira, P. Lauret, F. Díaz, M. David, Combining solar irradiance measurements, satellite-derived data and a numerical weather prediction model to improve intra-day solar forecasting, Renew. Energy 97 (2016) 599–610.

[9] E. Akarslan, F.O. Hocaoglu, R. Edizkan, Novel short term solar irradiance forecasting models, Renew. Energy 123 (2018) 58–66.
[10] F.F. Li, S.Y. Wang, J.H. Wei, Long term rolling prediction model for solar radiation combining empirical mode decomposition (EMD) and artificial neural network (ANN) techniques, J. Renew. Sustain. Energy 10 (1) (2018) 013704.
[11] G.M. Yagli, D. Yang, D. Srinivasan, Automatic hourly solar forecasting using machine learning models, Renew. Sust. Energ. Rev. 105 (2019) 487–498.
[12] H. Jiang, N. Lu, J. Qin, W. Tang, L. Yao, A deep learning algorithm to estimate hourly global solar radiation from geostationary satellite data, Renew. Sust. Energ. Rev. 114 (2019) 109327.
[13] S. Boubaker, M. Benghanem, A. Mellit, A. Lefza, A. Kahouli, L. Kolsi, Deep neural networks for predicting solar radiation at hail region, Saudi Arabia, IEEE Access (2021), https://doi.org/10.1109/ACCESS.2021.3062205.
[14] A. Mellit, A.M. Pavan, A 24-h forecast of solar irradiance using artificial neural network: Application for performance prediction of a grid-connected PV plant at Trieste, Italy, Sol. Energy 84 (5) (2010) 807–821.
[15] X. Qing, Y. Niu, Hourly day-ahead solar irradiance prediction using weather forecasts by LSTM, Energy 148 (2018) 461–468.
[16] G. Narvaez, L.F. Giraldo, M. Bressan, A. Pantoja, Machine learning for site-adaptation and solar radiation forecasting, Renew. Energy 167 (2021) 333–342.
[17] *https://power.larc.nasa.gov/data-access-viewer/*.
[18] M. Ghayekhloo, M. Ghofrani, M.B. Menhaj, R. Azimi, A novel clustering approach for short-term solar radiation forecasting, Sol. Energy 122 (2015) 1371–1383.
[19] P.F. Jiménez-Pérez, L. Mora-López, Modeling and forecasting hourly global solar radiation using clustering and classification techniques, Sol. Energy 135 (2016) 682–691.

Forecasting of photovoltaic output power using machine learning and deep learning algorithms

4

Chapter outline

4.1 Introduction

Millions of photovoltaic (PV) systems with different sizes (domestic, utility-scale, and commercial PV plans) have been installed worldwide [1]. The power produced by the PV plants depends mainly on two factors: meteorological, such as solar irradiance, surface air temperature, and surface wind velocity [2]; and environmental, such as tropical, arid, temperate, cold, and polar [3]; snow in the northern and sand in the southern regions are the most influential parameters. Other aspects such as failures in PV modules including degradation, aging, dust accumulation and hotspots [4,5] also influence the system performance.

Handbook of Artificial Intelligence Techniques in Photovoltaic Systems. https://doi.org/10.1016/B978-0-12-820641-6.00004-1
Copyright © 2022 Elsevier Ltd. All rights reserved.

PV output power forecasting has attracted, over the past two decades, the attention of many researchers and academics, and is currently one of the hottest topics in the area of renewable energy integration [6].

Most available methods were focused on the one-step ahead and short-term PV output power forecasting [7–9]; however, multistep and long-term forecasting stay a challenging issue in this area. In fact, accurate PV power forecasting can be beneficial for grid planning and scheduling, energy management, minimizing the operational costs, safe operation, quality and for balancing supply, effective dispatching decisions, demand, and avoiding penalties to plant managers due to deviations between the scheduled and the produced power [10].

Numerous PV power forecasting methods have been proposed in the literature. With reference to the forecast horizon, these methods or approaches can be grouped into four categories [11]:

(i) Very short-term forecasting with a time horizon ranging from a few seconds to some minutes: it could be applied for control and management of PV plants.
(ii) Short-term forecasting up to 48–72h ahead, mainly applied for control of power system operations, unit commitment, and economic dispatch.
(iii) Medium-term forecasting from a few days to 1 week ahead.
(iv) Long-term forecasting from a few months to a year or more, usually used for the maintenance and the planning of PV plants.

In addition to algorithm performance, time horizon, and time resolution, the availability and quality of the data play an important role in prediction accuracy [12].

This chapter covers some examples of the application of machine learning, ensemble, and deep learning methods for short-term PV output power forecasting:

(1) The first application is about forecasting one-step ahead PV output power, using machine learning algorithms (e.g., multi-linear regression, decision tree, random forest, support vector machine, K-nearest neighbor, multilayer perceptron neural network, and recurrent neural networks); in this application, only historical data (PV power measured in a given location) are required.
(2) The second application focuses on the use of artificial neural networks to forecast PV power output using the measured PV power and other meteorological parameters that influence the prediction accuracy such as solar irradiance, air temperature, relative humidity, and wind speed. The neural networks employed are: multilayer perceptron, recurrent neural network, and radial basis function network.
(3) The third application concerns the investigation of ensemble-learning algorithms (EL) for one-step ahead forecasting. Simple EL algorithms are used (e.g., MaxVoting, averaging, and weighted average).
(4) The fourth application deals with the utilization of deep learning natural networks for one-step and multistep ahead forecasting of output PV power. The algorithms investigated include deep learning neural networks such as, long short-term memory (LSTM), bidirectional LSTM, gated recurrent unit, convolutional neural network, and hybrid configurations.
(5) The fifth application shows how to qualify variation in the predicted results by calculating the confidence interval (CI) and the prediction interval (PI), which is very important for analyzing the uncertainties of the forecasters. As an example, the Bootstrap algorithm will be employed for the CI estimation and gradient boosting for PIs.

Eighteen examples have been described and discussed to help readers to understand how to apply machine learning, and deep learning in this topic. Four datasets have been used to develop the forecasters. These datasets were collected from different regions and for different timescales (1 min, 2 min, and 1 h). All models presented have been written and developed using Python programming language and Matlab® environment.

4.2 Forecasting of photovoltaic power based on data-driven methods

Due to the intermittent nature of solar energy, including the variation of solar irradiation, cloud, and air temperature, forecasting of the power produced by PV arrays is a crucial task and remains a challenging issue. PV power forecasting has been considered as a regression problem (time series prediction) [13]. Forecasting of PV power based on data-driven methods [14] can be achieved by three approaches [6]:

The first approach uses only historical records of PV output power (i.e., measured data in situ using data logger); so, the relationship between the input and the output data can be represented by the following expression:

$$\{p(t),p(t+1),...,p(t+k)\}=f\{p(t-n),p(t-n+1),p(t-n+2),...,p(t-1)\} \quad (4.1)$$

where $p(t)$ is the actual power, $p(t-n)$ is the previous power, $p(t+k)$ is the forecasted power at step k, and f is a functional dependency between past and future. $t\in\{1,..,n\}$, n is the length of the measurements.

Approaches rely upon forecasted meteorological parameters such as, air temperature, relative humidity, wind speed, pressure, solar irradiance, and cloud cover. These parameters can be obtained from satellite images, numerical weather prediction (NWP) models, or statistical models such as autoregressive (AR), autoregressive moving average (ARMA), autoregressive integrated moving average (ARIMA) , and others:

$$\{p(t),p(t+1),p(t+2),...,p(t+k)\}=f\{G(t+k),T_a(t+k),R_h(t+k),\\ W_s(t+k),C_c(t+k),...\} \quad (4.2)$$

where $G(t+k)$, $T_a(t+k)$, $W_s(t+k)$, $R_h(t+k)$, and $C_c(t+k)$ are the forecasted solar irradiance, air temperature, wind speed, relative humidity, and cloud cover, respectively.

Approaches that combine the use of historical power records with meteorological parameter forecasts:

$$\{p(t),p(t+1),...,p(t+k)\}=f\{p(t-n),p(t-n+1),...,p(t-1),\\ G(t+k),T_a(t+k),W_s(t+k),R_h(t+k),C_c(t+k),...\} \quad (4.3)$$

In most cases, forecasted solar irradiance is not available, so Eq. (4.3) becomes:

$$\{p(t),p(t+1),\ldots,p(t+k)\}=f\{p(t-n),p(t-n+1),\ldots,p(t-1),T_a(t+k),\\ W_s(t+k),R_h(t+k),C_c(t+k),\ldots\} \quad (4.4)$$

These methods can be used for both one-step and multistep ahead forecasting:

(a) One-step ahead *forecasting*

At time t, one-step ahead forecasting consists in estimating the future value of the PV output power expected at time $(t+1)$, based on the actual and previously measured data; this can be formulated by the following equation [4]:

$$\hat{p}_{t+1}=f(p_t,p_{t-1},\ldots,p_{t-n+1}) \quad (4.5)$$

where $t\in\{k,..,n-1\}$, $\{p_t,p_{t-1},\ldots,p_{t-n+1}\}$ are the actual and past values of the time series, $\hat{p}_{t+1}$. For example, with three inputs the relationship between the input and the output can be written as

$$\begin{bmatrix}\hat{p}_4\\ \hat{p}_5\\ \hat{p}_6\\ \hat{p}_7\\ \vdots\end{bmatrix}=f\begin{bmatrix}p_1 & p_2 & p_3\\ p_2 & p_3 & p_4\\ p_3 & p_4 & p_5\\ p_4 & p_5 & p_6\\ \vdots & \vdots & \vdots\end{bmatrix} \quad (4.6)$$

(b) Multistep ahead forecasting

Using a similar notation, the multistep ahead forecasting problem can be formulated by the following equation [4]:

$$\{\hat{p}_{t+1},\hat{p}_{t+2},\ldots,\hat{p}_{t+H}\}=f(p_t,p_{t-1},\ldots,p_{t-n+1}) \quad (4.7)$$

where H is the forecast horizon and $\{\hat{p}_{t+1},\hat{p}_{t+2},\ldots,\hat{p}_{t+H}\}$ is the forecasted time series. For example, in the case of two steps ahead ($H=2$) with five inputs , the relationship between the input and output can be written as follows:

$$\begin{bmatrix}\hat{p}_6 & \hat{p}_7\\ \hat{p}_8 & \hat{p}_9\\ \hat{p}_{10} & \hat{p}_{11}\\ \vdots & \vdots\end{bmatrix}=f\begin{bmatrix}p_1 & p_2 & p_3 & p_4 & p_5\\ p_3 & p_4 & p_5 & p_6 & p_7\\ p_5 & p_6 & p_7 & p_8 & p_9\\ \vdots & \vdots & \vdots & \vdots & \vdots\end{bmatrix} \quad (4.8)$$

4.3 Datasets description and preparation

The datasets used in this chapter for the development of various models are described as follows.

4.3.1 Dataset 4.1

The first dataset consists of historical records of PV power of a polycrystalline PV module (55 W). The data were collected at Marmara University, Istanbul (Turkey), for a period of 1 year (the sampling time is 2 min). Fig. 4.1A shows the collected P_{pv} (262,804 samples). Correlation between $P_{pv}(t)$ and $P_{pv}(t+1)$, $t \in \{0,N-1\}$, is shown in Fig. 4.1B.

4.3.2 Dataset 4.2

The second database contains measured data for T_a, T_C, G, and PV. The PV power is produced by a medium-scale PV plant of 500 kW, Bari (Italy). Fig. 4.2 shows a plot of

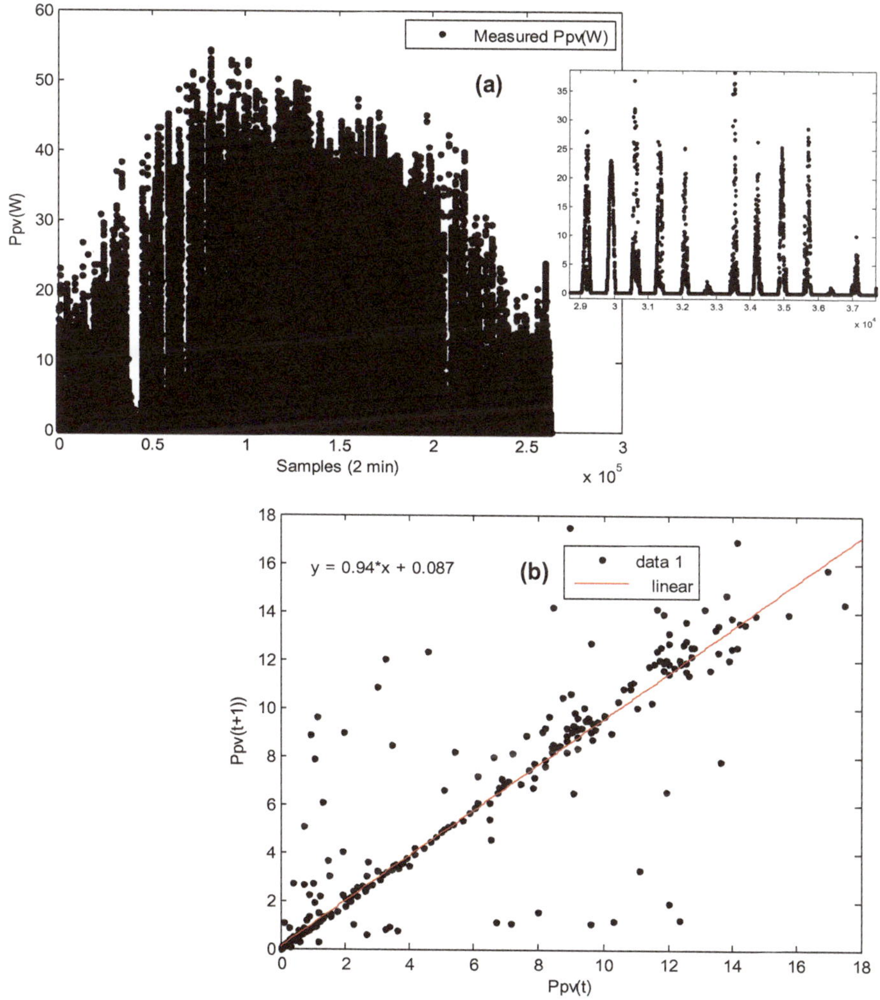

Fig. 4.1 Dataset 4.1: (A) measured PV power (262,804 samples) and (B) correlation between $P_{pv}(t)$ and $P_{pv}(t+1)$.

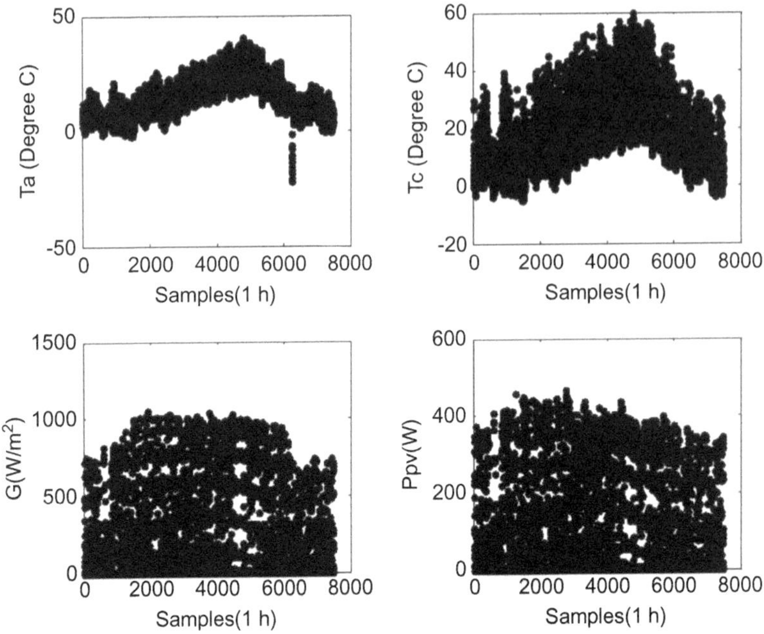

Fig. 4.2 Dataset 4.2: collected T_a, T_C, G, and P_{pv} (7445 samples).

the measured data corresponding to 323 days. The data was collected each hour; so, the size of the dataset is 7445 samples.

The correlation between these parameters (G, T_a, and T_C) is plotted in Fig. 4.3. As can be seen from Fig. 4.3, the correlation between G and T_a is not good ($r = 54\%$); acceptable correlation is obtained between G and T_C, and P_{pv} and T_C, respectively, while very good correlation is obtained by G and P_{pv} ($r = 98\%$).

4.3.3 Dataset 4.3

The third dataset consists of forecasted meteorological parameters such as: T_a, W_s, P_r, and R_h, and measured G and PV power of a small-scale PV system of 3.5 kW. As an example, Fig. 4.4 shows a plot of different forecasted meteorological parameters and measured G and P_{pv}. The data were collected (each hour) at Tamanrasset region, south of Algeria.

4.3.4 Dataset 4.4

The fourth dataset comprises historically measured PV power; the data were produced by a small-scale PV system of 4 kW, installed at Trieste University, Italy. The data were collected every 1 min and only daytime values ($P_{pv} > 0$ W) are considered; the size of the dataset is 224,172 samples (see Fig 4.5A). Fig. 4.5B shows the produced DC PV power by the system for 3 days (1500 samples).

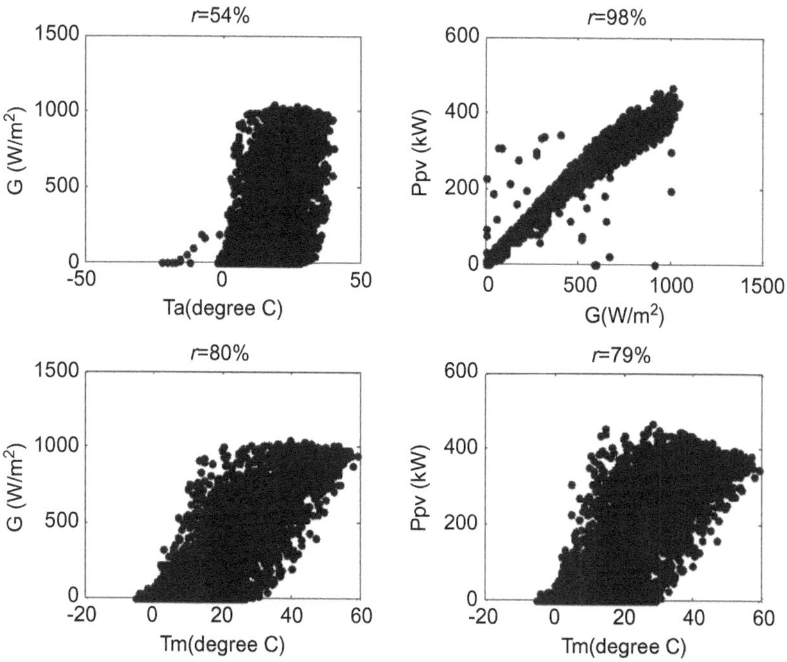

Fig. 4.3 Correlation (r) between parameters T_a, T_C, and G.

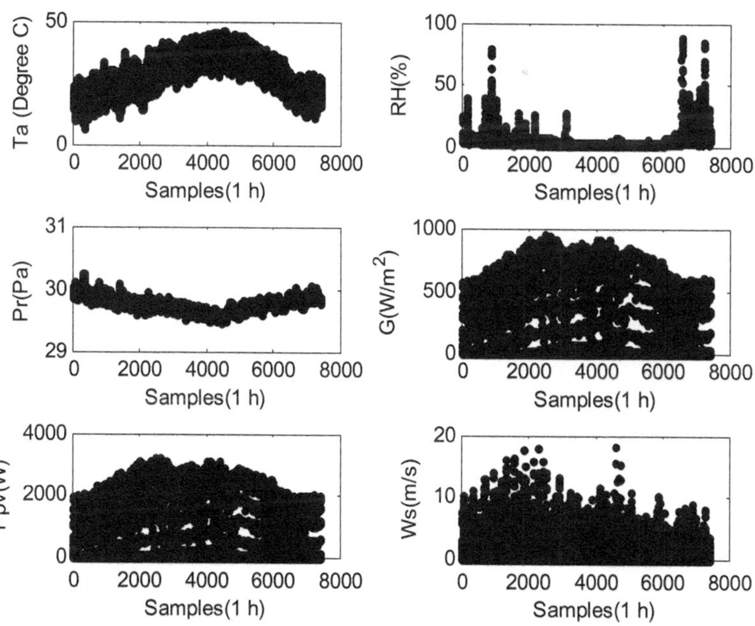

Fig. 4.4 Dataset 4.3: meteorological data and in situ solar irradiance and PV power (7445 samples).

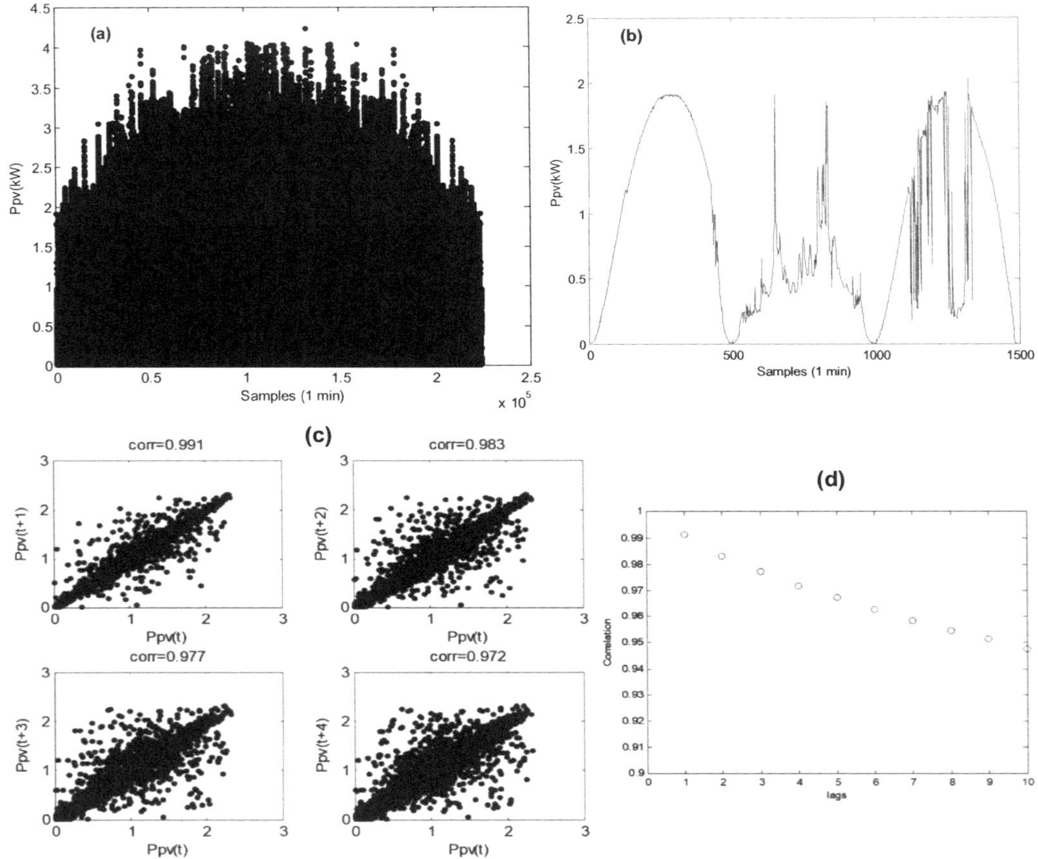

Fig. 4.5 Dataset 4.4: (A) collected P_{pv} (224,172 samples), (B) plot produced P_{pv} during a period of 3 days (sunny and cloudy), (C) Scatter curves of the GHI vs different lags ($y(t+1)$, $y(t+2)$, $y(t+3)$ and $y(t+4)$) and (D) correlation evolution.

Fig. 4.5C shows correlation between different lags $P_{pv}(t+1)$, $P_{pv}(t+2)$, $P_{pv}(t+3)$, and $P_{pv}(t+4)$. Correlation is decreased slowly when increasing the steps ($t+1$, $t+2$, ...) as shown in Fig. 4.5D.

4.4 Machine learning for forecasting of photovoltaic output power

In this section, the first database is used to develop and evaluate models for one-step ahead forecasting of the PV power. The algorithms investigated in this section are described in Chapter 2 (e.g., LR, SVM, K-NN, DT, RF, MLP, and RNN). In these examples, 85% of data have been used to train and validate the models and the rest 15% is for testing. *MinMaxScaler* and *train_test_split* functions are used. The function (*train_test_split_Ppv*) permits a user to vary the number of inputs $\{P_{pv}(t-1), P_{pv}(t-2), \ldots P_{pv}(t)\}$, while the output is set to $P_{pv}(t+1)$. For example, seq_len = 10 means that 10 samples are available $\{P_{pv}(1), P_{pv}(2), P_{pv}(3)\ldots, P_{pv}(10)\}$, and the purpose is to try to forecast the next value, $P_{pv}(11)$. Fig. 4.6A shows an example of the adopted process for P_{pv} one-step ahead forecasting, and Fig. 4.6B reports an example of the matrix created for three inputs and one output.

The procedure is summarized in the flowchart shown in Fig. 4.7.

Example 4.1: Multi-linear regression (MLR).

In this example a multiple-linear regression algorithm is applied, named *{LinearRegression()}*. The main command lines are (**Listing 4.1**):

```
#Listing 4.1 MLR
LR = LinearRegression()
LR.fit(x_train, y_train)
y_pred = LR.predict(x_test)
```

Example 4.2: Decision tree (DT)

Here, the decision tree algorithm is developed by using *DecisionTreeRegressor()* function. The command lines are shown in the **Listing 4.2**.

```
#Listing 4.2 DT model
DT = DecisionTreeRegressor(random_state = 0)
DT.fit(x_train, y_train)
y_pred = DT.predict(x_test)
```

Example 4.3: Random forest (RF)

In this example, we investigate the random forest algorithm; the *RandomForestRegressor()* function is used. The essential command lines are reported in **Listing 4.3:**

(a)

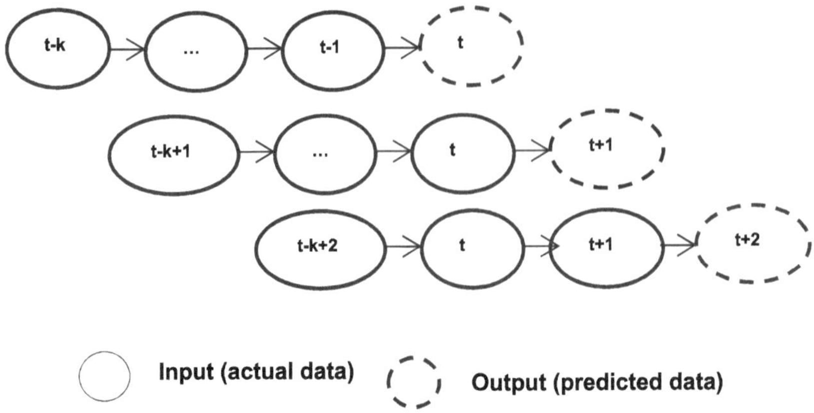

(b)

	Output	Input matrix		
	y(t+3)	y(t)	y(t+1)	y(t+2)
1000	3.159	2.664	2.812	3.002
1001	3.280	2.812	3.002	3.159
1002	3.444	3.002	3.159	3.280
1003	3.570	3.159	3.280	3.444
1004	3.741	3.280	3.444	3.570
...	...	...	...	...

Fig. 4.6 (A) Example of input (measured) and output (forecasted) sequences and (B) example of 3×1 (three inputs and one output) samples.

```
#Listing 4.3 RF model
RF= RandomForestRegressor(n_estimators=100,max_features=1,
  oob_score=True)
RF.fit(x_train, y_train)
y_pred = RF.predict(x_test)
```

Example 4.4: Support vector machine (SVM)

In this example, a support vector machine algorithm is developed by using the *svm.SVR()* function. Various kernel functions like, the *rbf, linear, ploy,* and *sigmoid*, have been evaluated. The main command lines are shown in **Listing 4.4**:

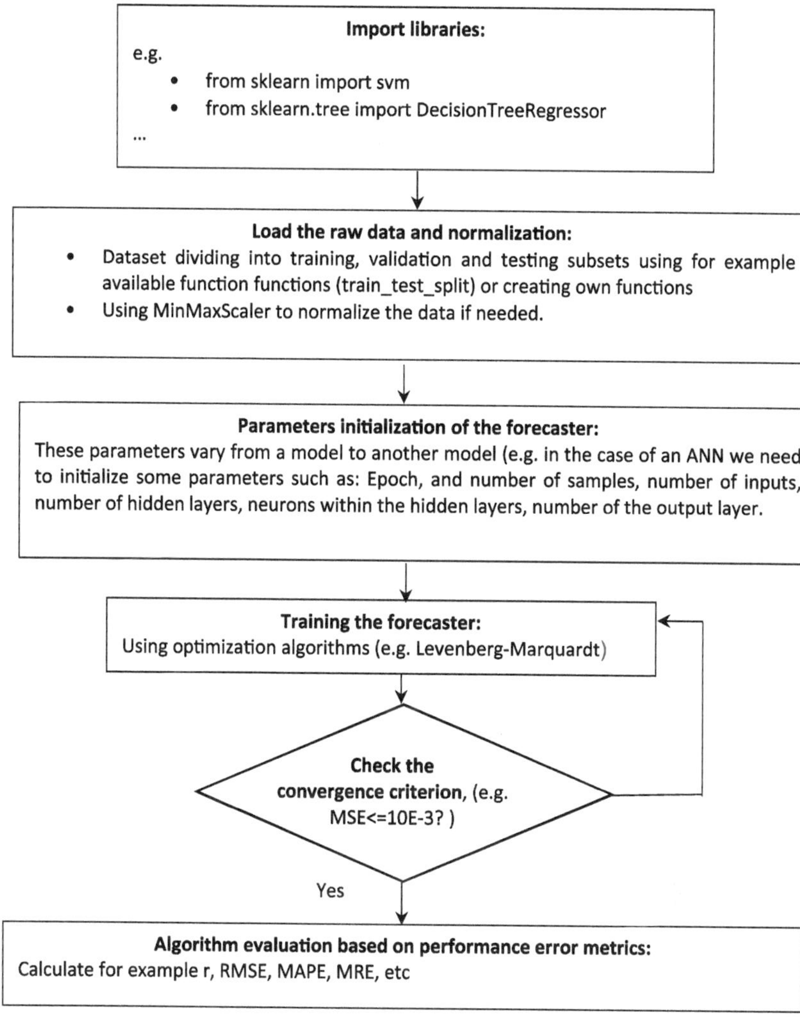

Fig. 4.7 Flowchart for the training and assessment of ML algorithms (SVM, LR, kNN, DT RF, and ANNs).

```
#Listing 4.4SVMmodel
SVM=svm.SVR(kernel='rbf', C=10)
SVM.fit(x_train, y_train)
y_pred=SVM.predict(x_test)
```

Example 4.5: K-nearest neighbors (K-NN)

In this example, the K-nearest neighbors algorithm, named {*KNeighborsRegressor ()*} function, is used to develop the model. The important parameter here is the choice of K, and for this the K between 1 and 50 were used and it has been found that the optimal one is for K = 10. The code is given in the following listing:

```
#Listing 4.5: k-NNmodel
Knn=KNeighborsRegressor(n_neighbors=10, leaf_size=10,p=3)
Knn.fit(x_train, y_train)
y_pred=knn.predict(x_test)
```

Example 4.6: Multilayer perceptron (MLP) neural networks

MLP neural network function, named {*MLPRegressor()*}, is used to develop the forecaster. Various activation functions have been examined. The main command lines are presented in **Listing 4.6**.

```
#Listing4.6: MLPmodel
MLP=MLPRegressor(hidden_layer_sizes=(3,2),activation="logistic",
 random_state=1, max_iter=1000)
MLP.fit(x_train, y_train)
y_pred=MLP.predict(x_test)
```

Example 4.7: Recurrent neural networks (RNNs)

In this example, a simple recurrent neural newel (Elman) is used based on the implemented function {`SimpleRNN()`}. Various configurations have been tested by modifying the number of nodes in the hidden layer and the type of the activation functions. The key command lines are shown in **Listing 4.7**.

```
#Listing 4.7: RNN model
RNN= Sequential()
RNN.add(SimpleRNN ( units =5, activation ="tanh", input_shape =
  (seq_len-1 , 1)))
RNN.add( Dense ( units =1, activation ='linear'))
sgd = SGD (lr =0.0001 , momentum =0.97 ,nesterov = True )
RNN.compile ( loss ='mean_squared_error',optimizer = sgd )
RNN.fit( x_train , y_train , batch_size =32 ,epochs =500)
```

The results obtained are summarized in Table 4.1, which shows the calculated error metrics between measured and one-step ahead forecasted PV power.

As can be seen from Table 4.1, the correlation coefficient r ranges between 90% and 95% and the lower value of RMSE is obtained by the recurrent neural network forecaster. For all modes investigated, the MAE do not exceed 0.31 W, and the maximum value of wMAPE is less than 0.251% provided by the decision tree algorithm. The lower value of MRPE is obtained by the RNN, MLP, and SVM forecasters. These three forecasters have good correlation coefficients of 95.1%, 94.5%, and 94%, respectively. As an example, Fig. 4.8A shows a comparison between the measured and forecasted P_{pv} by the RF algorithm, and Fig. 4.8B depicts the corresponding scatter curve.

Table 4.1 Error metrics for the investigated machine learning algorithms for one-step ahead forecasting of PV power.

Model parameters	r (%)	RMSE (W)	wMAPE (%)	MAE (W)	MRPE (W)
MLR forecaster NI = 9	94.4	1.103	0.241	0.296	4.939
DT forecaster copy_X = True, fit_intercept = True, n_jobs = None, normalize = False, NI = 7	89.9	1.463	0.251	0.309	10.542
RF forecaster n_estimators = 100, max_features = 1, NI = 15	94.1	1.134	0.208	0.256	4.950
SVM forecaster kernel = 'rbf', C = 100, NI = 5	93.9	1.147	0.192	0.236	3.530
K-NN forecaster K = 10, leaf_size = 50, $p = 3$, NI = 7	93.3	1.203	0.201	0.246	4.149
MLP forecaster hidden_layer_sizes = (7,5), activation = "logistic", NI = 3, epoch = 100	94.5	1.098	0.200	0.246	1.041
RNN forecaster units = 5, activation = "tanh", optimizer = sgd, los = mse, epoch = 500	95.1	1.001	0.180	0.240	1.022

It is worth noting that the machine learning algorithms investigated are simple from the structure point of view and can be easily implemented. However, the accuracy could be improved by tuning the parameters carefully, such as optimization algorithm, activation function, number of hidden layer, number of inputs, kernel, etc.

4.5 Neural networks for forecasting of photovoltaic output power from meteorological parameters and historical power

There are mainly three approaches to forecast the PV power using meteorological parameters (such as T_a, R_h, W_s, and P_r). The first one consists of forecasting the PV power (e.g., $t+1$) using as input meteorological parameters and PV power at time (t). The second approach aims to forecast the PV power (e.g., $t+1$) from forecasted

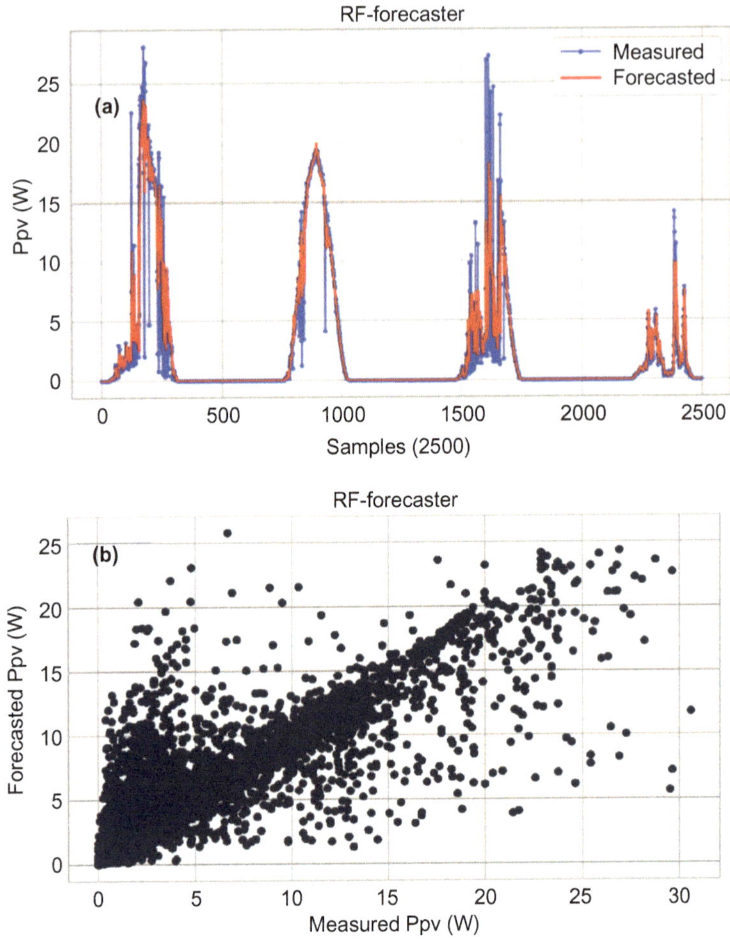

Fig. 4.8 Measured vs forecasted P_{pv} using RF machine learning algorithm: (A) superposition curves for 2500 samples and (B) scatter curve between both series.

meteorological parameters $(t+1)$, for example, by NWP models and historical P_{pv} at time (t). The last one aims to estimate the PV power at time (t) from meteorological parameters at time (t) (similar to model-based approximation function), then PV power could be forecasted by using forecasted parameters. In the first and the third approaches, the forecasted meteorological parameters are required to realize the forecasting. Forecasted meteorological parameters could be obtained from NWP models or using ML algorithms (e.g., RNNs).

In this section, two examples will be presented; the first one is about the forecasting of PV output power from a PV array of 500 kW based on meteorological measured data and historical PV power in situ (first approach), while the second example aims to forecast PV power by a small-scale PV system of 3.2 kW, based on the forecasted meteorological data, for example, by numerical weather prediction models such as T_a, W_s, R_h, and P_r, and measured values of solar irradiance (G) and PV power (second approach). The aim is to show the influence of the forecasted meteorological data on the accuracy of the neural network-based models. To do this, a Matlab code is developed and described below for one-step ahead forecasting, including the required functions.

Example 4.8: MLP forecaster (using measured meteorological parameters)

In this example, the second database 4.2 presented in Section 4.3 is used to develop an MLP forecaster. With regard to the correlation calculated in Section 4.3, it can be concluded that the most significant parameters are G and T_C, which means that these parameters are indispensable and helping to reach good performance accuracy.

To do this, a simple type of NNs is used which is a feed-forward neural network (ffnn). The developed models are formulated by the following equations [11]:

$$P_{pv}(t+1) = f_{ffnn}\{T_a(t), T_C(t), G(t), P_{pv}(t)\} \tag{4.9}$$

$$P_{pv}(t+1) = f_{ffnn}\{T_a(t), T_C(t), G(t)\} \tag{4.10}$$

$$P_{pv}(t+1) = f_{ffnn}\{T_C(t), G(t)\} \tag{4.11}$$

$$P_{pv}(t+1) = f_{ffnn}\{T_C(t), G(t), P_{pv}(t)\} \tag{4.12}$$

$$P_{pv}(t+1) = f_{ffnn}\{G(t), P_{pv}(t)\} \tag{4.13}$$

Thus, as indicated in Eqs. (4.9)–(4.13) the user tries to forecast the next value $(t+1)$ of P_{pv} based on the present values (t) of meteorological parameters and P_{pv}. The developed Matlab code is given below (see **Listing 4.8a**).

```
%Listing 4.8a: MLP forecaster
clear all; clc, close all
%loading data
v=xlsread('e:\Database2cha4.xlsx')

v=[v(:,1),v(:,2),v(:,3),v(:,4)];
N=length(v(:,1));Ntr=0.85*N;Lin=1:Ntr-1; Lout=2:Ntr;
Input=[v(Lin,1), v(Lin,2), v(Lin,3),v(Lin,4)];
Output=[v(Lout,4)];

%Network configuration
net=newff(minmax(Input),[12,7,1],
  {'tansig','tansig','purelin'},'trainlm');
net.trainParam.show =10;
net.trainParam.epochs =1000;
net.trainParam.goal =1e-2;
net.performFcn = 'msereg';
[net,tr]=train(net, Input, Output);
```

```
% Simulation step
Lin1=length(Ntr):1:N-1;
Lout1=length(Ntr)+1:1:N;
Input_sim=[v(Lin1,1), v(Lin1,2), v(Lin1,3), v(Lin1,4)];
Target_sim=[v(Lout1,4)];
rs=sim(net, Input_sim);
```

The *MinMax* function is used to normalize the input vector *Input*. To train the network, 85% samples (6352 samples corresponding to 264 days) have been used and 15% samples (1122 samples corresponding to 46 days) are used to test the network. When the model reaches a satisfactory performance (e.g., msereg <= 10E-1), the weights, bias, and the structure of network are saved using the following command lines (**Listing 4.8b**).

```
Listing 4.8b: MLP forecaster
Weights and bias
w1=net.IW{1};w2=net.LW{2};b1=net.b{1};b2=net.b{2};
ffnn1=net; save E:\Book_Cha4\ffnn1
```

It should be noted that the training process stops when any of these conditions are met: (a) the maximum number of epochs is reached, (b) performance has been minimized to the goal, and (c) the maximum amount of time has been exceeded.

To view and generate the Simulink block we can use the following commands:

```
View(net)
Genism(net)
```

Fig. 4.9 depicts the structure employed for the FFNN.

The following code is used to evaluate the models' performances. The function *errors_test* is used in order to assess models' accuracy. The error metrics used are RMSE, MRPE, MAE and *r*. A cumulative distribution function is also used (see **Listing 4.8c**).

```
%Listing 4.8c: MLP forecaster
% Performance
Per=perfom(net,ts,rs)

%Statistical test and verification
[mrpe,r,rmse,mae]=errors_test(ts,rs);
fprintf('RMSE: %0.3f | MRPE: %0.3f | R2: %0.3f | MAE: %0.3f\n',rmse,
  mrpe,r(1,2),mae);

%Kolmogorove cumulative s=distribution function Fx
```

```
l=length(ts); x=min(ts):0.01:max(ts);
[F_s1,F_r1,DIF1]=Kolmogorov(l,ts,rs,x);
figure, plot(x,F_r1,'k-',x,F_s1,'b:')
ylabel('Cumulative distribution function Fx');
xlabel('x');legend('Measured','Forecasted')

% saving results
result=[ts;rs];xlswrite('e:\model#1.xlsx',result)
```

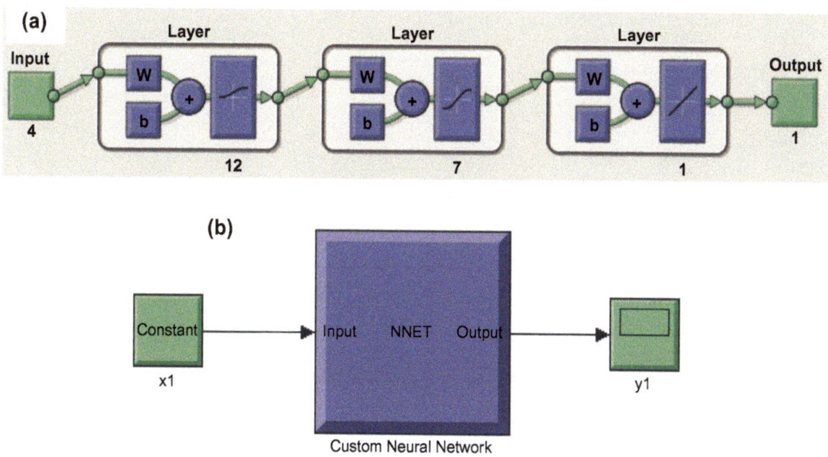

Fig. 4.9 (A) Displaying of the network using the *view* function and (B) creating a Simulink block using the *genism* function.

The *errors_test* function is depicted in **Listing 4.8d.**

```
%Listing 4.8d: error test function
function [mrpe, r, rmse, mae]=errors_test(vm,vs),
% Coefficient of correlation
r=corrcoef(vm,vs);
% Variance and covarinace Test
cov1=cov(abs(vm),abs(vs));
var1=var(abs(vm),abs(vs));
% Mean test
mm=mean(vm);ms=mean(vs);
% RMSE
n=length(vm);mse=0;
for i=1:n,
    mse=mse+(vs(i)-vm(i))o2;
end,
rmse=sqrt((mse)/n);
% Mean absolute Error
```

```
s=0;
for i=1:n,
    s=s+abs(vm(i)-vs(i));
end,
mae=s/n;
% Mean Percentage Error
s=0;
for i=1:n,
    s=s+((vs(i)-vm(i))/vm(i));
end,
mpe=(s/n)*100;
% Mean Absolate Percentage Error
s=0;
for i=1:n,
    s=s+abs((vm(i)-vs(i))/vm(i));
end,
mape=(s/n)*100;
%mean relative percent absulate error
mrpe=(((ms-mm)/mm)*100);
```

The *Kolmogorov* function (see **Listing 4.8e**).

```
%Listing 4.8e: Kolmogorov test function
function [F_s,F_r,DIF]=Kolmogorov(l,v1,v2,x)
cum=[];cum1=[];
for i=1:length(x)
   xc=find(v1<x(i));
   xc1=find(v2<x(i));
   Nbr_x=length(xc);
   X_c=Nbr_x/l;
   Nbr_x1=length(xc1);
   X_c1=Nbr_x1/l;
   cum=[cum;X_c];
   cum1=[cum1;X_c1];
end;
F_s=cum; F_r=cum1; DIF=F_r-F_s;
```

Eventually in order to use the model, the network P_{pv} values estimated using the *sim* function need to be loaded. As an example, see the following command lines.

```
Load E:\Book_Cha4\ffnn1
Input=[27, 33 800 4000]
Result=sim(ffnn1,input)
```

Table 4.2 Error metrics between measured and forecasted hourly PV power.

Model	RMSE (kW)	MRPE (%)	r (%)	MAE (kW)
Structure In × 15 × 9 × 1, In is the input variables Optimization algorithm: Levenberg-Marquardt (Trainlm) Performance function: Mean squared error/reg performance function (*msereg*) Transfer function:(*Tansig*) Number of epoch = 1000 Error = 10E-10				
Model 1 (Inputs: T_a, T_C, G, and P_{pv})	4.591	1.298	93.2	2.743
Model 2 (Inputs: T_a, T_C, and G)	4.620	1.016	93.1	2.630
Model 3 (Inputs: T_C and G)	4.824	1.286	92.5	2.907
Model 4 (Inputs: T_C, G, and P_{pv})	4.558	1.761	93.0	2.703
Model 5 (Inputs: G and P_{pv})	4.569	0.716	92.4	2.957

The obtained simulation results are listed in Table 4.2.

With reference to Table 4.2, the best values of r have been obtained by the first, the fourth, and the fifth models, respectively. For all structures investigated, MRPE does not exceed 2% and the value of the MAE is about 3 kW, while the RMSE ranges between 4.42 and 4.98 kW. As example, Fig. 4.10 shows the simulation results between measured and simulated P_{pv} by model 3 for 7 days.

Other functions and optimization algorithms could be used to make a comparative study. For example, one can modify the transfer function (e.g., *logsig*), the optimization algorithm (e.g., *trainbr*, *traingdx*, *traingdm*, etc.), and the performance functions (e.g., *rms*, *mae*, or *sse*).

Available NNs in the Matlab toolbox could be also investigated to assess the forecasting accuracy, for example, generalized regression neural network (GRNN) (*newgrnn* function), radial basis function network (*newrbe* function), and recurrent neural network (*newelm* function). See the following command lines.

```
net=newgrnn(p,t,spread);
net=newelm(p,t,[20],{'tansig','purelin'},'traingdx');
net=newrbe(P,T,spread)
```

Example 4.9: MLP forecaster (using forecasted meteorological data)

In this example, the third database is used. The same codes and functions are used in this example (see example #4.8). The goal is to forecast the PV power based on forecasted meteorological parameters such as T_a, R_h, P_r, W_s, and G, as well as to check the prediction accuracy. Three ffnn-based models have been developed; the relationship between the inputs and the output are given as

$$P_{pv}(t+1) = f_{ffnn}\{T_a(t+1), R_h(t+1), P_r(t+1), W_s(t+1), G(t), P_{pv}(t)\} \quad (4.14)$$

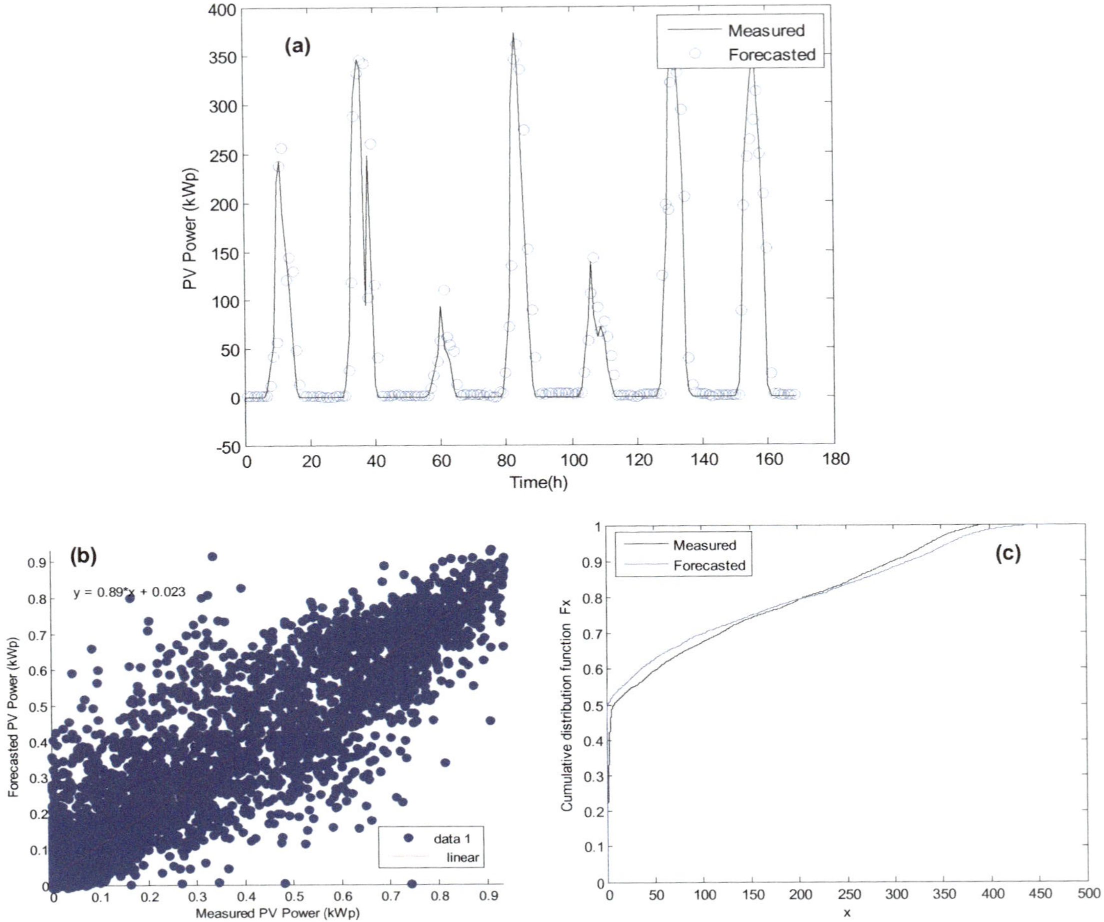

Fig. 4.10 (A) Measured versus forecasted PV power using the third model for 7 days, (B) scatter curve, and (C) cumulative distribution function between measured and forecasted PV power.

$$P_{\text{pv}}(t+1)=f_{\text{ffnn}}\{T_a(t+1),R_h(t+1),P_r(t+1),W_s(t+1),G(t)\} \tag{4.15}$$

$$P_{\text{pv}}(t+1)=f_{\text{ffnn}}\{T_a(t+1),R_h(t+1),P_r(t+1),W_s(t+1)\} \tag{4.16}$$

From Eqs. (4.14)–(4.16), historical values of P_{pv} is not always considered as input of the models. A comparison between measured and forecasted P_{pv} by the first model is depicted in Fig. 4.11A. Scatter curves and the cumulative distribution plots show clearly good correlation between both series (see Fig. 4.11B).

The calculated error metrics are reported in Table 4.3.

Based on the calculated error metrics reported in Table 4.3, it can be observed that good results are obtained by the first model ($r=98\%$) with a lower value of MRPE. The second model presents also satisfactory results ($r=95\%$). The third model presents lower accuracy ($r=80\%$).

As can be concluded from the results obtained in this section, the availability of the forecasted meteorological parameters with the historical PV power and solar irradiance play a very important role in the prediction accuracy. However, forecasted meteorological parameters without using historical values of G and P_{pv} do not perform well. To compare the results, other ML algorithms could be investigated by changing the configuration, adjusting parameters, etc.

4.6 Ensemble learning for forecasting of photovoltaic power for one-step ahead

In this section, ensemble learning (EL) algorithms is applied to forecast P_{pv} for one-step ahead. To do this, the first database is used. The examined EL algorithms are simple but powerful: (1) MaxVoting, (2) averaging, and (3) weighted average. These algorithms are described in Chapter 2; three predictors named DT,K-NN, and LR are used and then the three predictors will be used. For example, the *VotingRegressor* function can be used to estimate the final prediction, which is a simple majority vote among the predicted output [15,16].

Example 4.10: EL algorithm (MaxVoting)

A basic code of an EL algorithm (DT, K-NN, and LR) is given in **Listing 4.10**.

```
#Listing 4.10 EL MaxVoting method
# Loading libraries
from sklearn.tree import DecisionTreeRegressor
from sklearn.neighbors import KNeighborsRegressor
from sklearn.linear_model import LinearRegression
from sklearn.ensemble import VotingRegressor
# Models development separately
model1 = DecisionTreeRegressor()
model2 = KNeighborsRegressor(n_neighbors=10,leaf_size=50,p=3)
model3= LinearRegression()
```

```
# Models fitting
model1.fit(x_train,y_train)
model2.fit(x_train,y_train)
model3.fit(x_train,y_train)
# Models prediction
pred1=model1.predict(x_test)
pred2=model2.predict(x_test)
pred3=model3.predict(x_test)
# Maxvoting regressor
model = VotingRegressor([('DT', model1), ('kNN', model2), ('lr',
  model3)])
model.fit(x_train,y_train)
model.score(x_test,y_test)
```

The following code shows the model parameters:

```
VotingRegressor (estimators=[('DT',
                      DecisionTreeRegressor (ccp_alpha=0.0,
                                             criterion='mse',
                                             max_depth=None,
                                             max_features=None,
                                             max_leaf_nodes=None,
                                             min_impurity_decrease=0.0,
                                             min_impurity_split=None,
                                             min_samples_leaf=1,
                                             min_samples_split=2,
                                             min_weight_fraction_leaf=0.0,
                                             presort='deprecated',
                                             random_state=None,
                                             splitter='best')),
                           ('kNN',
                      KNeighborsRegressor (algorithm='auto', leaf_size=50,
                                             metric='minkowski',
                                             metric_params=None,
                                             n_jobs=None, n_neighbors=10,
                                             p=3, weights='uniform')),
                           ('lr',
                     LinearRegression (copy_X=True, fit_intercept=True,
                                       n_jobs=None, normalize=False))],
                 n_jobs=None, weights=None)
```

The correlation is 92.7% obtained by MaxVoting ML algorithm. Fig. 4.12A shows the variation of DL, K-NN, LR, and MaxVoting algorithm whereas Fig. 4.12B reports a comparison between the measured PV power and the one predicted by using the three EL methods (MaxVoting, averaging, and weighted average). From Fig. 4.12B, it is seen that acceptable correlation is obtained by the three EL algorithms.

EL algorithms however, presented here are simple, but other advanced EL algorithms could be used, such as Boosting and Bagging methods, including AdaBoost, gradient tree boosting, random forest, and other [16,17].

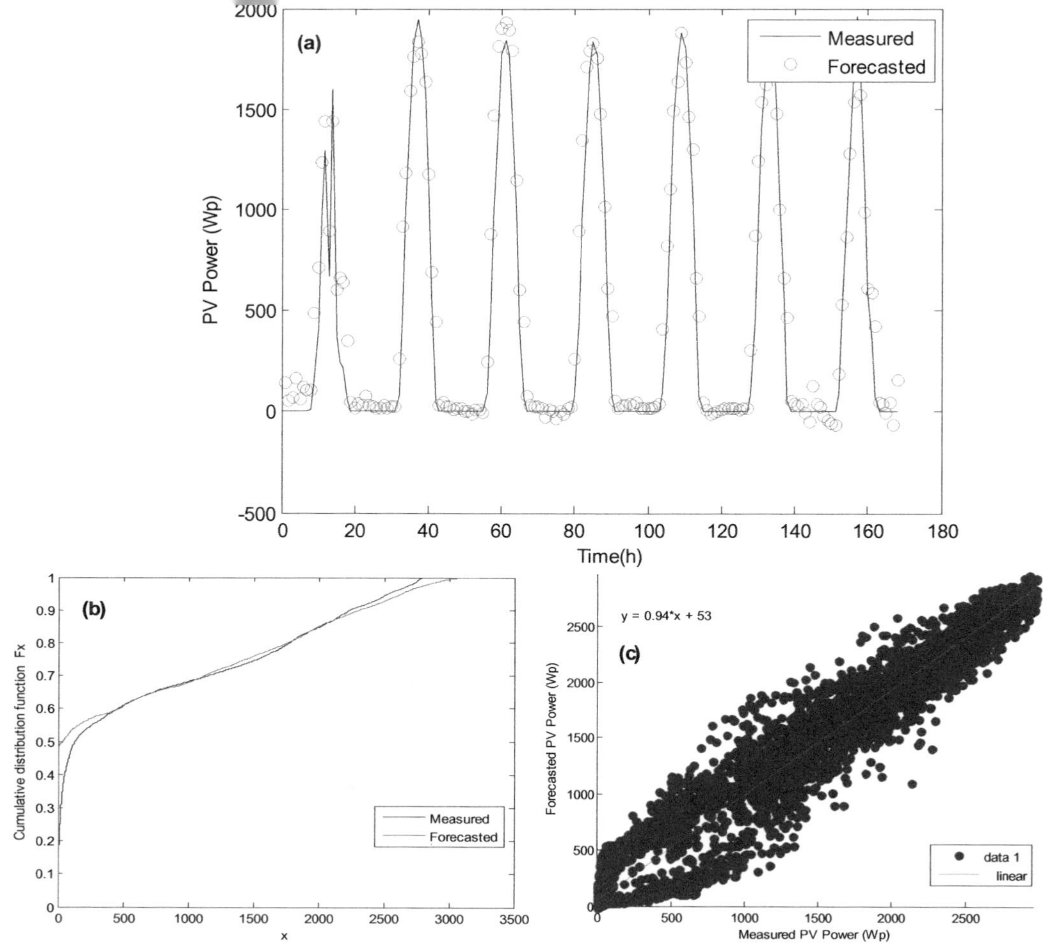

Fig. 4.11 (A) Measured versus forecasted PV power using the first model for 7 days, (B) cumulative distribution function between measured and forecasted P_{pv}, and (C) scatter curve.

Table 4.3 Error metrics between measured and forecasted PV power.

Model	RMSE (W)	MRPE (%)	r (%)	MAE (W)
Structure {In × 7 × 5 × 1}, In is the input variables Optimization algorithm: Levenberg-Marquardt (*Trainlm*) Performance function: mean squared error w/reg performance function (*msereg*) Transfer function:(*Tansig*) Number of epoch = 500 Error = 10E-2				
Model 1 (Inputs: T_a, R_h, P_r, W_s, G, and P_{pv})	18.97	1.380	98	12.87
Model 2 (Inputs: T_a, R_h, P_r, W_s, G, and P_{pv})	42.95	1.174	95	18.60
Model 3 (Inputs: T_a, R_h, P_r, and W_s)	59.29	2.097	80	42.20

4.7 Deep learning for forecasting of photovoltaic power (one-step and multistep ahead)

In this section, the application of deep learning algorithms to forecast PV power for one-step and multistep ahead is shown. The investigated DL algorithms are: LSTM, stacked LSTM, GRU, Bidirectional LSTM, CNN, and a hybrid CNN-LSTM. The fourth dataset is used; it consists of 224,172 measured samples. These algorithms are described in detail in Chapter 2. A total of 85% (190,546 samples) of the data were used to train the models and the rest 15% (33,625 samples) were used for testing the models. The data (historical values of PV power) are normalized used the *MinMaxScaler* function. The codes below are written and developed in Python language and related packages, more specifically the DL libraries [18] are used here.

4.7.1 One-step ahead forecasting

The following functions permit to split the database into training and testing portions (**Listing 4.11**).

```
#Listing4.11: Splitfunction
Input_Samples=4
def creat_sequences(data, seq_len):
    vect=[]
    for index in range(len(data) - seq_len):
        vect.append(data[index: index+seq_len])
    return np.array(vect)
```

```
def datasplit(data_raw, seq_len, train_split):
    data=creat_sequences(data_raw, seq_len)
    num_train=int(train_split*data.shape[0])
    X_train=data[:num_train, :-1, :]
    y_train=data[:num_train, -1, :]
    X_test=data[num_train:, :-1, :]
    y_test=data[num_train:, -1, :]
    return X_train, y_train, X_test, y_test
#call the function
X_train, y_train, X_test, y_test=datasplit(Norm_data, Input_Samples,
  train_split=0.80)
```

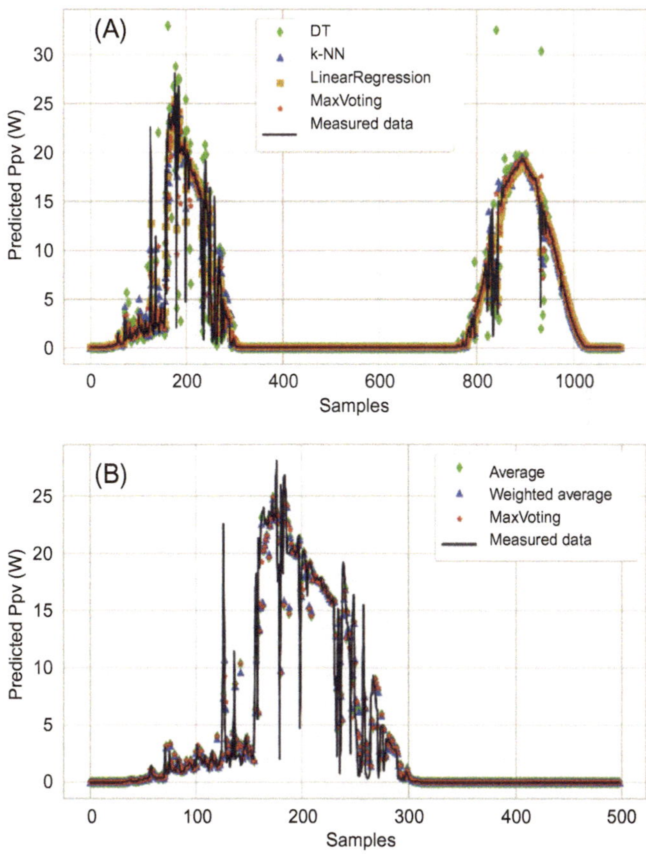

Fig. 4.12 (A) Measured vs forecasted results using the MaxVoting EL algorithm for 1100 samples and (B) comparison between the three EL algorithms (averaging, MaxVoting, and weighted average).

Example 4.11: LSTM forecaster model

The main line commands for developing a LSTM model are summarized as follows (**Listing 4.12**).

```
#Listing 4.12: LSTMmodel
model=Sequential ()
model.add(LSTM(NU, activation='relu',input_shape=(X_train.shape
  [1], 1)))
model.add(Dense(1))
model.summary()
model.compile(loss='mean_squared_error', optimizer='adam')
Performance=model.fit( X_train, y_train, epochs=50, batch_size=128,
  shuffle=False,validation_split=0.15)
```

Example 4.12: Stacked LSTM model (SLSTM)

Similar to the LSTM, more layers are added, as shown in the following code (**Listing 4.13**).

```
#Listing 4.13: SLSTM model
 S model
model=Sequential()
model.add(LSTM(100, activation='relu', return_sequences=True,
  input_shape=(X_train.shape[1], 1)))
model.add(LSTM(50, activation='relu'))
model.add(Dense(1))
```

Example 4.13: Bidirectional LSTM model (BiLSTM)

Similar to LSTM, the LSTM is replaced only by the bidirectional LSTM, using the following command line:

```
# Bi LSTM model
model.add(Bidirectional(LSTM(units=100, activation='relu',
input_shape=(X_train.shape[1], X_train.shape[2]))))
```

Example 4.14: GRU model

Similar to the LSTM code, simply LSTM is replaced in the third line by GRU.

```
# GRUmodel
model.add(GRU(NU, activation='relu',input_shape=(X_train.shape
[1], 1)))
```

Example 4.15: CNN model

The CNN model consists of a Conv1d layer, MaxPooling layer, Flatten layer, Dense (100) layer, and the last Dense layer with one node. The corresponding code is shown in **Listing 4.14**:

```
#Listing 4.14: CNNmodel
cnn=Sequential()
cnn.add(Conv1D(filters=64, kernel_size=2, activation='relu',
  input_shape=(X_train.shape[1], 1)))
cnn.add(MaxPooling1D(pool_size=2))
cnn.add(Flatten())
cnn.add(Dense(100, activation='relu'))
cnn.add(Dropout(0.2))
cnn.add(Dense(1))
```

Table 4.4 Statistical errors for the forecasted PV powers.

Model	RMSE (kW)	MRPE (%)	r (%)	MAE (kW)
LSTM $(3 \times 50 \times 1)$	0.023	1.032	99.2	0.009
BiLSTM $(3 \times 50 \times 1)$	0.024	2.784	99.0	0.011
Stacked LSTM$(3 \times 50 \times 25 \times 1)$	0.024	1.443	99.2	0.010
GRU$(3 \times 50 \times 1)$	0.024	2.715	99.2	0.011
BiGRU $(5 \times 50 \times 1)$	0.023	1.485	99.2	0.010
CNN $(7 \times 100 \times 1)$	0.028	5.577	98.8	0.017

The number of inputs is set to 3, {i.e., $P_{pv}(t)$, $P_{pv}(t+1)$, and $P_{pv}(t+3)$} for all models investigated, and the future value {i.e., $P_{pv}(t+4)$} is predicted. The following parameters are also kept the same as before: NU = 50, Epoch = 100, BS = 128. The loss function is the MSE, the optimization algorithm is *Adam* and the activation function is ReLU; these functions can be chosen by the user. With reference to Table 4.4, the correlation coefficient r is in the range (98.5%–99.2%) revealing a good correlation between the measured and the forecasted powers for all the tested models. Globally, in terms of RMSE, MAPE, and MAE all models perform well, while CNN provides relatively lower accuracy.

This is an example of the model summary using the command line:

```
model.summary()
Layer (type)                 Output Shape              Param #
=================================================================
lstm_3 (LSTM)                (None, 50)                10400
dense_3 (Dense)              (None, 1)                 51
=================================================================
```

```
Total params: 10,451
Trainable params: 10,451
Non-trainable params: 0
```

Fig. 4.13A shows a comparison between measured and forecaster one-step ahead PV power by the LSTM forecaster. Fig. 4.13B reveals that the correlation between the power forecasted using LSTM forecaster.

4.7.2 Multistep ahead forecasting

This subsection aims to develop a number of multistep forecasters using the LSTM network. Seven forecasters were trained using the fourth database and can predict 2, 3, 4, 5, 10, and 20 steps ahead (i.e., 2, 3, 4, 5, 10, and 20 min).

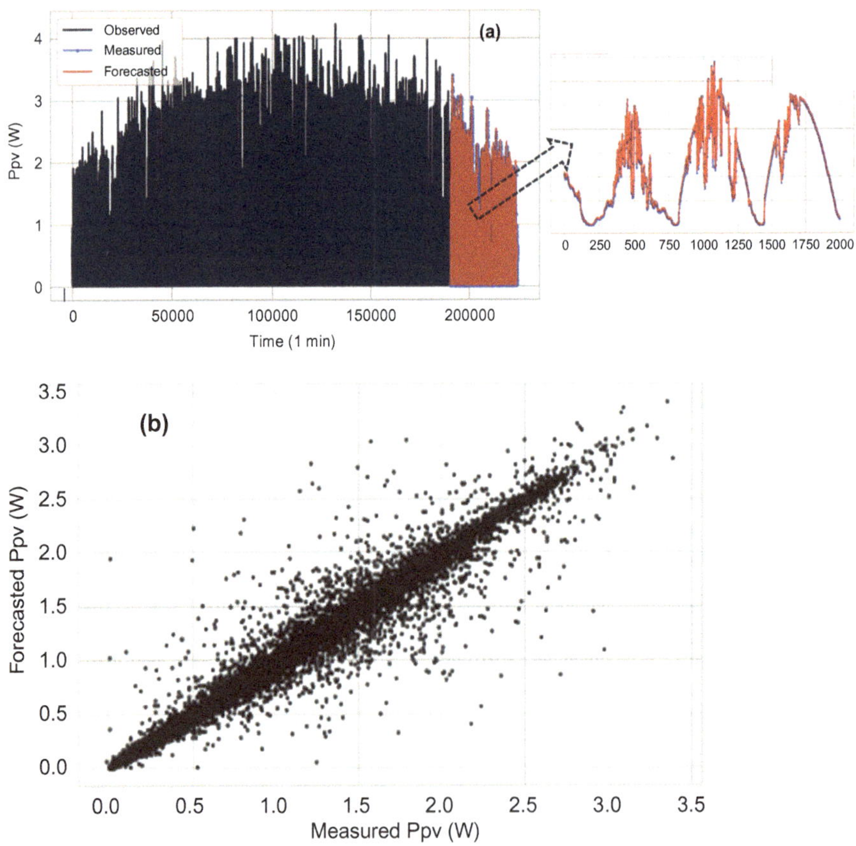

Fig. 4.13 (A) Measured versus forecasted P_{pv} using LSTM forester and (B) correlation between the measured and forecaster P_{pv}.

	y(t)	y(t+1)	y(t+2)	y(t+4)
0	0.004730	0.018029	0.023845	0.006196
1	0.049905	0.146209	0.146051	0.144734
2	0.138600	0.141763	0.144809	0.144486
3	0.134342	0.140282	0.140835	0.135812
4	0.124409	0.126204	0.132638	0.133086
...	...	...	...	...

_train Training matrix (input data)

	y(t+5)	y(t+6)
0	0.049905	0.146209
1	0.138600	0.141763
2	0.134342	0.140282
3	0.124409	0.126204
4	0.120624	0.118795
...	...	...

_train Training matrix (output data)

Fig. 4.14 Data preparation for training the LSTM network (4 inputs and 2 outputs).

Example 4.16: LSTM for multistep

The following function (*data_preparation*) allows to build the training and testing data (**Listing 4.15a**).

```
#Listing 4.15a: data preparation function
def data_preparation(n,m,data):
    x_train, y_train=
    for i in range(n,len(data)-m,n):
        x_train.append(data[i-n:i])
        y_train.append(data[i:i+n-m])
    return x_train, y_train
n, m=50,30
x_train,y_train=data_preparation(n,m,data_train)
```

where *n* and *n-m* in the above listing are the number of input and the number of the output respectively; in this case $n=50$ and the number of the steps ahead is $n\text{-}m=20$ steps ahead, which means 20min ahead. An example of the prepared input and output training matrix (x_train and y_train) for $n=4$ and $m=2$ is shown in Fig. 4.14.

The following function (*normalization*) permits to normalize the data (**Listing 4.15b**):

```
#Listing4.15b : normalisationfunction
def normalisation (x_train,y_train):
    x_scaler=MinMaxScaler()
    x_train=x_scaler.fit_transform(x_train)
    y_scaler=MinMaxScaler()
    y_train=y_scaler.fit_transform(y_train)
    return x_train,y_train
x_train,y_train=normalisation(x_train,y_train)
```

The main code is given in **Listing 4.15c**.

```
#Listing 4.15c: sequential LSTMmodel
model=Sequential()
model.add(LSTM(50, activation='relu'))
model.add(Dense(n-m))
model.compile(loss='mean_squared_error', optimizer='adam')
history=model.fit( x_train, y_train, epochs=50, batch_size=128,
  verbose=1)
```

The configurations of the forecasters together with the statistical errors are listed in Table 4.5. The best MAE, RMSE, and MAPE correspond to the first configuration (two-steps ahead), and the convergence time increases with the number of inputs and outputs.

Fig. 4.15 shows as an example the comparison between the measured and the forecasted power values in the case of configuration #2 ($10 \times 50 \times 5$) corresponding to a five-steps ahead prediction.

4.8 Uncertainty quantification and interval prediction

Quantification of uncertainties associated with photovoltaic power forecasts is essential for PV plants (e.g., optimization, management, and control). A confidence interval

Table 4.5 Statistical errors and training times for different LSTM-based multi-step forecasters.

LSTM model parameters Number of unit = 50, batch size S = 128, epoch = 100	**RMSE (kW)**	**MAPE (%)**	***r* (%)**	**MAE (kW)**	**Training time (s)**
Configuration 1 ($4 \times 50 \times 2$)	0.129	0.644	98.6	0.063	352
Configuration 2 ($10 \times 10 \times 5$)	0.156	0.132	97.9	0.079	499
Configuration 3 ($30 \times 50 \times 10$)	0.186	0.139	96.7	0.106	626
Configuration 4 ($50 \times 50 \times 20$)	0.228	0.148	95.3	0.124	834
Configuration 5 ($180 \times 50 \times 30$)	0.338	1.230	79.4	0.281	2089

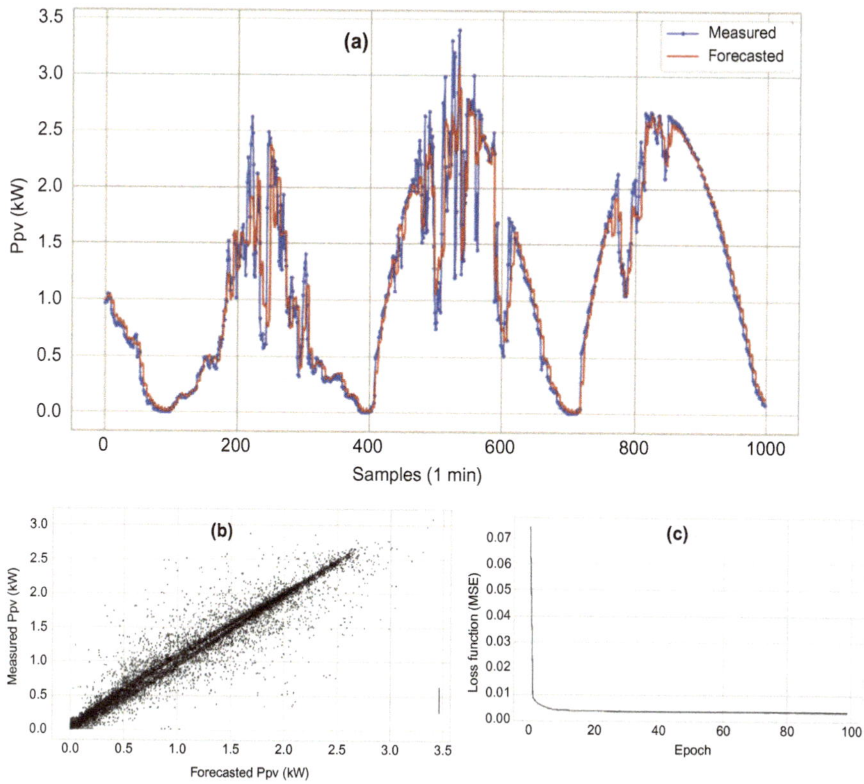

Fig. 4.15 LSTM-based multistep forecaster, configuration 2 ($10 \times 50 \times 5$): (A) supposition curve between measured and forecasted, (B) scatter curve, and (C) the loss function (MSE).

(CI) quantifies the uncertainty on an estimated population variable, such as the mean or standard deviation. However, a prediction interval (PI) quantifies the uncertainty on a single observation estimated from the population [19]. A CI can be used to quantify the uncertainty of the estimated ability of a model in forecasting, while a PI can be used to quantify the uncertainty of a single forecast.

4.8.1 Uncertainty quantification

Here, we use the Bootstrap CI [20] in order to evaluate the uncertainty of the power forecasted.

Example 4.17: Uncertainty quantification using Bootstrap algorithm

For example, the forecasted data by the LSTM (Configuration 1) case of multistep forecasting (two-steps ahead) are used. The Boostrap CI procedure is summarized as follows:

Step #1: Draw N samples from the original sample with replacement ($N = 10{,}000$).
Step #2: Find the median for each samples.
Step #3: Arrange these sample medians in order of magnitude.
Step #4: Calculate the middle 95% of the medians in order to get a 95% confidence percentile.

CIs deal with the accuracy of our LSTM forecaster model, i.e., of the mean of the target probability distribution [20]. The bootstrap is widely used to find and plot the sampling distribution of a statistic (e.g., mean). The basic bootstrap function is given in **Listing 4.16.**

```
#Listing 4.16: bootstrap function
def bootstrap(data, n=10000, func=np.mean):
    simulations=list()
    sample_size=len(data)
    xbar_init=np.mean(data)
    for c in range(n):
        itersample=np.random.choice(data, size=sample_size,
        replace=True)
    simulations.append(func(itersample))
  simulations.sort()
  def ci(p):
      u_pval=(1+p)/2.
      l_pval=(1-u_pval)
      l_indx=int(np.floor(n*l_pval))
      u_indx=int(np.floor(n*u_pval))
      return(simulations[l_indx],simulations[u_indx])
 return(ci,simulations)
```

The procedure was implemented in Python using the *percentile*, *resample,* and *accuracy_score* functions considering 5500 samples of the forecasted PV power by the LSTM model. As an example, Fig. 4.16 shows the distribution function for 10,000 bootstrap resampling; the mean value is 1.51 kW.

Table 4.6 presents the calculated bootstrap CI (lower and upper intervals) at different confidence percentiles (80%, 85%, 90%, and 95%).

As shown in Table 4.6, for all confidence percentiles, the mean forecasted power (1.51 kW) is included within the confidence interval.

4.8.2 Interval prediction

Prediction intervals consider the accuracy with which the targets themselves can be predicted, i.e., they are based on the estimates of the distribution [21]. PIs describe

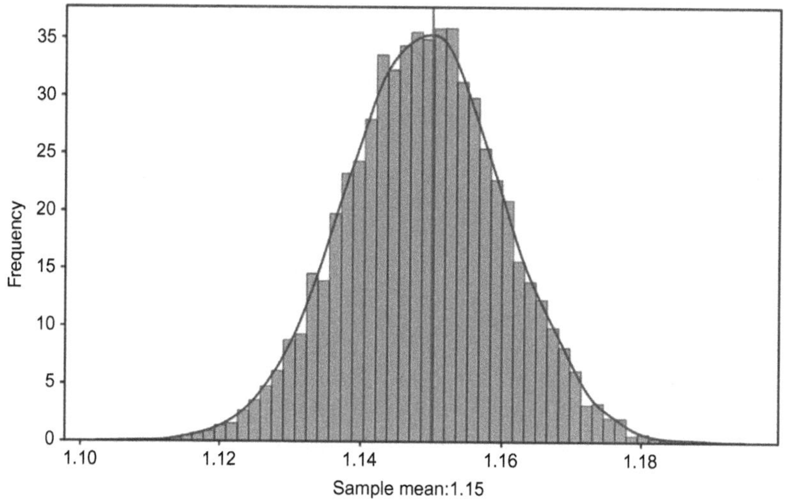

Fig. 4.16 The distribution function of the sample means.

Table 4.6 CI for different confidence percentiles.

Confidence percentiles (%)	Bootstrap CI		Mean (kW)	Standard error $\frac{\sigma}{\sqrt{n}}$
	Lower (kW)	Upper (kW)	1.51	0.011
80	1.724	1.756		
85	1.710	1.760		
90	1.706	1.762		
95	1.701	1.768		

the uncertainty for a single specific outcome. A PI is often larger than the confidence interval as it must take the confidence interval and the variance in the output variable being predicted, into account.

Example 4.18: Prediction interval using gradient boosting algorithm

In this example, a gradient boosting algorithm (*GradientBoostingRegressor)* is applied to estimate and plotting the interval predictions. For this, the first dataset 4.1 is used for one-step ahead PV power forecasting. The following code permits calculating the PI using the function *GradientBoostingRegressor* (see **Listing 4.17**).

```
#Listing4.17: gradientboostingalgorithm
from sklearn.ensemble import GradientBoostingRegressor
UPPER_ALPHA=0.9
LOWER_ALPHA=0.1
```

```
mid_model=GradientBoostingRegressor(loss="ls")
lower_model=GradientBoostingRegressor(loss="quantile",
  alpha=LOWER_ALPHA)
upper_model=GradientBoostingRegressor(loss="quantile",
  alpha=UPPER_ALPHA)
lower_model.fit(x_train, y_train)
mid_model.fit(x_train, y_train)
upper_model.fit(x_train, y_train)

pred=pd.DataFrame(y_test)
pred['mid']=mid_model.predict(x_test)
pred['lower']=lower_model.predict(x_test)
pred['upper']=upper_model.predict(x_test)

v=pred.values
real=v[:,0]
lower1=v[:,1]
mid1=v[:,2]
upper1=v[:,3]
L=400
x=list(range(1, L))
fig=plt.figure()
x1=real[1:L]
x2=mid1[1:L]
x3=upper1[1:L]
x4=lower1[1:L]
plt.plot(x, x1, 'b.', markersize=8, label=u'Measured')
plt.plot(x, x2, 'r', label=u'Forecasted')
plt.plot(x, x3, 'k')
plt.plot(x, x4, 'k')
plt.fill(np.concatenate([x, x[::-1]]),np.concatenate([x3, x4[::-
  1]]),alpha=0.5, fc='gray', ec='None', label='95%
  predictioninterval')
plt.xlabel('Samples')
plt.title('Predictionintervals')
plt.ylabel('PVPowerPpv(W)')
plt.legend()
plt.show()
```

As an example, Fig. 4.17 shows the calculated interval prediction (alpha = 95%) for the forecasted PV power. As can be seen in this plot, the variation of the forecasted PV power does not exceed the interval prediction limit (shaded area in zoomed portion).

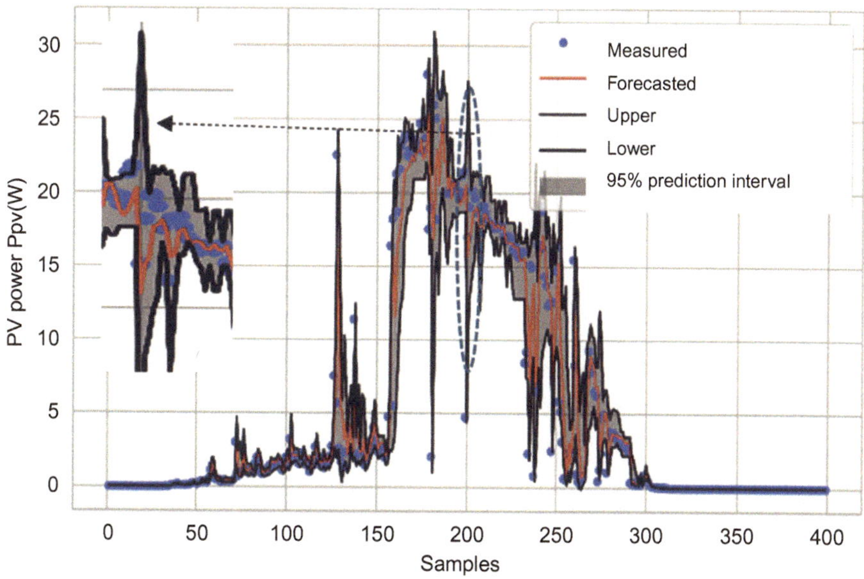

Fig. 4.17 Example of the prediction interval for 1 day (400 samples) using a gradient boosting algorithm.

4.9 Summary

A total of 18 examples, including codes with explanations, have been described and discussed. The codes have been developed using Python language and Matlab environment for one-step and multistep forecasting. Four databases with different timescales (1 min, 2 min, and 1 h) measured in four regions have also been used for this purpose. PV power prediction interval and confidence interval are also considered in this chapter.

Supervised machine learning algorithms are the most used in this topic, while very limited works are related to the application of unsupervised machine learning algorithms. Long-term PV power forecasting is not considered in this chapter, but it could be achieved by using NWP models. Machine learning-based forecasters, on the other hand, have shown their capability for PV power forecasting as well as deep learning-based models. However, deep learning-based models are more suitable in the case of large databases, unlike machine learning-based models.

References

[1] IRENA, World Adds Record New Renewable Energy Capacity in 2020. https://www.irena.org/newsroom/pressreleases/2021/Apr/World-Adds-Record-New-Renewable-Energy-Capacity-in-2020. (Accessed 5 April 2021).

[2] S. Jerez, I. Tobin, R. Vautard, J.P. Montávez, J.M. López-Romero, F. Thais, B. Bartok, O. B. Christensen, A. Colette, M. Déqué, G. Nikulin, The impact of climate change on photovoltaic power generation in Europe, Nat. Commun. 6 (1) (2015) 1–8.
[3] M.C. Peel, B.L. Finlayson, T.A. McMahon, Updated world map of the Köppen-Geiger climate classification, Hydrol. Earth Syst. Sci. 11 (5) (2007) 1633–1644.
[4] A. Livera, M. Theristis, G. Makrides, G.E. Georghiou, Recent advances in failure diagnosis techniques based on performance data analysis for grid-connected photovoltaic systems, Renew. Energy 133 (2019) 126–143.
[5] A. Mellit, G.M. Tina, S.A. Kalogirou, Fault detection and diagnosis methods for photovoltaic systems: a review, Renew. Sust. Energ. Rev. 91 (2018) 1–17.
[6] A. Mellit, A.M. Pavan, V. Lughi, Deep learning neural networks for short-term photovoltaic power forecasting, Renew. Energy 172 (2021) 276–288.
[7] A.M. Nobre, C.A. Severiano Jr., S. Karthik, M. Kubis, L. Zhao, F.R. Martins, E.B. Pereira, R. Rüther, T. Reindl, PV power conversion and short-term forecasting in a tropical, densely-built environment in Singapore, Renew. Energy 94 (2016) 496–509.
[8] A. Mellit, A.M. Pavan, V. Lughi, Short-term forecasting of power production in a large-scale photovoltaic plant, Sol. Energy 105 (2014) 401–413.
[9] W. Yin, Y. Han, H. Zhou, M. Ma, L. Li, H. Zhu, A novel non-iterative correction method for short-term photovoltaic power forecasting, Renew. Energy 159 (2020) 23–32.
[10] J. Antonanzas, N. Osorio, R. Escobar, R. Urraca, F.J. Martinez-de-Pison, F. Antonanzas-Torres, Review of photovoltaic power forecasting, Sol. Energy 136 (2016) 78–111.
[11] A. Mellit, A. Massi Pavan, E. Ogliari, S. Leva, V. Lughi, Advanced methods for photovoltaic output power forecasting: a review, Appl. Sci. 10 (2) (2020) 487.
[12] E. Ogliari, A. Nespoli, Photovoltaic plant output power forecast by means of hybrid artificial neural networks, in: A Practical Guide for Advanced Methods in Solar Photovoltaic Systems, Springer, Cham, 2020, pp. 203–222.
[13] F. Lazzeri, Machine Learning for Time Series Forecasting With Python, John Wiley & Sons, 2020.
[14] J.Á.G. Ordiano, S. Waczowicz, M. Reischl, R. Mikut, V. Hagenmeyer, Photovoltaic power forecasting using simple data-driven models without weather data, Comput. Sci. Res. Dev. 32 (1–2) (2017) 237–246.
[15] I.K. Nti, A.F. Adekoya, B.A. Weyori, A comprehensive evaluation of ensemble learning for stock-market prediction, J. Big Data 7 (1) (2020) 1–40.
[16] S. Raschka, Python Machine Learning, Packt Publishing, 2015.
[17] A.C. Müller, S. Guido, Introduction to Machine Learning With Python: A Guide for Data Scientists, O'Reilly Media, 2016.
[18] https://keras.io/.
[19] D.P. Foster, R.A. Stine, R.P. Waterman, Prediction and confidence intervals in regression, in: Business Analysis Using Regression, Springer, New York, NY, 1998, pp. 85–104.
[20] B. Efron, R.J. Tibshirani, An Introduction to the Bootstrap, Chapman and Hall, New York, NY, 1993, https://doi.org/10.1007/978-1-4899-4541-9.
[21] T.M. Heskes, W.A.J.J. Wiegerinck, H.J. Kappen, Practical confidence and prediction intervals for prediction tasks, in: Progress in Neural Processing, World Scientific, 1997, pp. 128–135.

Optimization of photovoltaic systems based on artificial intelligence techniques

5

Chapter outline

5.1 Introduction

Optimization of photovoltaic (PV) array is an interesting step to capture the maximum yield by considering climate change effects, environmental issues, and other factors that lower the efficiency of photovoltaic devices [1]. The PV system should produce the maximum power during its lifetime, under different climatic conditions and loads. Optimization of the PV power produced by PV plants can be mainly achieved by two methods: (1) tracking the maximum power point of the PV array and (2) rearrangement of photovoltaic modules in an array in order to avoid shading effect. Both methods aim to reduce output power losses and are based on the development of an algorithm that can be basically implemented in digital circuits, such as microcontroller, dSPACE, and reconfigurable field programmable gate array (FPGA), for real-time applications.

This chapter covers the application of optimization methods based on artificial intelligence techniques for maximizing the photovoltaic output power under uniform and non-uniform solar conditions. The material presented is mainly focused on the application of fuzzy logic, neural networks, neuro-fuzzy, particle swarm optimization, and various hybrid combinations of these techniques. There are mainly two ways for maximizing the PV output power: (1) by tracking the maximum power from a photovoltaic array using maximum power tracking algorithms and (2) by photovoltaic array reconfiguration (static or dynamic techniques), which are globally used to prevent shading effects. Eight examples are presented in detail using Matlab®/Simulink.

Handbook of Artificial Intelligence Techniques in Photovoltaic Systems. https://doi.org/10.1016/B978-0-12-820641-6.00005-3
Copyright © 2022 Elsevier Ltd. All rights reserved.

5.2 Maximum power point tracking methods

Maximum power point tracking (MPPT) methods were designed to extract the maximum power of PV modules (string or array). A large number of MPPT methods were developed based on classical approaches, advanced methods, and other hybrid combinations [2]. Among classical methods, hill climbing (HC), perturb and observe (P&O), incremental conduction (InCond), and other improved versions are the most used ones. These methods are simple and can be implemented easily, however, they have shown lower efficiency particularly under a rapid variation of solar irradiance and partially shading conditions (PSCs) [3]. Methods based on artificial intelligence (AI) techniques, mainly using fuzzy logic (FL) and neural networks (NN) have proven their ability to solve this issue [4,5], such as tracking the maximum power under the fast transition of solar irradiance. Metaheuristic methods such as genetic algorithm (GA), particle swarm optimization (PSO), and other evolutionary algorithms [6] are the most suitable in the case of PSCs; however, their implementation for real-time applications is challenging as they need more skill and experience [7]. Other hybrid methods which combine one classical method and one advanced method have also shown their benefits to improve efficiency as well as reduce the implementation cost [8,9]. Numerous MPPTs have been verified experimentally at the laboratory scale [10,11]. To the best knowledge of the authors, conventional methods are the most commercialized techniques, and among them are the MPPT-based artificial intelligence methods.

A PV module has a highly non-linear I-V characteristic varying with the solar irradiance and air temperature. The low energy conversion efficiency of PV cells remains a barrier to the prolific growth of solar electricity. Fig. 5.1A and B shows the plot of real P-V curves measured under uniform and non-uniform solar irradiance. As can be seen under uniform solar irradiance, the P-V characteristic presents one unique point, the so-called global maximum power point (GMPP), at which the PV array operates with maximum efficiency and produces maximum output power.

As the P-V characteristics of a PV module vary with solar irradiation, under non-uniform solar irradiance, it is possible to have multiple local MPPs (LMPPs) and only one GMPP for the entire array. Under non-uniform conditions, the shaded solar cells (see Fig. 5.1C) become reverse biased and behave as load, discharging the power against the fully radiated cells. The non-uniform solar irradiance occurs quite frequently due to clouds and shadowing from various items such as leaves, electric poles, and neighboring buildings and trees.

When a PV module/string or array is connected directly to a load without any adaptation stage, the maximum power cannot be extracted from the PV module (see Fig. 5.2). The efficient operation of a PV array requires an optimum transfer of energy generated in the PV array to the load.

Thus, to improve the conversion efficiency of the electric power generation, an adaptive stage (DC-DC converter) with a MPPT algorithm should be integrated into the whole installation so that the PV module/string or array will be able to deliver the maximum power available under all possible system operation conditions, as shown schematically in Fig. 5.3.

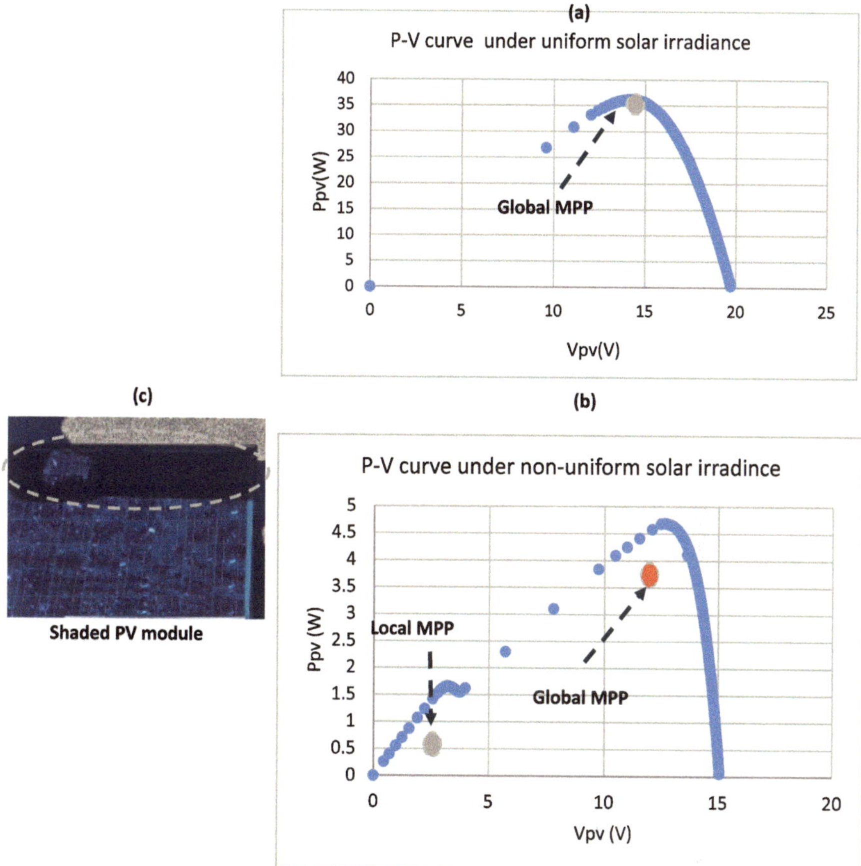

Fig. 5.1 Measured P-V curves of a PV module: (A) uniform solar irradiance, (B) non-uniform solar irradiance, and (C) PV module under non-uniform solar irradiance (shaded solar cells).

DC-DC converters used to track the MPPT can be classified into two categories: (i) non-isolated and (ii) isolated. The non-isolated category includes boost (step-up), buck (step-down), and buck-boost. The isolated category comprises flyback, forward, push-pull, half-bridge, and full-bridge [12].

Isolation describes the electrical separation between the input and output of a DC-DC converter. A non-isolated DC-DC converter has a DC path between its input and output (Fig. 5.4A), while an isolated DC-DC converter uses a transformer to eliminate the DC path between its input and output (Fig. 5.4B), mainly use or safety issue.

A considerable number of MPPT algorithms are available in the literature [2].

5.3 Photovoltaic array reconfiguration methods

Nowadays, fixed PV architecture is one of the most popular architectures employed in PV systems. This kind of architecture can be easily realized, with an exact and simple

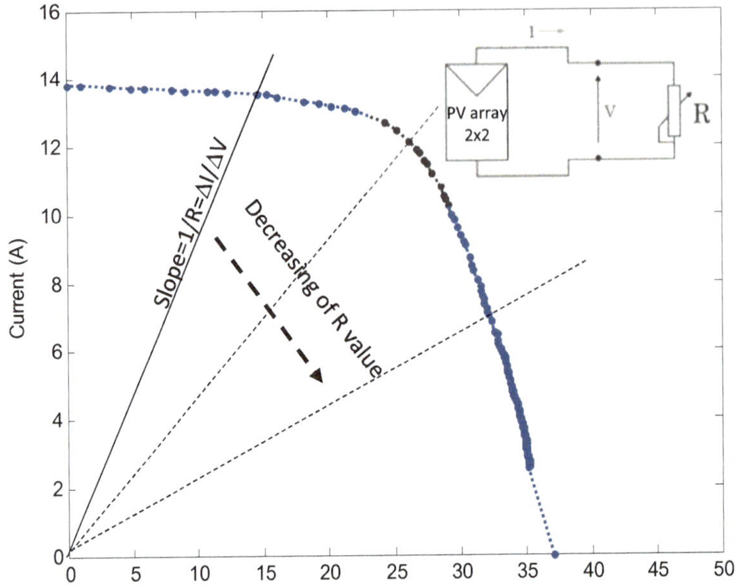

Fig. 5.2 I-V curve of a PV array module (4 PV modules connected in 2 series by 2 parallel) connected to a restive load.

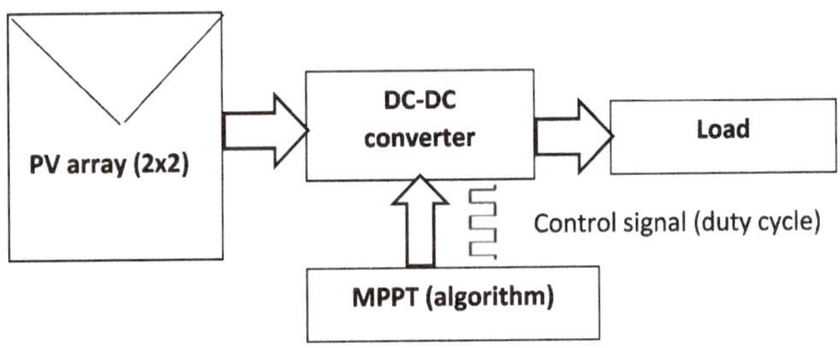

Fig. 5.3 Basic structure of a PV array connected with a DC-DC converter and MPPT algorithm.

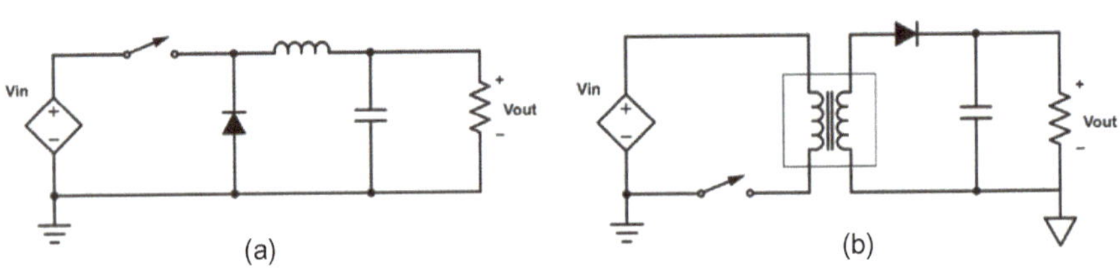

Fig. 5.4 DC-DC converters: (A) non-isolated DC converter and (B) isolated DC-DC converter using a transformer.

resizing of the PV array rows at a minimal cost. However, under partially shaded conditions (PSCs), this architecture cannot perform well due to performance degradation. Reconfigurable PV array architectures are more effective than the fixed ones in terms of the reduction of partial shading effect.

Methods based on PV array reconfiguration are basically designed in order to prevent the shading effect, by making a rearrangement of PV modules in an array. Numerous configurations are proposed in the literature [13]; they can be also classified into two main techniques: passive and active techniques [14].

For example, Fig. 5.5A–D shows a 4×4 partially shaded PV array with different configurations series-parallel (SP), total cross tied (TCT), bridged-linked (BL), and honey-comb (HC) [15]. The corresponding P-V curves are depicted in Fig. 5.5E.

Static techniques consist of relocating the shaded modules to the un-shaded positions and equalizing the row current difference, e.g., SuDoKu, Zig-Zag, Magic Square, Puzzle, non-symmetrical reconfiguration, and other techniques [13,14] can be employed. These techniques are relatively simple and do not require additional switches or sensors, but their main limitation is that they require skilled persons to

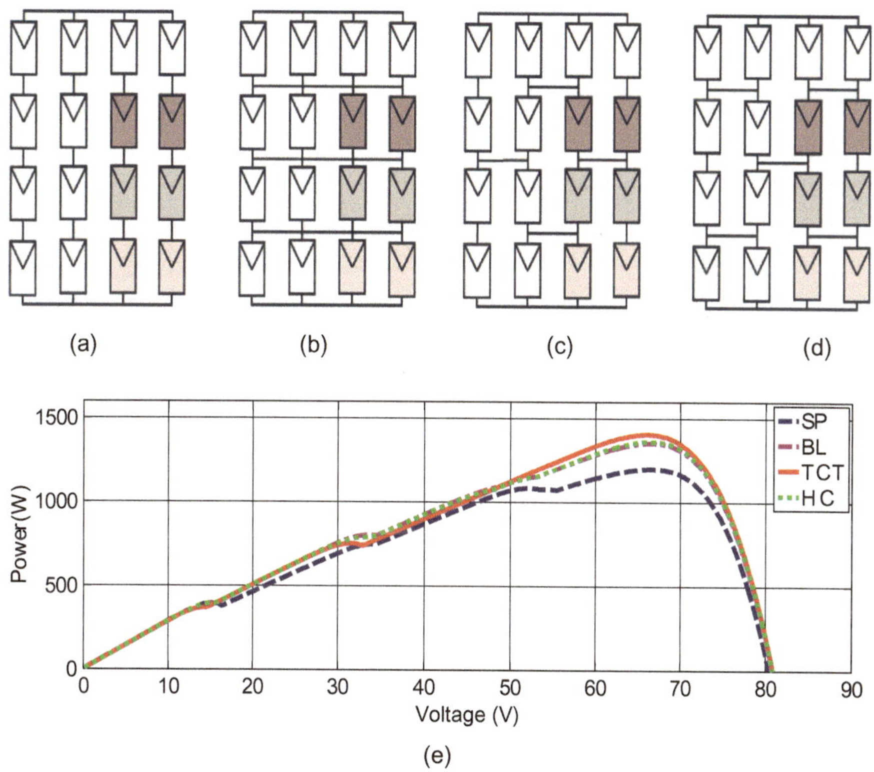

Fig. 5.5 PV array architectures, (A) SP, (B) TCT, (C) BL, (D) HC, and (E) the corresponding P-V curves.

perform physical changes in the installation with high wiring complexity and are mainly appropriate for small-array configurations.

Dynamic techniques are regrouped basically into three categories: (i) electrical array reconfiguration, (ii) irradiation equalization, and (iii) adaptive array reconfiguration. In these techniques, fuzzy logic and metaheuristic algorithms are the most used ones. These techniques are appropriate for large-scale PV arrays; they are usually implemented in electronic circuits which increase the cost of the overall system which are relatively expensive as additional equipment is required. The implementation of a control algorithm aims to manage the switching matrix.

A typical structure of a dynamic PV array reconfiguration method is shown in Fig. 5.6. It consists of a PV array under partially shaded conditions, switches, and a control algorithm which could be implemented on a low-cost microcontroller. The PV array is configured based on the following configuration: SP, TCT, HC, and BL [13]. The switching matrix generates the switching conditions to adjust the connections between PV modules to scatter the shade. During PSC, the control technique reconfigures the PV array dynamically by receiving the switching conditions from the switching matrix.

The control algorithm aims to find an optimal configuration of a PV array under different patterns of shading. Numerous algorithms have been developed in the

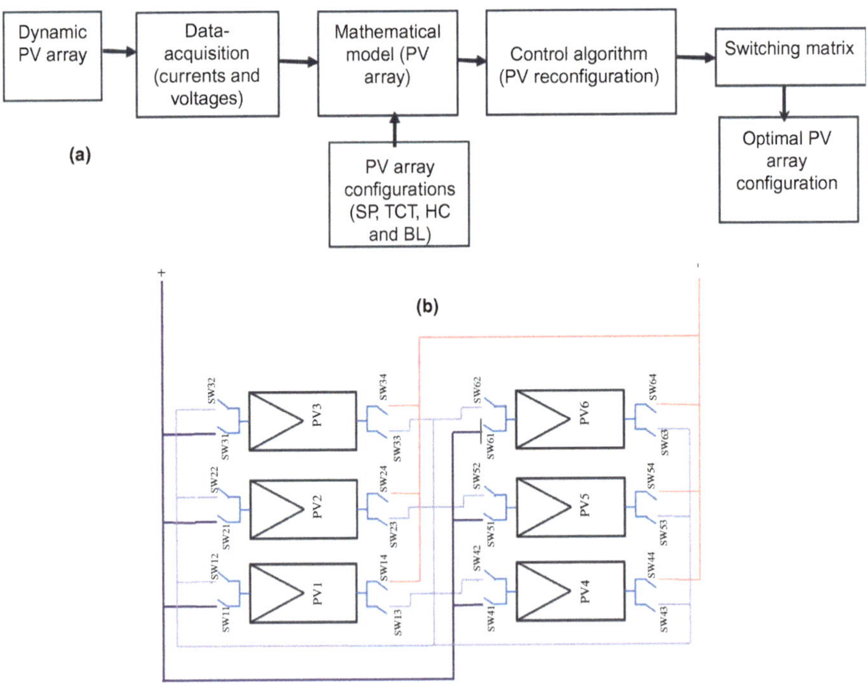

Fig. 5.6 (A) A basic structure of dynamic PV array with switching matrix and (B) an example of a dynamic PV array.

literature [14], such as electrical array configuration, irradiation equalization reconfiguration, adaptive array reconfiguration, metaheuristic optimization, and other approaches.

Generally, TCT is the most common topology in reconfigurable PV array architectures, especially for dynamic PV array reconfigurations, due to the large number of parallel strings included in this configuration. Despite the complexity of its interconnection scheme, the TCT topology offers good energy performance. TCT-based reconfiguration methods require a high number of sensors and switches and sometimes even complex control algorithms that in turn require a large computation process to determine when to turn on or off the switches.

5.4 Maximum power point tracking using artificial intelligence techniques

This section aims to provide some examples of the application of neural networks, fuzzy logic, and metaheuristic algorithms.

Example 5.1: Tracking the MPP using P&O algorithm and a PI controller

Fig. 5.7A shows a basic Matlab/Simulink model of the well-known perturb and observe (P&O) and proportional integral (PI) controllers. The power rating of the PV array is 100 kW which consists of 50 parallel strings; each string contains 10 PV modules connected in series. The pseudo-code of the P&O algorithm is summarized as follows [16]:

```
Read Vpv(n), Ipv(n), Vpv(n-1) and Ipv(n-1),
Initialize dD and
calculate dPpv(n)=Ppv(n)-Ppv(n-1); dVpv(n-1)=Vpv(n)-Vpv(n-1);
If dPpv>0 then
      If dVpv> 0 then D(n)← D(n)+dD
      Else D(n)← D(n)-dD
      Endif
Else
      If dVpv> 0 then D(n)← D(n)-dD
      Else D(n)← D(n)+dD
      Endif
Endif
Where dD is the variation step, D is duty cycle
```

The above P&O code is implemented as a Matlab function (P&O). The PI controller parameters have been estimated by trial-error tuning method, which is based on guess-and-check, or could be estimated using available tuning tool in MathWorks as shown in Fig. 5.7B [17].

The MPPT algorithm has been evaluated under different values of solar irradiation with the fast transition of solar irradiance as shown in Fig. 5.8.

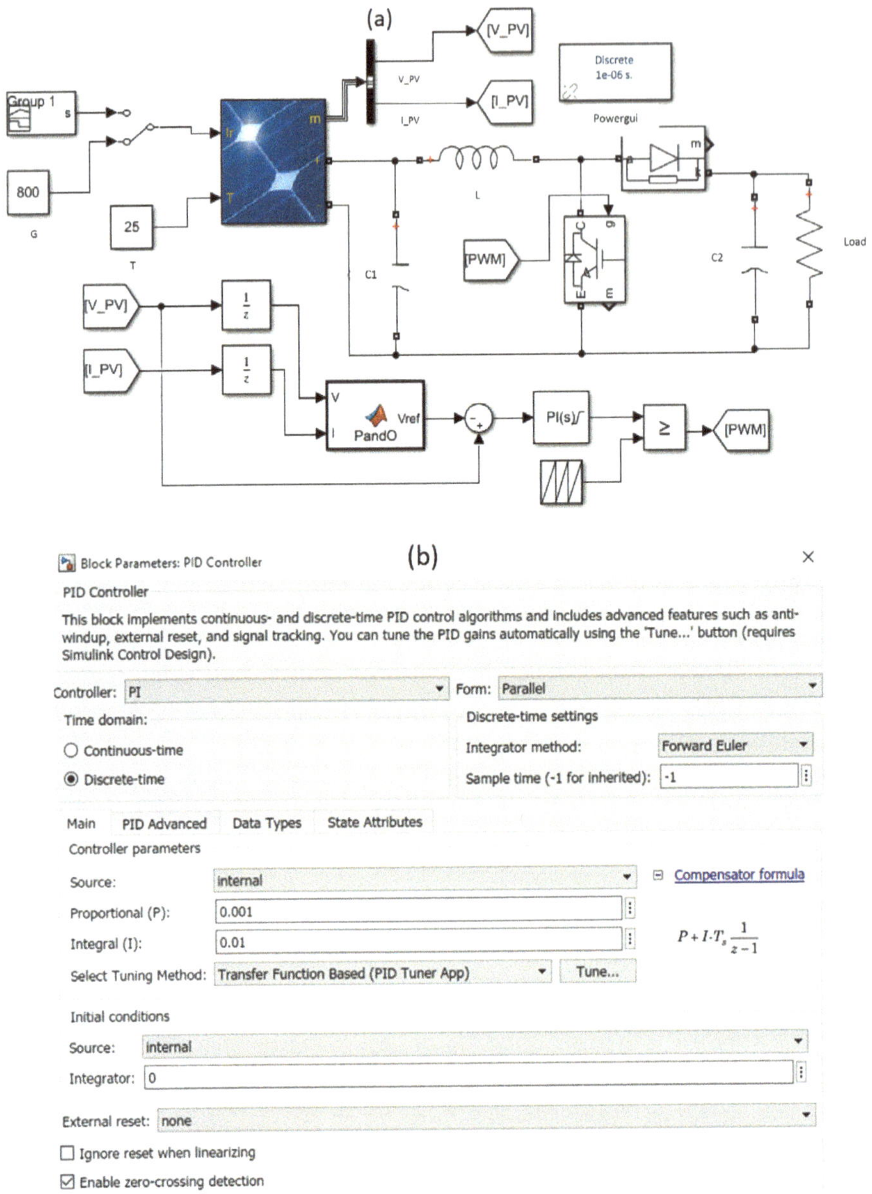

Fig. 5.7 (A) Simulink model of perturb and observe algorithm and (B) block parameters of PI controller.

As can be seen, the maximum power is reached after a few oscillations, however, in the case of rapid variation of solar irradiance, the algorithm presents more oscillations around the MPP which leads to a remarkable loss in power. Efficiency given in Eq. (5.1) is the indicator, mostly used, to evaluate the performance of the MPPT algorithms.

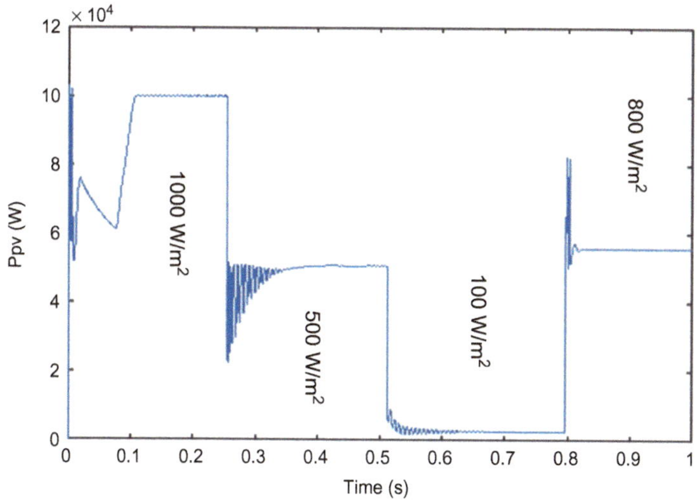

Fig. 5.8 MPPT-P&O tracking performance under a fast transition of solar irradiance.

$$\text{eff} = \frac{\int_0^t P_{mp}(t)dt}{\int_0^t P_{\max}(t)dt} \tag{5.1}$$

where P_{mp} represents the output power of the PV system with MPPT and P_{max} is the output power at true MPP.

5.4.1 Tracking the maximum power point under uniform conditions using AI techniques

MPPT algorithms based on AI are developed to improve the efficiency of the system and solve some issues presented by conventional algorithms (such as P&O, InCond, etc.). Thus, this section aims to present four examples of the application of AI techniques in tracking the maximum power, which includes fuzzy logic, optimized fuzzy logic, neural networks, and neuro-fuzzy controllers. The whole system consists of a PV module/string or array, current and voltage sensors, DC-DC boost converter, a load, and the main MPPT block as shown in Fig. 5.9. Matlab/Simulink is used to implement the different MPPTs [18].

The input of the MPPT block differs from one method to another, e.g., in the case of the fuzzy logic controller (FLC) only measured I_{pv} and V_{pv} are needed, while in the case of machine learning including, neural networks (NNs), support vector machine (SVM), K-nearest neighbor (K-NN), and other hybrid configurations, such as neuro-fuzzy, only the solar irradiance and air temperature are used. The output is the same (duty cycle D) for most MPPT techniques which are used to command the switch of

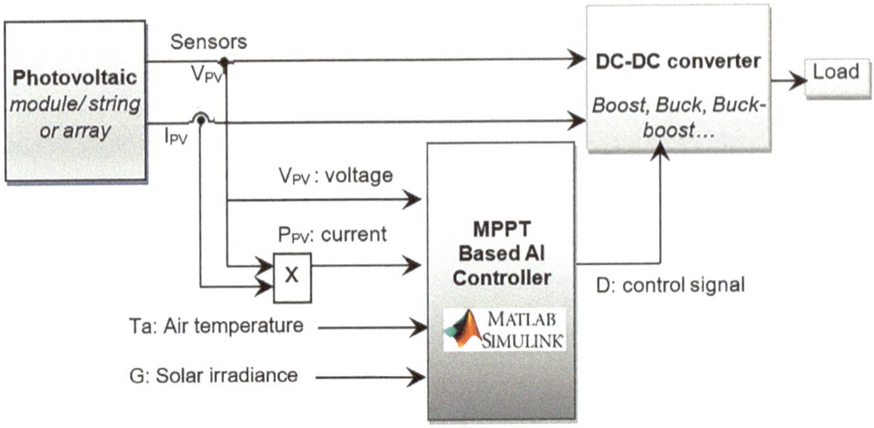

Fig. 5.9 Basic structure of MPPT-based AI controller using Matlab/Simulink.

the converter (MOSFET or IGBT power transistors). Matlab/Simulink is the main programming language used in this topic, due to the availability of various toolboxes, e.g., FL, NNs, machine learning and statistics, and SimScape toolboxes.

Example 5.2: Tracking the MPP using MPPT-FL controller (fast variation of solar irradiance)

The Simulink model of the MPPT-based FL controller is shown in Fig. 5.10A. The PV generator consists of four MSX-80 PV modules connected in series. The rated power is about 332 W at STC.

The FLC generates the control signal of the DC-DC boost converter that interfaces the PV source with the load. The FLC considered is based on Mamdani fuzzy inference system with two inputs and one output. The inputs are defined as the error (E) and the change of error (ΔE) such that [18]

$$E(n) = \frac{P_{pv}(n) - P_{pv}(n-1)}{V_{pv}(n) - V_{pv}(n-1)} \tag{5.2}$$

$$\Delta E(n) = E(n) - E(n-1) \tag{5.3}$$

The FLC output is the change of duty cycle, which is calculated according to the centroid method as follows:

$$\Delta D = \frac{\sum_{i=1}^{N} \mu_i dD_i}{\sum_{i=1}^{N} \mu_i} \tag{5.4}$$

where μ_i is the weighting factor and dD_i is the value that corresponds to the membership function of ΔD. The duty cycle value used to generate the control signal of the boost converter is updated according to

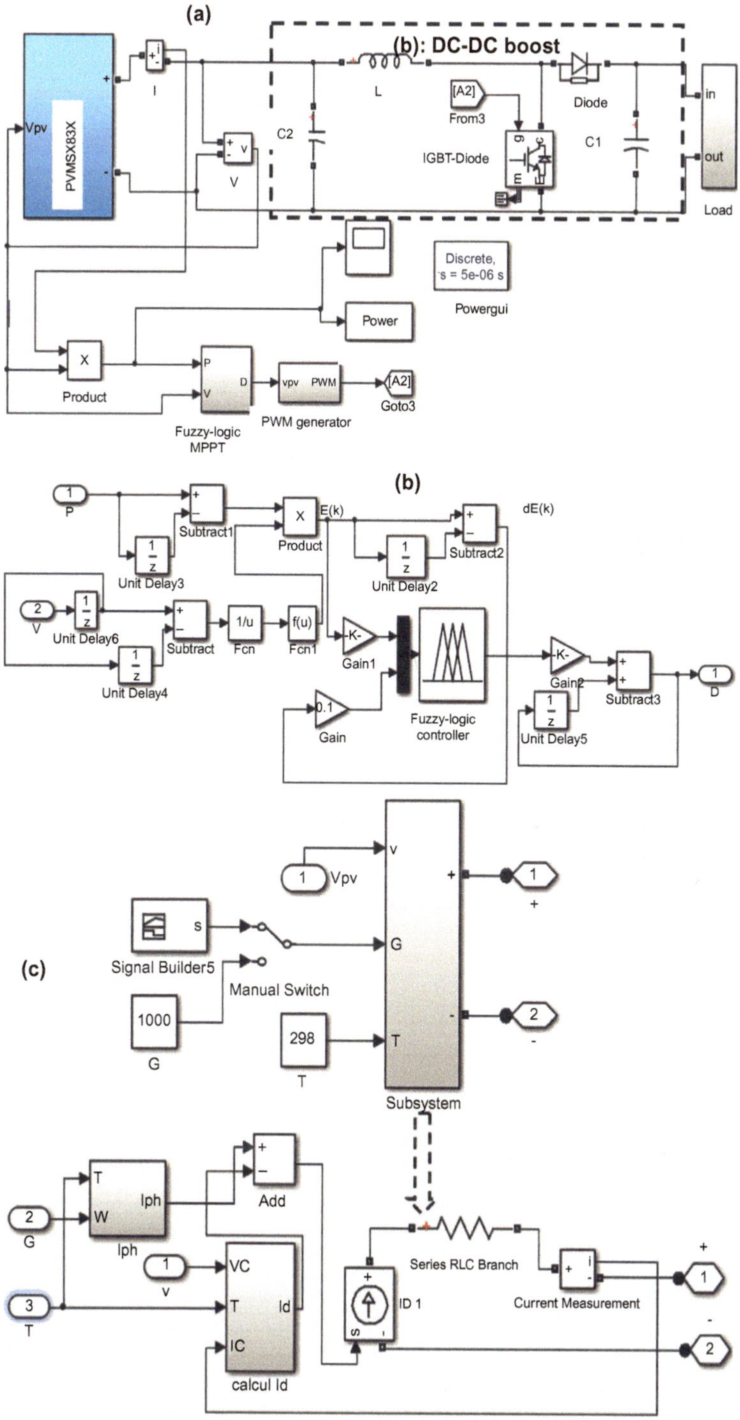

Fig. 5.10 (A) Simulink model of MPPT-based fuzzy logic controller, (B) MPPT-based fuzzy logic method, and (C) MSX80 PV array.

$$D(n) = D(n-1) + \Delta D(n) \tag{5.5}$$

Fuzzy tool is used to create the MPPT fuzzy rules as shown in Fig. 5.11A. Fuzzy surface is shown in Fig. 5.11B.

The Gaussian and generalized bell membership functions were chosen for the premises and consequents of FLC are illustrated in Fig. 5.12A–C.

There are five fuzzy sets of input and output variables represented by the linguistic labels NB (Negative Big), NM (Negative Medium), NS (Negative Small), ZE (Zero), PS (Positive Small), PM (Positive Medium), and PB (Positive Big). Hence, 25 fuzzy rules, which are given in Table 5.1, are used to decide the correct perturbation of the duty cycle. If the operating point of the PV generator moves in the right direction toward MPP, such that the power is increased, the same direction of perturbation of *D* is kept. Otherwise, the sense of perturbation is inversed. In order to check the static and dynamic performance of the FLC, the PV system under study is simulated for variable irradiation levels (1000, 500, 400, 200, and 100 W/m^2) while the air temperature is kept constant at 25°C.

The MPPT-based FLC steps are summarized as follows:

Step #1: Set an initial value for the duty ratio (D).
Step #2: Measure Vpv(n) and Ipv(n).
Step #3: Calculate Ppv(n) ← Vpv(n)*Ipv(n).
Step #4: Calculate Error E(n) ← [Ppv(n)-Ppv(n-1)]/[Vpv(n)-Vpv(n-1)] and change in error ΔE(n) = E(n)-E(n-1).
Step #5: Fuzzificatin and defuzzification based on the rules created.
Step #6: Calculate ΔD and update D ← D + ΔD.

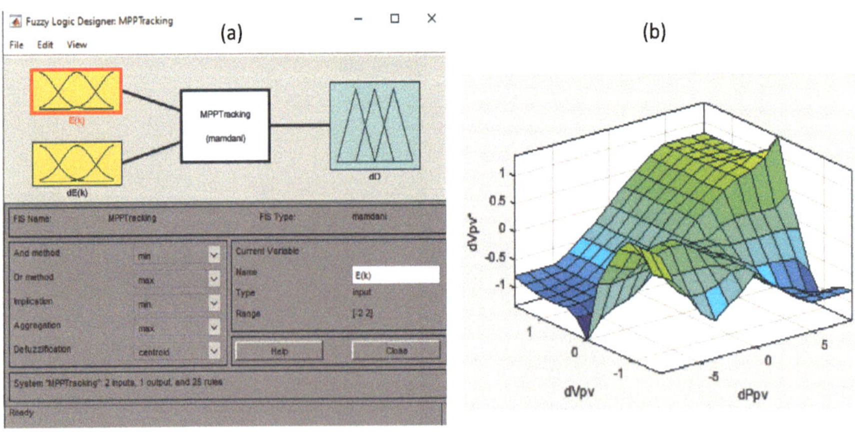

Fig. 5.11 (A) Fuzzy tools for designing the rules and (B) fuzzy surface of FLC-MPPT.

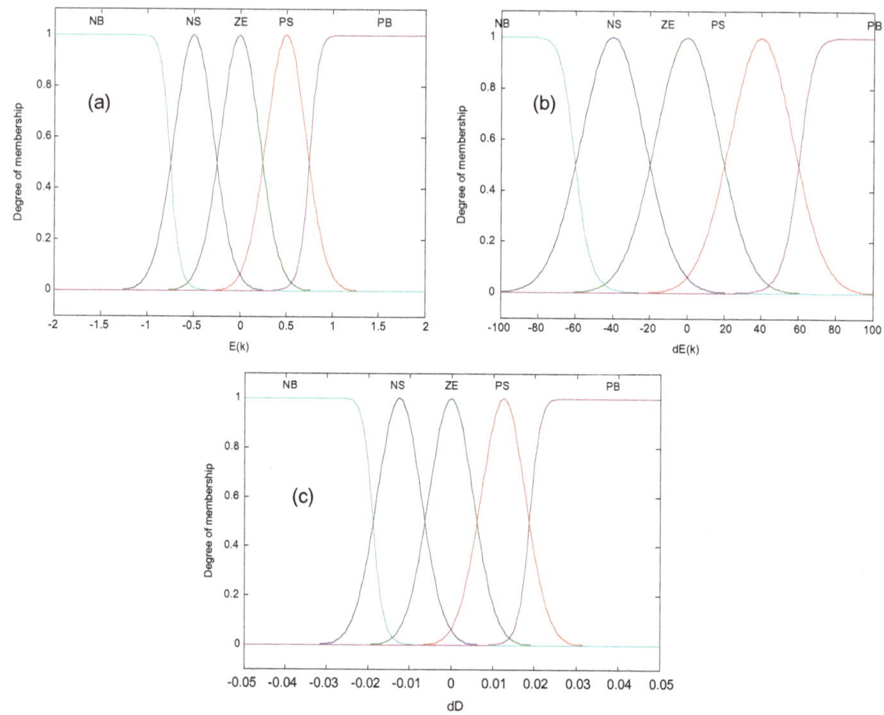

Fig. 5.12 The membership functions for the system variable: (A) E, (B) ΔE, and (C) dD.

Table 5.1 Adopted rules of the MPPT-based FLC.

E	ΔE	NB	NS	ZE	PS	PB
NB		PB	PB	PS	PB	PB
NS		PB	PS	PS	PS	PB
ZE		NS	NS	ZE	PS	PS
PS		NB	NS	NS	NS	NB
PB		NB	NB	NS	NB	NB

As can be observed from Fig. 5.13 the FLC adopted has a fast convergence speed and minimal static oscillations around the MPP.

Generally, the FLC provides acceptable results and can be easily implemented. Furthermore, this technique does not require the knowledge of the characteristics of the exact PV model used. However, the shortcoming of fuzzy computation concerns the ability to obtain the correct fuzzy rules and membership functions, which heavily rely on prior knowledge of the system and has an impact on the robustness and reliability of the method. Moreover, in the absence of expert knowledge, the design becomes slow and not optimized. The developed FLC is more suitable in the case

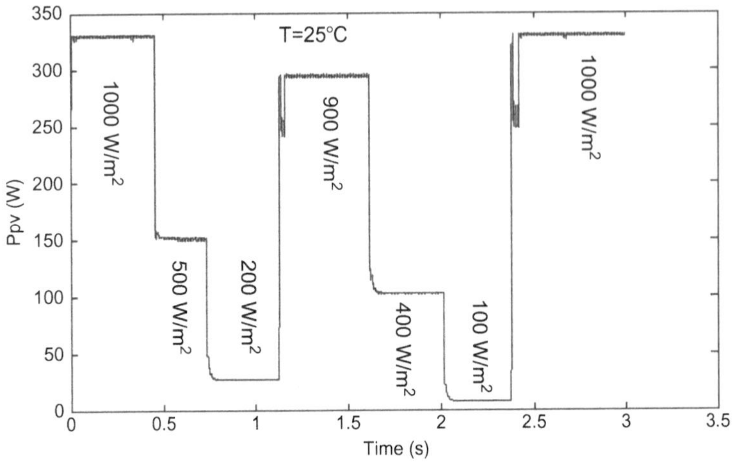

Fig. 5.13 MPPT-FLC tracking performance under a fast transition of solar irradiance.

of fast transition of solar irradiance, but will not be able to track the MPP under abnormal conditions (PSCs) [8].

Example 5.3: Tracking the MPP combining FL with one evolutionary algorithm

To optimize the FLC, tuning membership functions, and rule base parameters, the evolutionary algorithms have been developed [19] and proven their capability to improve the efficiency [20,21]. The most used evolutionary algorithms in this topic are the Genetic Algorithms (GA), PSO, and Cuckoo Search (CS) [22]. In this example CS with FL are employed. The same Simulink model shown in Fig. 5.8A is used, but the MPPT block is modified by introducing new optimized membership functions based on CS optimization and FL.

The optimization done by CS is based on the maximization of integral absolute error (IAE) given by

$$ISE = \int_0^{\infty} \left|P_{\text{pv}}(t)^* - P_{\text{pv}}(t)\right| dt \tag{5.6}$$

where $P_{\text{pv}}(t)^*$ is the power value calculated at STC, and $P_{\text{pv}}(t)$ is the power value measured at instant t.

As it stands, IAE definition cannot be used directly as a fitness parameter, since the value of fitness must increase with efficiency, thus, for evaluating the fitness of different nests, the following equation is used:

$$\text{Fitness} = 1000/(1 + \text{ISE}) \tag{5.7}$$

From which obviously, the more the fitness value tends toward the value 1000, the more the solution approaches the global optimum, which in this case is the maximum power point. CS is based on three idealized rules [23]:

- each cuckoo lays one egg at a time, and dumps it in a randomly chosen nest;
- the best nests with high quality of eggs (solutions) will be carried over to the next generations;
- the number of available host nests is fixed, and a host can discover an alien egg with probability pa ∈ [0 1]. In this case, the host bird can either throw the egg away or abandon the nest to build a completely new nest in a new location.

The pseudocode of CS algorithm is given by Ref. [23].

```
Objective function f(x) x=(x1, x2, ...xd)T
Generate initial population of n host nests xi(1,2,...,n)
While (t <MaxGeneration) or (stop criterion)
Get a cuckoo randomly by Levy flights
evaluate its quality / fitness Fi
Choose a nest among n (say .j) randomly
if (Fi >Fj) then
replace j by the new solution;
end if
A fraction pa of worst nests is abandoned and new is build;
Keep the best solutions (or nests with quality solutions;
Rank the solutions and find the current best;
End while
Post process result and visualization
```

The optimal values of the CS are: population size ($n=20$), probability of discovery (pa = 0.25), and maximum generation number (max_gen = 100). Fig. 5.14 shows the evolution of the highest and average fitness, for all research processes. The highest fitness value increases gradually during the research process, converging toward its maximum value 1000, corresponding to the optimal solution, reached during

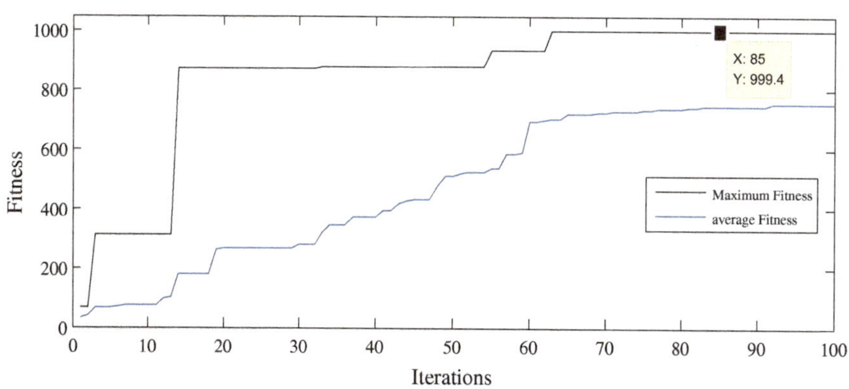

Fig. 5.14 Evolution of maximum and average fitness of the CS algorithm.

generation number 85. The average fitness curve shows that, during the research process, the solutions (nests) approach the best solution.

The shape of the membership function for the three variables (E, ΔE, and D) is shown in Fig. 5.15 and Table.5.2 reports the tuned rule base.

For abrupt changes of solar irradiance, the profile of Fig. 5.16A, was applied to a PV array with MPPT-based CS-FLC. The curves of Fig. 5.16B shows that the approach behaves as expected for different levels of solar irradiance change, with a transition time from one level to another, at about 0.19 s.

The simulation results shown in Fig. 5.16B, show that the examined MPPT based on CS-FLC presents a very good accuracy. This is achieved by optimizing the membership function shapes, the logic rules selection, and their weights simultaneously. The optimization of weights allows to minimize the logic rule number, which is very beneficial for hardware implementation.

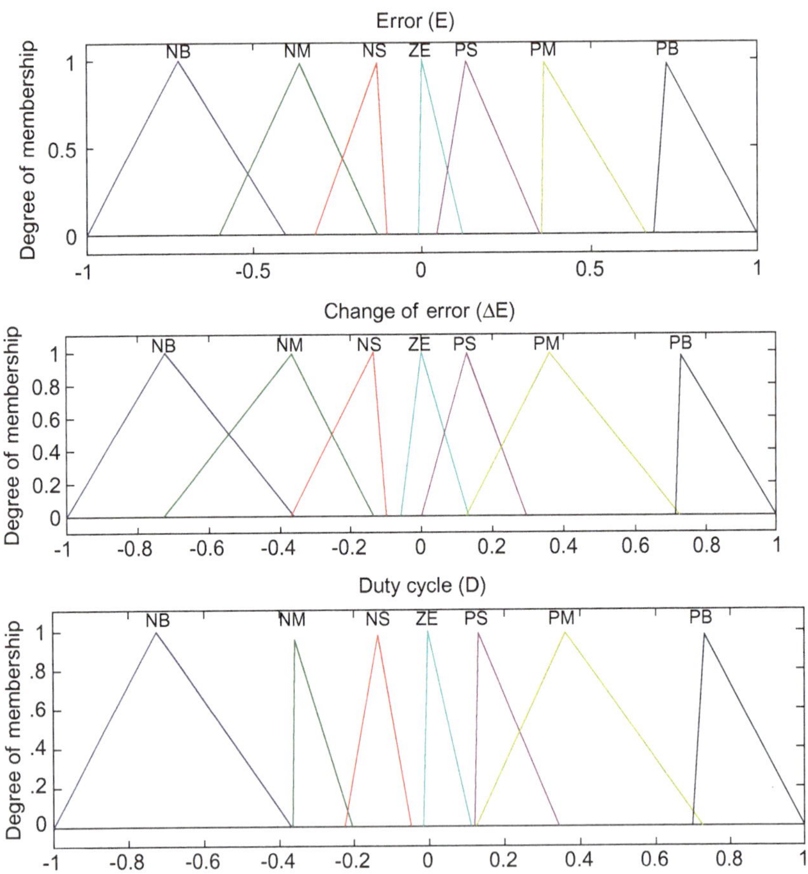

Fig. 5.15 Optimized membership functions obtained for CS-FLC variables.

Table 5.2 Optimized membership functions obtained for CS-FLC variables.

ΔE	**NB**	**NM**	**NS**	**ZE**	**PS**	**PM**	**PB**
NB	PM (0.125)	PS (0)	PM (0.064)	NM (0.89)	NS (0.192)	PB (1)	NB (0)
NM	PS (1)	NM (1)	NB (0.420)	PS (0.341)	NM (0.491)	NS (0.140)	NB (0.178)
NS	PS (0)	NB (0.608)	PB (1)	NB (0.553)	ZE (0.436)	PB (0.143)	PB (0.853)
ZE	PS (1)	NB (1)	NB (1)	NB (1)	NS (0.310)	NM (0.858)	PB (0.943)
PS	PB (1)	NB (1)	NB (0.676)	NB (0.105)	PS (0)	PM (0.096)	NB (0.090)
PM	NB (0.893)	PB (0.967)	ZE (0.566)	NS (0.609)	PB (0)	NB (0)	NM (1)
PB	PM (0)	PB (0.122)	PB (0)	NB (0.254)	ZE (1)	PS (0.350)	PB (0)

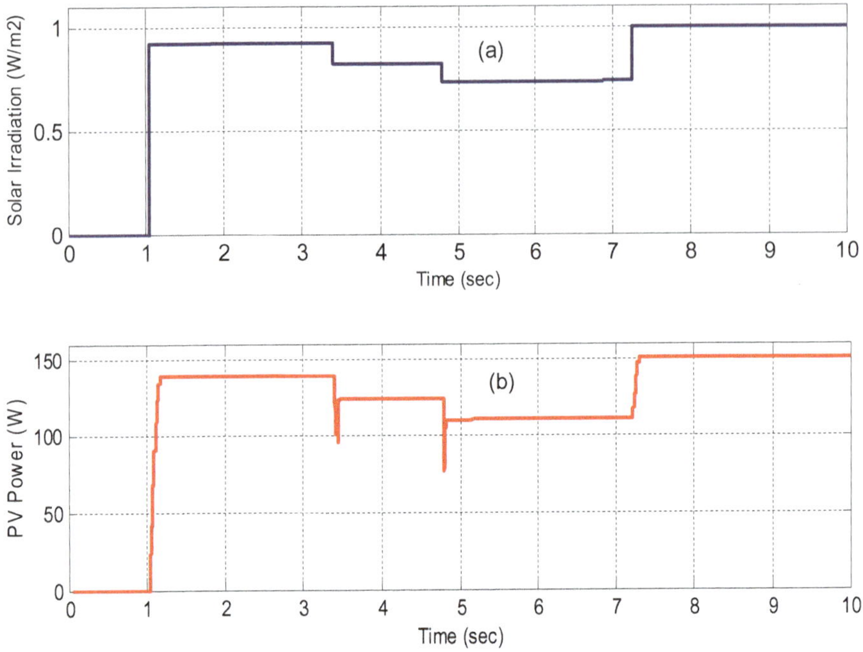

Fig. 5.16 MPPT-based CS-FLC response to an abrupt change of solar irradiance.

Example 5.4: Tracking the MPP using MPPT-NN controller (fast variation of solar irradiance)

Among the machine learning algorithms based on MPPT, neural network is mostly used to track the MPP [24]. For example, Fig. 5.17 depicts a Simulink model of a MPPT-based neural network.

The main task of this method is the prediction of V_{ref}, which can be used as a reference signal for the PI controller in order to command the switch of the DC-DC converter via a PWM signal. The V_{ref} can be estimated from solar irradiance and air temperature measurement using a simple multilayer perceptron NN. A dataset of solar irradiance, air temperature, and the corresponding voltage (V_{ref}) is used to train the NN (see Fig. 5.18). The MPPT-based NN is summarized as follows [24]:

Step #1: Data preprocessing and normalization.
Step #2: Neural networks parameters initialization (training algorithm, number of epoch, error rate, etc.).
Step #3: Training the neural network.
Step #4: Data preprocessing and normalization.
Step #5: Test and performance verification of the estimated Vref.

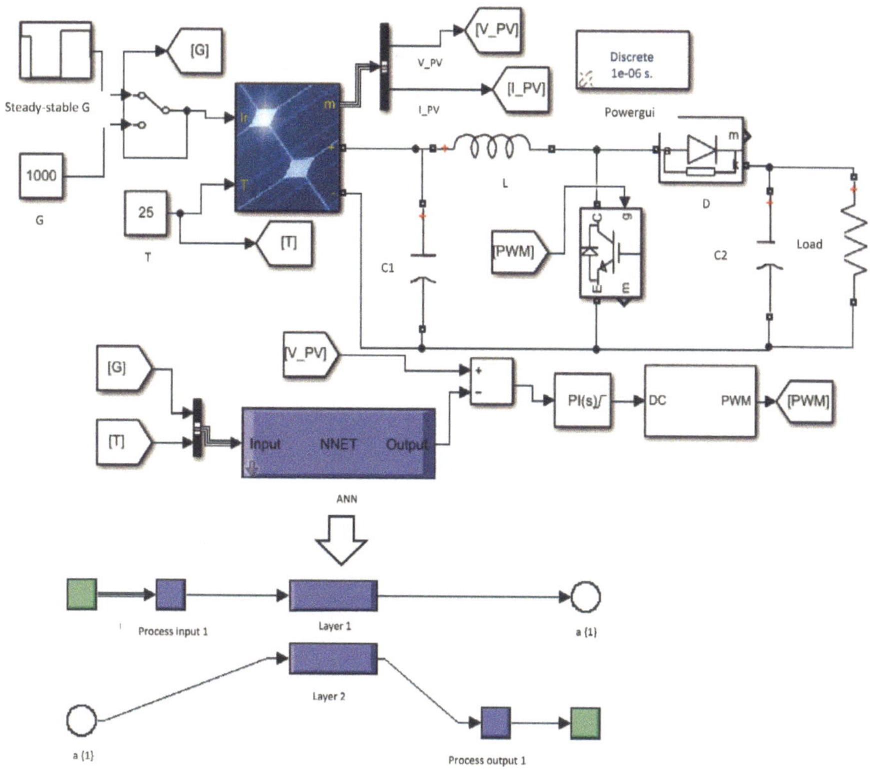

Fig. 5.17 Simulink model of MPPT-based neural network controller.

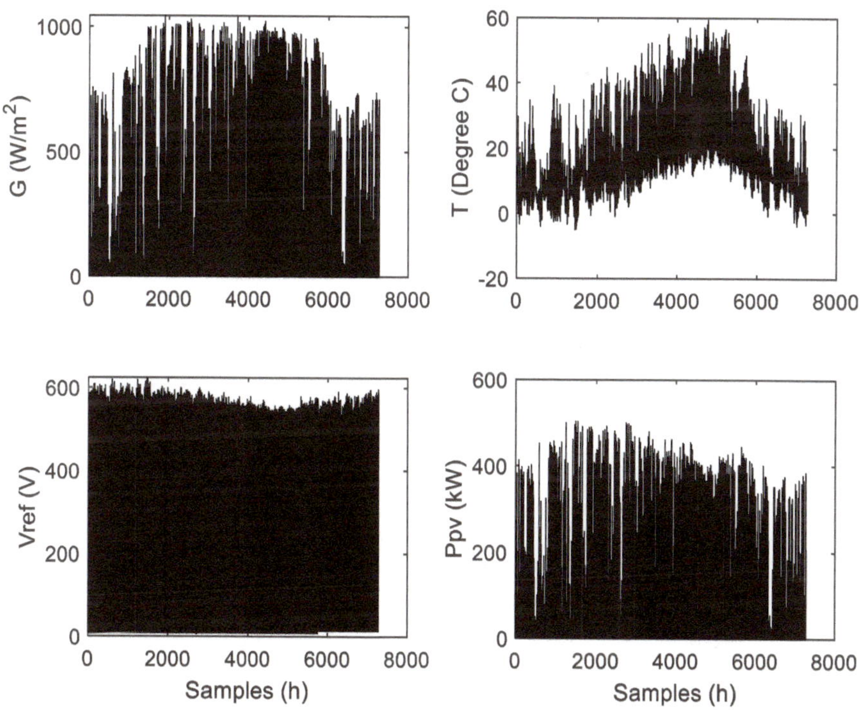

Fig. 5.18 Dataset: G, T, V_{ref}, and the produced power P_{pv}.

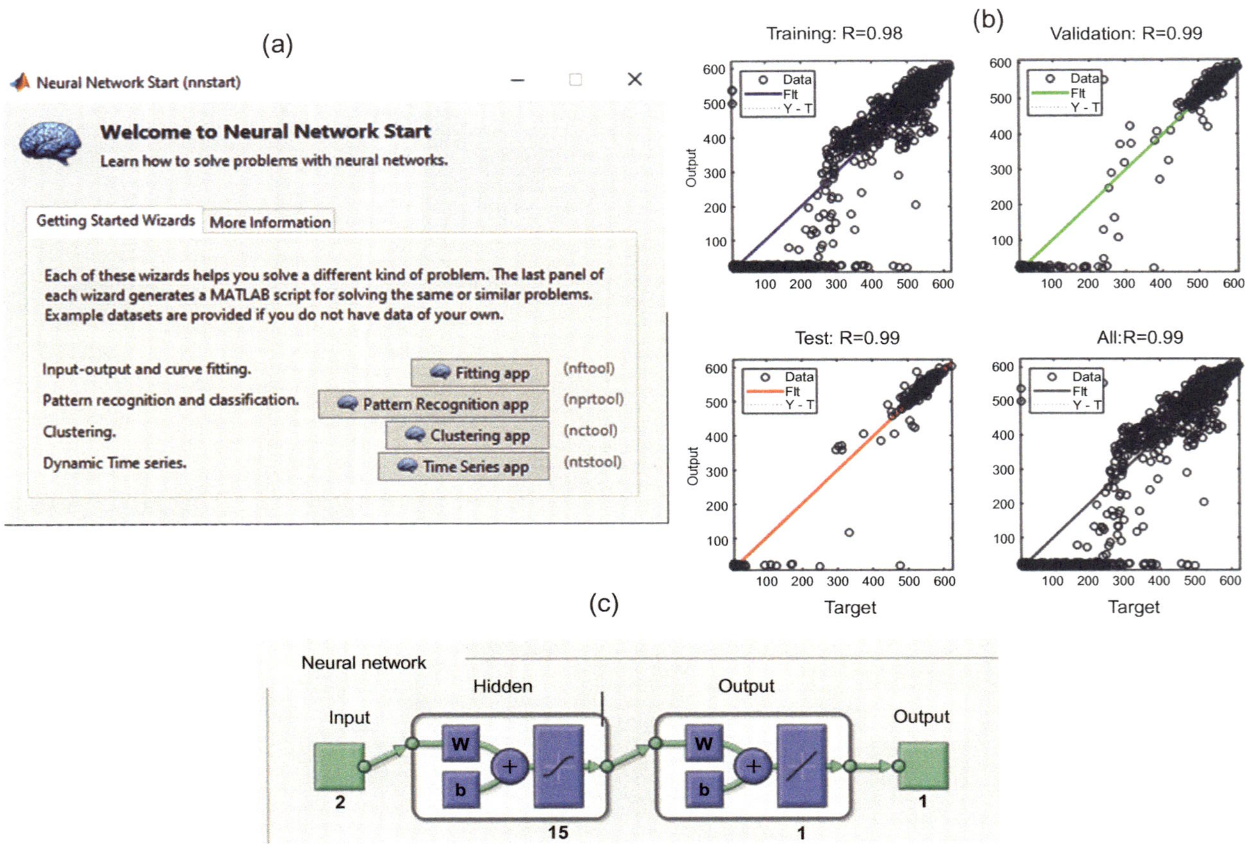

Fig. 5.19 (A) **Nnstart** tool of MathWorks, (B) correlation comparisons, and (C) MLP network for V_{ref} estimation.

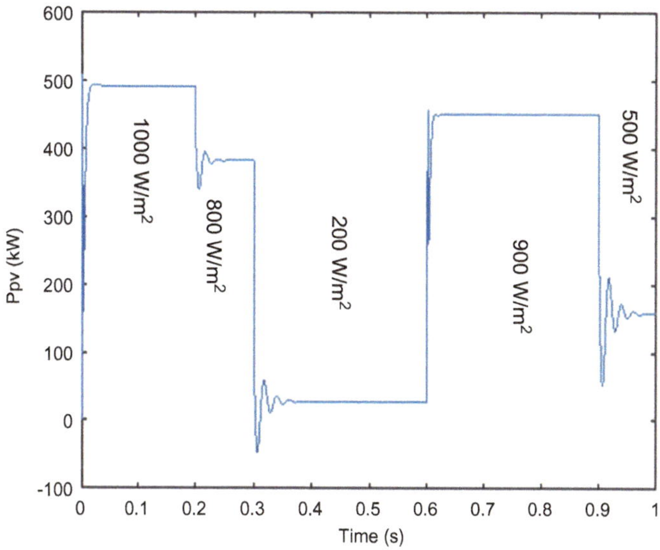

Fig. 5.20 MPPT-NN tracking performance under a fast transition of solar irradiance.

The available ***nnstart*** tool shown in Fig. 5.19A of MathWorks is used due to its simplicity. Fig. 5.19B shows the correlation during training, validation, and testing. In all cases, the calculated correlations are close to 1.

From Fig. 5.20, it can be observed that the algorithm performs well under fast variation of solar irradiance. The main advantages of a MPPT-based NN are that no detailed information about the system is required, which means that it is easy to be implemented and requires relatively less memory space. However, it needs a heuristic sense and it works as a black box. In addition, its robustness depends on the good training parameters. The main disadvantage of this technique is that it could fail when the PV modules start to be degraded by approximately 10%, in which case training with new data should be carried out periodically. Therefore, this is the most important point.

Example 5.5: Tracking the MPP using MPPT-ANFIS controller (fast variation of solar irradiance)

The same Simulink model is used here, just the neural network block is replaced by an Adaptive Neuro-Fuzzy Inference System (ANFIS) block as shown in Fig. 5.21. The model used here is the ANFIS.

The ANFIS uses a hybrid-learning rule combining back-propagation, gradient-descent, and least-squares algorithms to identify and optimize the Sugeno system's signals. The equivalent ANFIS architecture is a first-order Sugeno fuzzy model with 5×5 rules. The model has five layers and every node in a given layer has a similar function. The fuzzy IF-THEN rule set is used in which the outputs are linear combinations of their inputs [25]. The ***anfisedit*** tool (see Fig. 5.22A) of MathWorks is used to create the ANFIS model based on the dataset 5.1. The steps are shown as follows:

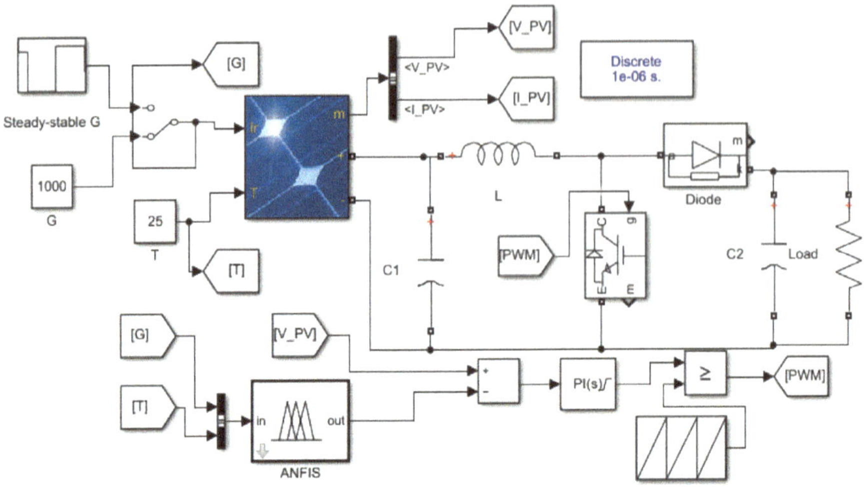

Fig. 5.21 Simulink model of MPPT-based ANFIS.

Step #1: Loading data and normalization.
Step #2: Implement the fuzzy system using for example grid-partition (select the number of rules 5 × 5, the type of the membership function).
Step #3: Training the ANFIS model (number of epoch, error rate, etc.).
Step #4: Verify the performance (error measured versus estimated).

The structure of the obtained ANFIS for V_{ref} estimation is depicted in Fig. 5.22B.

As shown in Fig. 5.23, the ANFIS controller exhibits good results. However, its robustness depends on the tuning of FL parameters and is relatively easy to be implemented compared to FLC.

It should be pointed that the examples presented here are not appropriate for abnormal conditions, e.g., partial shading effects, for which the model should be combined with other methods [8] or using adaptive or metaheuristics methods such as GA, PSO, CS, etc. [26,27]. It can be concluded that in terms of accuracy there is no big difference between NN and ANFIS.

5.4.2 Tracking the maximum power point under non-uniform conditions using hybrid methods

Shading effects on a PV array cause a huge amount of power loss. Partially shaded PV modules typically exhibit additional difficulties in tracking the MPP. In such circumstances, the power produced by PV systems is significantly reduced. Hybrid or adaptive methods have demonstrated their ability to track the global MPP under non-uniform solar irradiance. These methods combine mainly

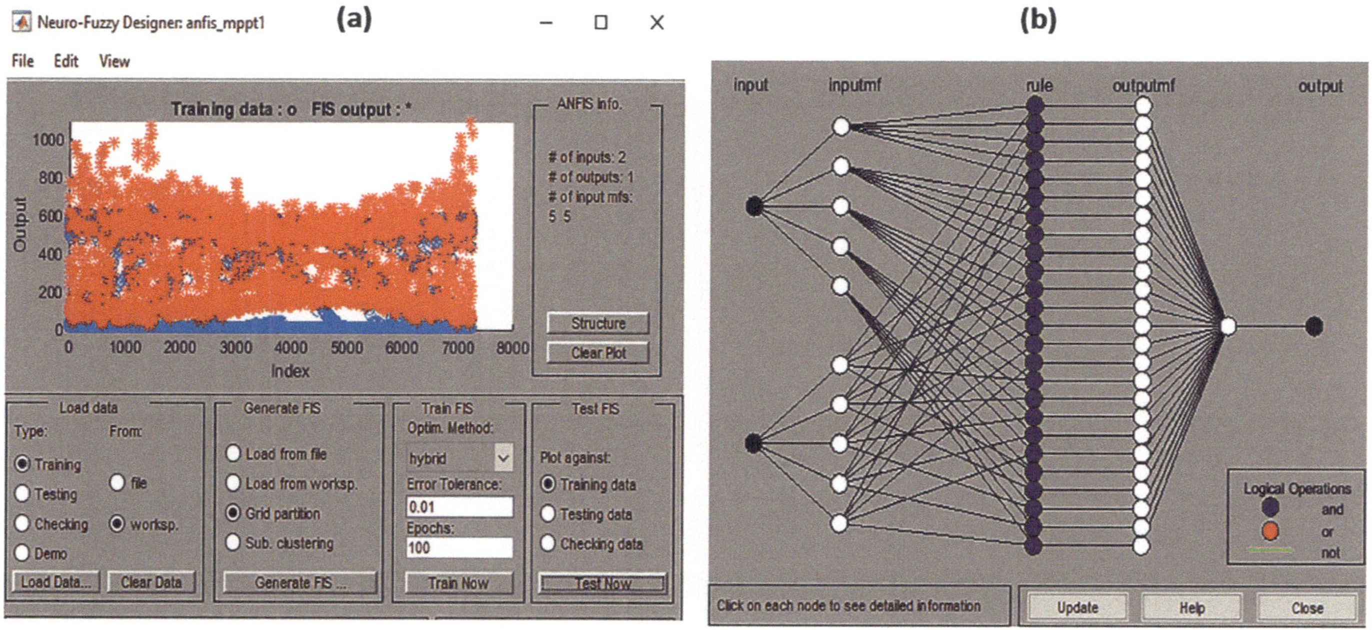

Fig. 5.22 (A) **Anfisedit** tool, (B) structure of the developed ANFIS for V_{ref} estimation.

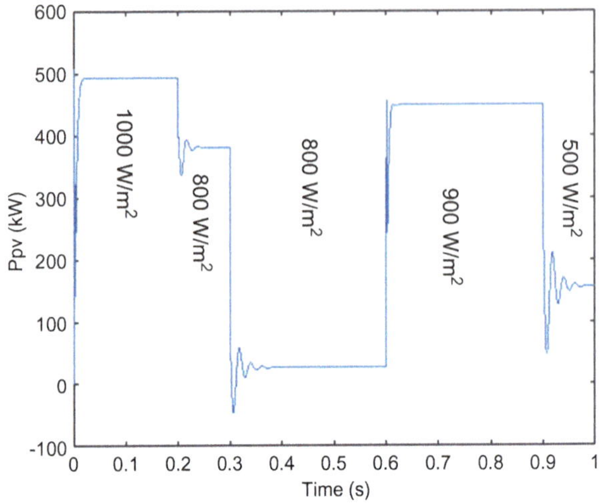

Fig. 5.23 MPPT-ANFIS tracking performance under fast transition of solar irradiance.

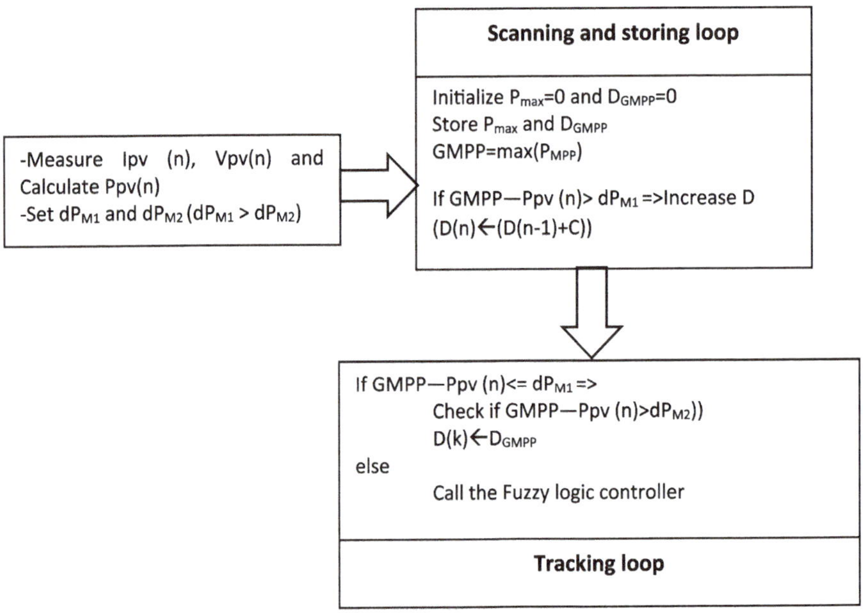

Fig. 5.24 Adaptive method for tracking the MPP under PSCs.

conventional (P&O, IncCond, and other improved versions) and advanced MPPT (e.g., FL, NN, etc.) [28].

Example 5.6: Tracking the GMPP under PSCs using an adaptive method

The flowchart of an adaptive method for tracking the GMPP under PSCs is shown in Fig. 5.24 and consists of two procedures [8]: (1) scanning and storing and (2) tracking loop.

Initially, $P_{\max}$ and D are set to 0. Then, $I_{pv}(n)$ $V_{pv}(n)$ for the nth instant are measured. Thus, the instant power output $P_{pv}(n)$ is calculated. ΔP_{M1} and ΔP_{M2} are two prefixed values that represent the allowable power difference between the GMPP and the current measured power. Moreover, ΔP_{M1} must be chosen greater than ΔP_{M2}.

The **scanning** and **storing** procedures are carried out during varying weather conditions or initial conditions, subsequently, the controller scans the PV power output to identify and store the GMPP and the corresponding D_{GMPP}. The scanning technique is based on the study and the critical observations of P-V characteristic curves under PSCs reported in Ref. [29], which shows that the peaks in the P-V characteristic curve occur roughly at multiples of 80% of open-circuit voltage. In addition, the amplitude of the peaks increases before the GMPP and displays the trend of decreasing as the peaks are longer away from the GMPP. The scanning procedure increases the duty cycle from its minimum value D_{min} to its maximum value D_{max} with a fixed step size (C). During this loop, the output power generated by the PV array must be analyzed to recognize the GMPP and its corresponding duty cycle D_{GMPP}.

The **tracking loop** aims to compare the GMPP and the actual power $P_{pv}(n)$, which permits the assurance that the MPPT is not trapped by local MPP, so, if the difference ($GMPP - P_{pv}(n)$) is greater than the prefixed value ΔP_{M1}, the duty cycle is increased. On the contrary, the tracking loop is carried out by examining if the difference ($GMPP - P_{pv}(n)$) is greater than the prefixed value ΔP_{M2}, thus D_{GMPP} is applied to move the operating power of the PV array as fast and as close as possible to the GMPP, otherwise, the FLC is carried out to reach the GMPP with high efficiency and a minimum of ripples.

More details on the process of the scanning procedure are reported in Ref. [29], while the pseudo algorithm of the procedure is as shown below.

Step #1: to obtain the GMPP, the stored maximum power Pmax must be compared with the actual power Ppv, and the GMPP is identified as the maximum value of Pmax.
Pmax = max (Ppv(d), Pmax).
GMPP = max (Pmax).
Step #2: To identify the DGMPP.
Step #3: Calculzate the power variation ΔPpv and the variation between the maximum power stored Pmax and the actual power Ppv, knowing that d is the actual duty cycle and C is the fixed step size of duty cycle.
ΔPpv(d) = Ppv(d) − Ppv(d − C).
ΔPmax(d) = Pmax(d) − Ppv(d).
Step #4: identify the DGMPP.
If ΔPmax(d) ≤ 0 **and** ΔPpv(d) ≥ 0 **then** DGMPP = d
Else DGMPP = DGMPP_old
end

The Simulink model of the adaptive method is shown in Fig. 5.25.

To verify the performance of the adaptive MPPT algorithm during non-uniform solar irradiance, the PV system is tested under different distributed solar irradiance

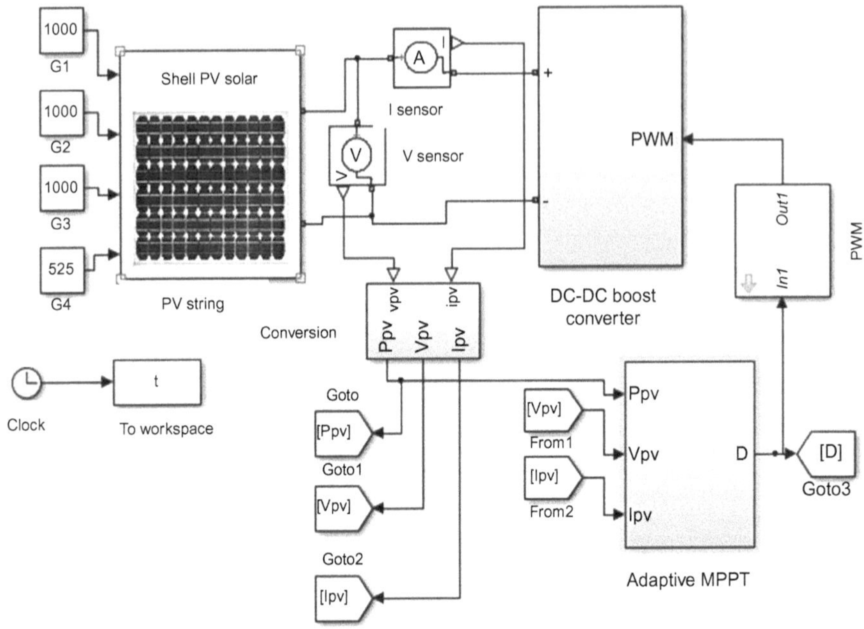

Fig. 5.25 Simulink model of adaptive MPPT.

levels. Three un-shaded PV modules receive 1000 W/m^2 whereas the fourth one is partially shaded with a solar irradiance of 525 W/m^2. It can be seen from Fig. 5.26 that the algorithm tracks the global maximum power point which is around 450 W at 1000 W/m^2, while the local MPPT is around 360 W. This method takes much more time to stabilize at the global MPP due to the scanning process.

5.4.3 Tracking the MPP under PSCs using evolutionary algorithms

Evolutionary algorithms, including GA, PSO, CS, and other versions, have shown their ability to track the global MPP under partial shading conditions, i.e., under non-uniform irradiance [23,30–32]. These techniques vary in terms of input parameters, scanning process, and have relatively the same accuracy.

Example 5.7: PSO-based tracking of MPP under non-uniform irradiance on a PV array

In this example, an application of a PSO algorithm to track the MPP under PSCs is presented. The Simulink model is shown in Fig. 5.27. The PV array consists of 9 PV modules of 3 × 3 (three strings connected in parallel series), the devolved power at STC is 1080 W/m^2.

The main equations of the PSO-based MPPT algorithm are written as [33]

$$D_i(t+1) = D_i(t) + \Delta D_i(t+1)\, i \in \{1, \dots, n_p\} \tag{5.8}$$

$$\Delta D_i(t+1) = w\Delta D_i(t) + c_1 r_1 \left(D_{op\_i} - D_i(t)\right) + c_2 r_2 \left(D_{op\text{-}g} - D_i(t)\right) \tag{5.9}$$

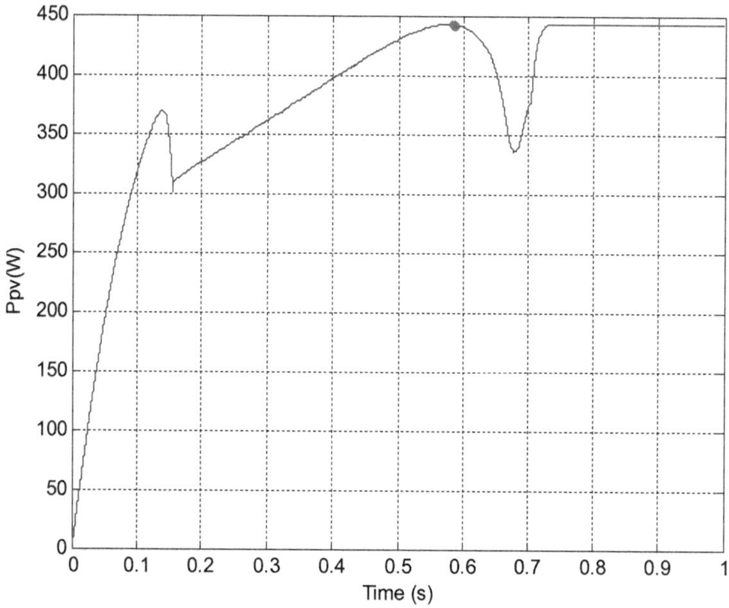

Fig. 5.26 The P_{pv} during the non-uniform irradiance (1000 and 525 W/m^2).

where $D_i(t)$ is the duty cycle of particle i at the instant t, $D_i(t+1)$ is the duty cycle of particle i at the instant $t+1$, $\Delta D_i(t+1)$ is the variation of the D of particle i at $t+1$, $\Delta D_i(t)$ is the variation of the D of particle i at t. D_{op_i} is the best D of particle i and D_{op_i} is the best D of all particles of the swarm, c_1 and c_2 are speed coefficients, r_1 and r_2 are random variables [0,1], and w is the inertia coefficient [33]. The pseudocode of the implemented PSO under Matlab is shown below:

Step #1 PSO parameters initialization (r_1, r_2, c_1, c_2, np, and w).
Step #2 Swarm optimization.
Step #3 PSO parameters initialization.
Step #4 Update the local and global optimal position.
Step #5 Update the speed and position of each particle.

As shown in Fig. 5.28, the MPPT-based PSO tracks the global maximum power which is around 720 W. But it takes much time to converge to the global maximum.

This method is suitable for PSCs, however, it exhibits significant algorithmic complexity, which increases the implementation cost of the global MPP control system.

The advantages of the intelligent MPPT methods are: they offer an alternative approach to conventional MPPT algorithms; they exhibit a faster converging speed, good performance, efficiency, less oscillation around the MPP under steady-state

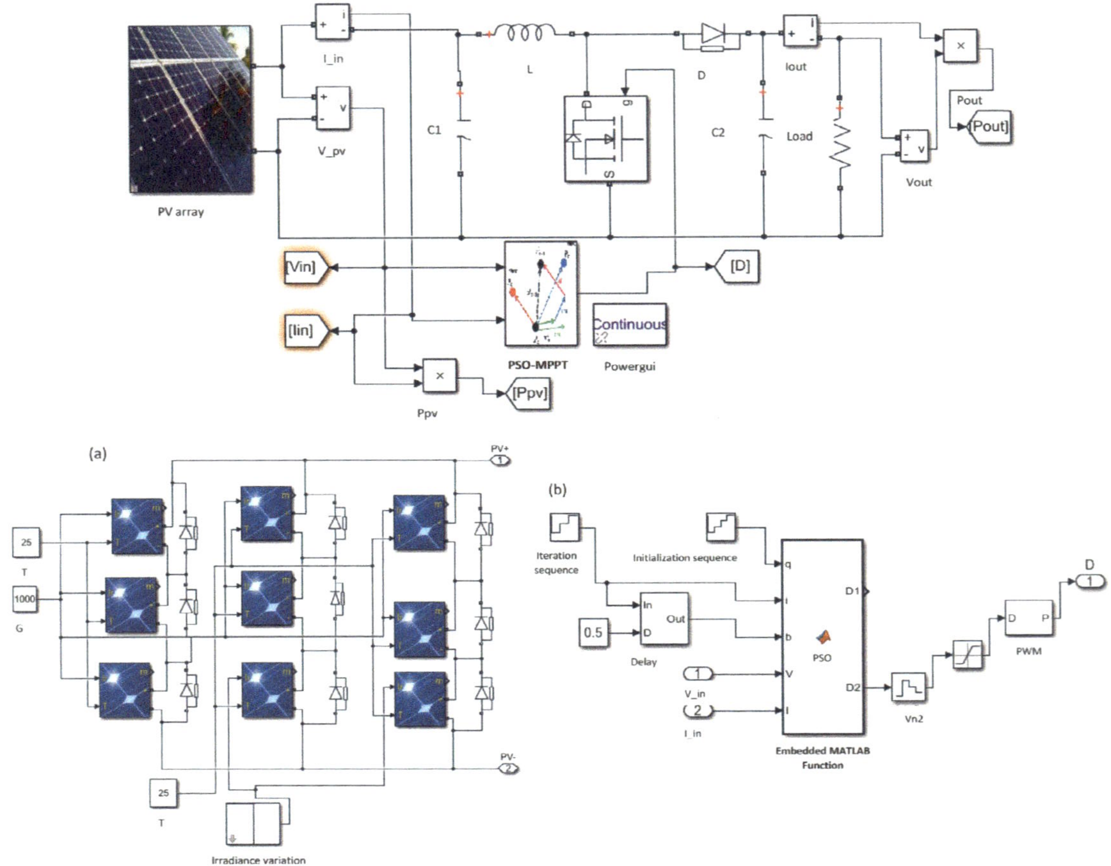

Fig. 5.27 Simulink model of MPPT-based PSO: (A) PV array and (B) PSO block.

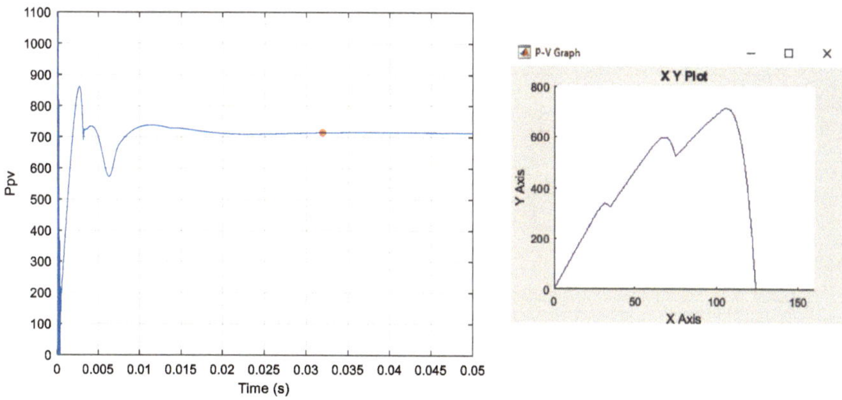

Fig. 5.28 The P_{pv} during the non-uniform irradiance (1000 and 525 W/m^2).

conditions, and no divergence from the MPP during varying weather conditions. However, their complexity implementation varies from one method to another.

5.5 PV module reconfiguration based on dynamic techniques and AI techniques

The fixed TCT, static, and dynamic PV array reconfiguration principles are shown in Fig. 5.29.

Example 5.8: Dynamic PV array reconfiguration using FL and IE

In this section, an example of the application of FL for dynamic PV array reconfiguration is presented. The Simulink model of a dynamic PV array method is shown in Fig. 5.30.

The system consists of three main parts:

(i) PV array with a switch network to change the electrical connections of PV modules in TCT configuration.

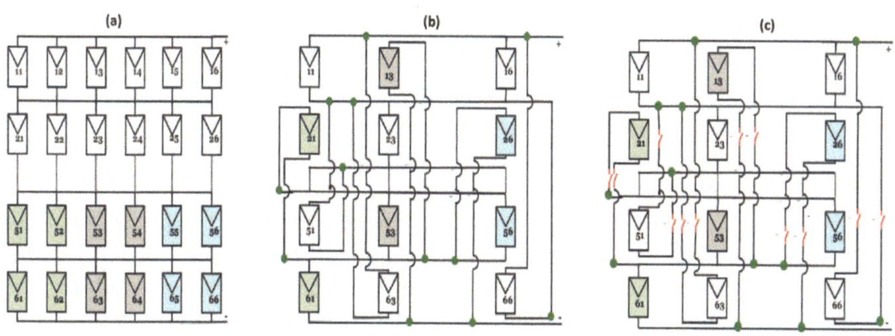

Fig. 5.29 (A) Fixed TCT, (B) static PV array, and (C) dynamic PV array reconfigurations.

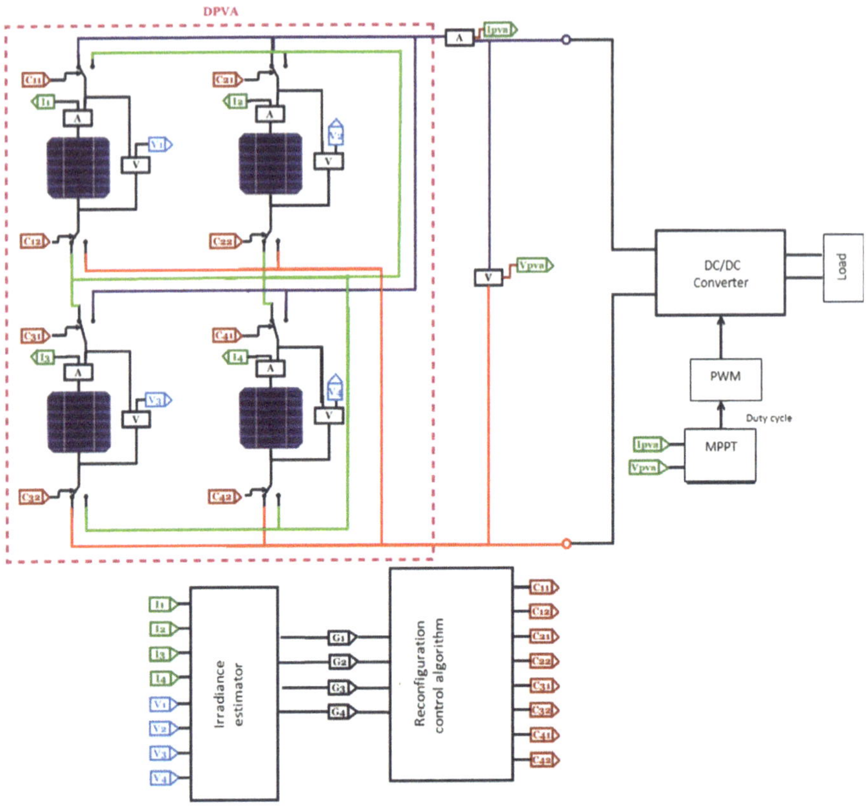

Fig. 5.30 Simulink model of a dynamic PV array method based on FL and IE.

(ii) The control part which acquires current and voltage measurements to estimate the irradiance value for each PV module generates the appropriate driven signals of switches and the suitable duty cycle from the MPPT algorithm.

(iii) Boost converter interposed between the PV array and the load.

The reconfiguration process is based on irradiance equalization (IE) approach [34]. The interconnections of the PV modules are reconfigured so that each row in the series string receives the same amount of irradiance [35]. The FL controller provides appropriate signals to drive the switch according to the suitable PV array configuration. The equalization goal is achieved by keeping the same number of PV modules per row or by forming a PV array with an unequal number of PV modules per row.

A PV array (4 strings with 4 PV modules in each string) simulation is used to evaluate the FL method [36]. The performance of the FL method is assessed across numerous partial shading patterns. The 16 PV modules implemented in the Matlab/Simulink environment have the same electrical characteristics ($P_{mp} = 300\,W$, $V_{mp} = 31.9\,V$, $I_{mp} = 9.40\,A$, $V_{oc} = 39.8\,V$, $I_{sc} = 9.75\,A$). The shading dispersion diagrams of the fixed TCT configuration and the reconfigurable configuration based on FL are shown in Fig. 5.31. As can be seen, the FL reconfiguration method disperses the shading more uniformly.

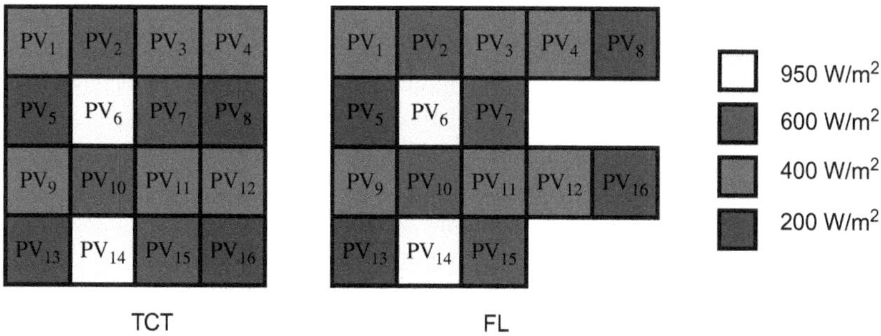

Fig. 5.31 Shade dispersion diagram under random shading.

Table 5.3 Calculated parameters under PSCs for both configurations.

Array configuration	V_{mp} (V)	I_{mp}(A)	P_{mp} (W)	V_{oc} (V)	I_{sc} (A)	Mismatch loss (%)
TCT	89.29	25	2265	102.5	32. 9	20
FL	87.29	27.8	2426	102.5	30.02	3.1

The mismatch loss calculation is reported in Table 5.3. The fixed TCT reconfiguration method operates with 20% of mismatch loss, while the FL method operates with 3.1% of mismatch loss, which indicates that the uniform dispersion of the shading and the reduction of a mismatch loss is around 17% [36].

The P-V characteristics of both configurations are shown in Fig. 5.32. As can be observed, the fixed TCT configuration has more peaks on the P-V curve, while this is reduced to one peak by the FL reconfiguration method. Due to the change in the number of modules per row by the FL configuration, I_{sc} is reduced in the new reconfiguration [36,37].

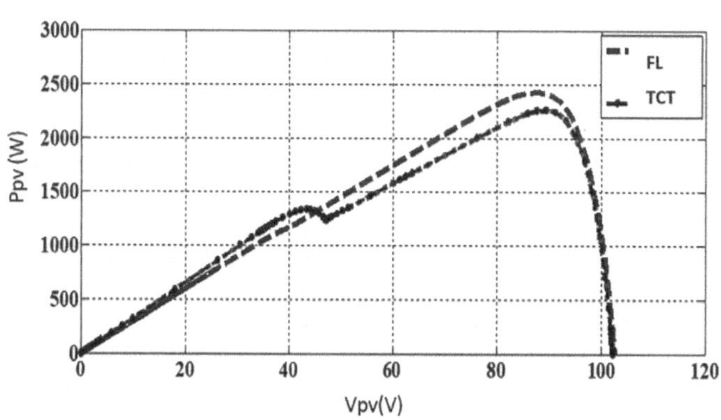

Fig. 5.32 P-V characteristic for fixed TCT, and FL reconfiguration methods.

Two main issues are involved in dynamic PV array architectures: (1) the topology choice and (2) the algorithm complexity. The number of required switches is conditioned by the topology choice, while the sensors number, optimal configuration, and preservation of the switches lifetime are conditioned by algorithm optimization. Thus, an optimal reconfigurable PV array architecture that is able to guarantee a maximum power at a minimum cost and fits the application requests can only be achieved if there is a good compromise between the topology choice and the reconfiguration algorithm is ensured.

5.6 Summary

This chapter provides and discusses various examples of the application of artificial intelligence techniques and metaheuristic methods to optimize and maximize the produced PV output power. The models presented have been developed and tested under Matlab/Simulink environment.

Despite the various contributions to track the maximum power and obtain optimal PV array configurations, the real-time implementation of a dynamic PV array configuration and intelligent MPPT methods is still complicated and involves considerable investment cost as numerous sensors and switches are required. In addition, advanced skills and knowledge are needed.

References

[1] S. Jerez, I. Tobin, R. Vautard, J.P. Montávez, J.M. López-Romero, F. Thais, B. Bartok, O. B. Christensen, A. Colette, M. Déqué, G. Nikulin, The impact of climate change on photovoltaic power generation in Europe, Nat. Commun. 6 (1) (2015) 1–8.

[2] M. Mao, L. Cui, Q. Zhang, K. Guo, L. Zhou, H. Huang, Classification and summarization of solar photovoltaic MPPT techniques: a review based on traditional and intelligent control strategies, Energy Rep. 6 (2020) 1312–1327.

[3] R. Boukenoui, R. Bradai, A. Mellit, M. Ghanes, H. Salhi, Comparative analysis of P&O, modified hill climbing-FLC, and adaptive P&O-FLC MPPTs for microgrid standalone PV system, in: 2015 International Conference on Renewable Energy Research and Applications (ICRERA), IEEE, 2015, pp. 1095–1099.

[4] X. Li, H. Wen, Y. Hu, L. Jiang, A novel beta parameter based fuzzy-logic controller for photovoltaic MPPT application, Renew. Energy 130 (2019) 416–427.

[5] A.B.G. Bahgat, N.H. Helwa, G.E. Ahmad, E.T. El Shenawy, Maximum power point tracking controller for PV systems using neural networks, Renew. Energy 30 (8) (2005) 1257–1268.

[6] M.A. Mohamed, A.A.Z. Diab, H. Rezk, Partial shading mitigation of PV systems via different meta-heuristic techniques, Renew. Energy 130 (2019) 1159–1175.

[7] R. Boukenoui, M. Ghanes, J.P. Barbot, R. Bradai, A. Mellit, H. Salhi, Experimental assessment of maximum power point tracking methods for photovoltaic systems, Energy 132 (2017) 324–340.

[8] R. Boukenoui, H. Salhi, R. Bradai, A. Mellit, A new intelligent MPPT method for standalone photovoltaic systems operating under fast transient variations of shading patterns, Sol. Energy 124 (2016) 124–142.

[9] K. Sundareswaran, S. Palani, Application of a combined particle swarm optimization and perturb and observe method for MPPT in PV systems under partial shading conditions, Renew. Energy 75 (2015) 308–317.
[10] R. Bradai, R. Boukenoui, A. Kheldoun, H. Salhi, M. Ghanes, J.P. Barbot, A. Mellit, Experimental assessment of new fast MPPT algorithm for PV systems under non-uniform irradiance conditions, Appl. Energy 199 (2017) 416–429.
[11] P. Soulatiantork, L. Cristaldi, M. Faifer, C. Laurano, R. Ottoboni, S. Toscani, A tool for performance evaluation of MPPT algorithms for photovoltaic systems, Measurement 128 (2018) 537–544.
[12] M.H. Taghvaee, M.A.M. Radzi, S.M. Moosavain, H. Hizam, M.H. Marhaban, A current and future study on non-isolated DC–DC converters for photovoltaic applications, Renew. Sust. Energ. Rev. 17 (2013) 216–227.
[13] G.S. Krishna, T. Moger, Reconfiguration strategies for reducing partial shading effects in photovoltaic arrays: state of the art, Sol. Energy 182 (2019) 429–452.
[14] A.M. Ajmal, T.S. Babu, V.K. Ramachandaramurthy, D. Yousri, J.B. Ekanayake, Static and dynamic reconfiguration approaches for mitigation of partial shading influence in photovoltaic arrays, Sustainable Energy Technol. Assess. 40 (2020), 100738.
[15] S. Malathy, R. Ramaprabha, Reconfiguration strategies to extract maximum power from photovoltaic array under partially shaded conditions, Renew. Sust. Energ. Rev. 81 (2018) 2922–2934.
[16] M. Seyedmahmoudian, B. Horan, T.K. Soon, R. Rahmani, A.M.T. Oo, S. Mekhilef, A. Stojcevski, State of the art artificial intelligence-based MPPT techniques for mitigating partial shading effects on PV systems—a review, Renew. Sust. Energ. Rev. 64 (2016) 435–455.
[17] S. Li, Q. Jiang, Study on pid parameters tuning method based on Matlab/Simulink, in: 2011 IEEE 3rd International Conference on Communication Software and Networks, IEEE, 2011, pp. 408–411.
[18] F. Chekired, A. Mellit, S.A. Kalogirou, C. Larbes, Intelligent maximum power point trackers for photovoltaic applications using FPGA chip: a comparative study, Sol. Energy 101 (2014) 83–99.
[19] O. Cordón, A historical review of evolutionary learning methods for Mamdani-type fuzzy rule-based systems: designing interpretable genetic fuzzy systems, Int. J. Approx. Reason. 52 (6) (2011) 894–913.
[20] C. Larbes, S.A. Cheikh, T. Obeidi, A. Zerguerras, Genetic algorithms optimized fuzzy logic control for the maximum power point tracking in photovoltaic system, Renew. Energy 34 (10) (2009) 2093–2100.
[21] A. Messai, A. Mellit, A. Guessoum, S.A. Kalogirou, Maximum power point tracking using a GA optimized fuzzy logic controller and its FPGA implementation, Sol. Energy 85 (2) (2011) 265–277.
[22] X.S. Yang, S. Deb, Cuckoo search via Lévy flights, in: 2009 World Congress on Nature & Biologically Inspired Computing (NaBIC), IEEE, 2009, pp. 210–214.
[23] S. Daraban, D. Petreus, C. Morel, A novel MPPT (maximum power point tracking) algorithm based on a modified genetic algorithm specialized on tracking the global maximum power point in photovoltaic systems affected by partial shading, Energy 74 (2014) 374–388.
[24] A. Mellit, S.A. Kalogirou, MPPT-based artificial intelligence techniques for photovoltaic systems and its implementation into field programmable gate array chips: review of current status and future perspectives, Energy 70 (2014) 1–21.

[25] A. Mellit, S.A. Kalogirou, ANFIS-based modelling for photovoltaic power supply system: a case study, Renew. Energy 36 (1) (2011) 250–258.
[26] A. Bouilouta, A. Mellit, S.A. Kalogirou, New MPPT method for stand-alone photovoltaic systems operating under partially shaded conditions, Energy 55 (2013) 1172–1185.
[27] Z. Salam, J. Ahmed, B.S. Merugu, The application of soft computing methods for MPPT of PV system: a technological and status review, Appl. Energy 107 (2013) 135–148.
[28] Ö. Çelik, A. Teke, A hybrid MPPT method for grid connected photovoltaic systems under rapidly changing atmospheric conditions, Electr. Power Syst. Res. 152 (2017) 194–210.
[29] H. Patel, V. Agarwal, Maximum power point tracking scheme for PV systems operating under partially shaded conditions, IEEE Trans. Ind. Electron. 55 (4) (2008) 1689–1698.
[30] H. Rezk, A. Fathy, A.Y. Abdelaziz, A comparison of different global MPPT techniques based on meta-heuristic algorithms for photovoltaic system subjected to partial shading conditions, Renew. Sust. Energ. Rev. 74 (2017) 377–386.
[31] A. Mohapatra, B. Nayak, P. Das, K.B. Mohanty, A review on MPPT techniques of PV system under partial shading condition, Renew. Sust. Energ. Rev. 80 (2017) 854–867.
[32] J. Ahmed, Z. Salam, A maximum power point tracking (MPPT) for PV system using cuckoo search with partial shading capability, Appl. Energy 119 (2014) 118–130.
[33] J. Shi, W. Zhang, Y. Zhang, F. Xue, T. Yang, MPPT for PV systems based on a dormant PSO algorithm, Electr. Power Syst. Res. 123 (2015) 100–107.
[34] L. Bouselham, A. Rabhi, B. Hajji, New reconfiguration method based on logic gates for small dynamic photovoltaic Array, in: A Practical Guide for Advanced Methods in Solar Photovoltaic Systems, Springer, Cham, 2020, pp. 123–138.
[35] M. Drif, A. Mellit, J. Aguilera, P.J. Pérez, A comprehensive method for estimating energy losses due to shading of GC-BIPV systems using monitoring data, Sol. Energy 86 (9) (2012) 2397–2404.
[36] L. Bouselham, A. Rabhi, B. Hajji, A. Mellit, Photovoltaic array reconfiguration method based on fuzzy logic and recursive least squares: an experimental validation, Energy (2021), 121107.
[37] L. Bouselham, A. Rabhi, B. Hajji, A. Mellit, C. El Fouas, An intelligent irradiance equalization approach based on fuzzy logic for small reconfigurable PV architecture, in: 2019 7th International Renewable and Sustainable Energy Conference (IRSEC), IEEE, 2019, pp. 1–6.

Machine learning and deep learning algorithms for fault diagnosis of photovoltaic systems

6

Chapter outline

6.1 Introduction

Photovoltaic (PV) systems (stand-alone, grid-connected, or hybrid configurations) are subject to many faults that have a negative impact on the system's efficiency, reliability, and safety. Thus, the operation and maintenance activities of such plants are important for the safety of the professionals working in this sector. Keeping solar PV plants at their maximum efficiency is a challenging task and around 2% of the PV modules are predicted to fail after 11–12 years.

Currently, with the rapid development of the PV industry, fault protection, detection, and diagnosis has become a very important issue for maintaining the safe and

Handbook of Artificial Intelligence Techniques in Photovoltaic Systems. https://doi.org/10.1016/B978-0-12-820641-6.00006-5
Copyright © 2022 Elsevier Ltd. All rights reserved.

stable operation of PV plants [1]. Safety and reliability are the key challenges facing the rapid growth of this kind of plant. Early fault detection and isolation enables the user to avoid the risk of damage to such systems. In this regard, many efforts have been made to address these issues using various methods [2].

Generally, fault detection and diagnosis methods comprise three major steps:

- *Fault detection step*: This can be achieved in a simple way, by comparing the power measured and calculated based on an accurate model [3,4]. It is also possible to use machine learning (ML) and deep learning (DL) frameworks to detect faults [5,6].
- *Fault classification and identification step*: This aims to classify and identify the causes of the fault. This step can be performed by developing appropriate algorithms. The complexity of the algorithm depends mainly on the nature of the faults to be identified. In some cases, PV faults exhibit the same symptoms while the causes are different, in this circumstance ML or DL algorithms are the most suitable methods to address this issue [7,8].
- *Fault localization and isolation step*: This is the most crucial and challenging step, particularly for large-scale PV systems, as it requires sophisticated equipment and expert knowledge. It can be achieved, e.g., by using aerial thermal images with deep learning to localize the overheated PV modules (hot spot problem) [9].

It is worth noting that some faults can be identified easily using simple methods [10], while other emerging faults that have the same symptoms or multiple faults need an effective algorithm (such as ML or deep DL) to identify and correctly classify the faults [11,12], and make the right decision before the fault is developed, thus avoiding catastrophic results.

This chapter intends to present only faults that occur on PV arrays, including PV modules and cells. The PV cell/module is the most sensible and frequently affected element as it is subject to long-time exposure to different climatic conditions, such as temperature, ultraviolet radiation, humidity, and wind. This chapter is focused more on the application of machine learning, ensemble learning, and deep learning methods to detect, classify, and identify certain faults in PV arrays. A total of 15 examples with 9 listings are presented in detail using a free ML library (Scikit-learn) for Python language.

6.2 Type of faults in photovoltaic arrays

Major faults in a PV array can be classified into two groups: (1) permanent faults and (2) temporal faults. Permanent faults include disconnected cables, failure of PV modules such as bubbles, sink moisture, delamination, snail trails, yellowing, invisible cell cracks and burnt cells, scratches, and aging modules [13]. These kinds of faults can be fixed by changing the defective PV modules. Temporal faults such as dust accumulation, snow, and artificial shading can be removed by operators without replacing the affected PV module. These faults contribute significantly to the reduction of the efficiency of PV modules sometimes up to 80%.

The cause of the fault could be external or internal, and can lead to a decrease in both the produced power and the whole system efficiency and reliability. In some cases, such as overheating solar cells (hot spot problem), if the fault is not detected early it can lead to hazard risk, thus in this circumstance, the system should be correctly protected. The main faults, affected elements, causes, and effects are summarized in Table 6.1 [2]. As can be observed from Table 6.1, hot spot, arc fault, line-to-line fault, and ground faults are the most catastrophic faults that can lead to the risk of fire [10].

Table 6.1 Summary of different type of faults on PV arrays with connected information: affected components, causes (internal or external), effects, and safety issue.

Type of fault	Components and visible fault	Causes		Effects and safety issue
		External	Internal	
Hot spot fault	Solar cell (no visualization)	Soiling, dust, snow, and shadow	Fragmentation of cells Current mismatch between cells High resistance or "cold" solder points Aging and degradation of cells	Damage of solar cells, open circuits, and fire may be caused.
Diode fault	Bypass diode or blocking diode (possible visualization)	Partially shaded cells and overheating	–	Damage diodes, short-circuited diode, open circuited or shunted diode
Junction box fault	Junction box (possible visualization)	Fretting corrosion, and oxidation	–	Damage and risk of fire
PV module fault or PV array fault	PV module (delamination, discoloration, burn marks, snail trails and classes breakage)	Connector failure, (disconnection), isolated from ground, encapsulation, stress high exposure to ultraviolet radiation, humidity and heat, loss in adhesion between glass, encapsulate and back cover, failure in solder bonds, local heating and ribbon breakage, sink moisture in the module	Induced on microcracks along edges, thermal stresses, impacts during transportation and installation, stress induced on microcracks along edges.	Shunted module, short circuit within a module, short-circuit between PV modules, leakage currents within a PV module, damage PV module, corrosion of solar cells, moisture penetration, corrosion and lower irradiance, accelerated degradation, yellowing and browning, oxidation of PV module, electric shock physical danger and failure may cause fire

Continued

Table 6.1 Continued

Type of fault	Components and visible fault	Causes		Effects and safety issue
		External	Internal	
Ground fault	PV string or PV array (possible visualization)	Insulation failure of cables, incidental short circuit between normal conductor and ground, cable insulation damage during the installation degraded sealant and water ingress, accidental short circuit inside the PV combiner box.	–	Damage of PV modules, and risk of fire
Arc fault	PV string	A short break is created in a conductor Two conductors of widely differing voltage are placed near one another Degradation in solder joints, wiring or connections inside the insulation damage due to mechanical destruction, aging, or wild life junction box, loosening of screws.	–	Damage of PV modules, and risk of fire
Line-to-line fault	PV array	An unintentional low impedance current path between two points, insulation failure of cables, and incidental short circuit between current carrying conductors	–	Damage PV modules, conductors, and risk of fire

6.3 Protection devices for photovoltaic systems

To protect and isolate faults in PV systems, various standards and codes were developed [14–16]. These standards [14–16] are used to protect systems from anti-islanding, lighting, surge, overvoltage, and other faults. Fig. 6.1 illustrates a simplified schematic representation of a grid-connected PV plant with the main protection devices. A ground fault detection and interruption device is used to protect the PV array during grounding faults and an over current protection device is used to protect from line-to-line faults. Arc fault circuit interrupter is a standard protection device generally used against series arc fault. Different commercial mitigation techniques are developed to prevent fire in grounded and ungrounded PV arrays. However, commercial products are related to the fact that these are usually closed source and thus are unlikely able to adapt to different location and operation conditions.

6.4 Fault detection and diagnosis methods

Globally, available fault detection and diagnosis methods can be classified into two major categories [2]: (1) visual inspection and thermal images and (2) electrical methods.

6.4.1 Visual inspection and thermography methods

These methods can be used for small-scale PV plants. The procedure is simple, but it requires some knowledge and skills. If a problem is observed, sophisticated equipment is needed such as an I-V tracer curve, and in some cases, a thermal camera is indispensable to identify the faulty PV strings or modules [17]. In most cases and through the diagnostic process, a PV string or module should be put off, i.e., disconnected from the main PV array. These methods can detect various PV module defects, such as dust accumulation, browning, discoloration, hot spot, snail trails, breaking class, and delamination. Advanced image processing algorithms are needed to identify invisible faults such as microcrack cells [18].

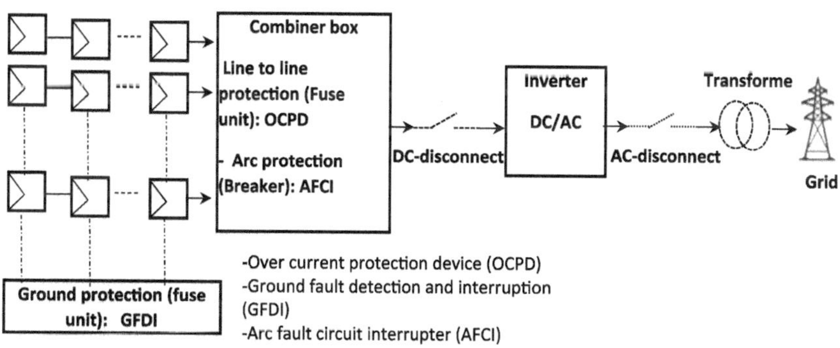

Fig. 6.1 Grid-connected PV system with protection devices.

6.4.2 Electrical methods

These include mainly methods based on electrical measurement of I-V curves, current, voltage, power, and other environmental parameters. Two main groups are found:

(a) *Statistical and signal processing methods*: These include time domain refractory [19], Earth capacitance measurement [20], and other improved versions [21]. These techniques can be used to detect and localize a fault particularly for small-scale PV plants. Here, during the testing process, the considered element (PV module or string) should be disconnected from the whole system. Signal processing methods are used to diagnose the faults based on the measured signal (voltage, current, and power) [22].

(b) *Data-driven methods*: In these methods, a data-acquisition system is indispensable to collect and prepare a dataset for eventual utilization. Data collected could include measurements of current, voltage, cell temperature, PV images (infrared or electroluminescence), and environmental parameters such as air temperature and solar irradiance. In this category, artificial intelligence techniques such as the ML is the most used method to detect and classify faults in PV arrays. Major ML algorithms used in this area are K-nearest neighbors (K-NN), decision tree (DT), support vector machine (SVM), random forest (RF), ensemble learning (EL), and neural networks (NNs) [1]. Computer vision and deep convolutional neural networks (ResUNet, the VGG-16, MobileNet, and other GAN versions) have been recently applied to identify and classify faults from thermal PV images using unmanned aerial vehicles [17,23].

6.5 Datasets description and preparation

Datasets are indispensable for the development of fault detection and diagnosis methods based on ML or DL algorithms. Two datasets are described in the next subsections. The first dataset comprises I-V curves measured in normal and abnormal working conditions. The second one contains I-V curves provided by a PV emulator. Both datasets will be used to develop fault detection and classification models based on some ML algorithms.

6.5.1 Dataset 6.1

The first dataset was prepared at the Renewable Energy Laboratory (RELab), University of Jijel, Algeria. It consists of 1500 I-V curves measured for a PV string of 10 modules (2 strings, each string has 5 PV modules connected in series). The Prova 1011 solar tracer instrument was used to collect the I-V curves under normal and abnormal working conditions. The type of PV module is BP Solar MSX120. Fig. 6.2 shows an example of normal and faulty measured I-V and P-V curves. The faults considered in this dataset are open-circuit, short-circuit, line-to-line, and mismatch effect.

The main features that can be extracted from the I-V curve are calculated as follows:

$$I_{sc}=I(V=0), V_{oc}=I(I=0), P_{mp}=(\max(P)), I_{mp}=I(y), V_{mp}=V(x), \text{and FF} = P_{mp}/(I_{sc}*V_{oc}).$$

where x, y is the coordinate values corresponding to the maximum power point in the P-V curve.

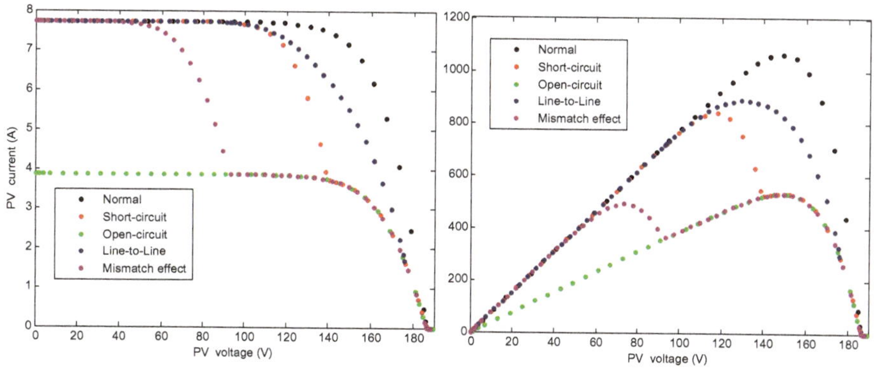

Fig. 6.2 Example of I-V and P-V curves under normal and faulty conditions (short-circuit, open-circuit, line-to-line, and mismatch).

Table 6.2 The main features extracted for normal and faulty I-V curves (I_{sc}, V_{oc}, I_{mp}, V_{mp}, P_{mp} and FF).

I-V curves with fault	Parameters					
	I_{sc}	V_{oc}	I_{mp}	V_{mp}	P_{mp}	FF
Normal I-V curve	7.84	187.1811	7.1592	143.9819	1030.8	0.71
Open circuit (one PV module)	3.8700	189.3846	3.5267	145.9453	514.7015	0.70
Short circuit (one PV module)	7.7400	188.1035	7.1374	112.8125	805.1873	0.55
Mismatch (partial shading)	7.7398	188.1362	6.6834	73.5238	491.3868	0.34
Line-to-line	7.7400	175.9195	6.8583	129.4528	887.8230	0.65

The calculated features of the curves plotted in Fig. 6.2 are listed in Table 6.2.

Numerous experiments have been carried out to build and prepare the dataset under various working conditions. The main features are then normalized as follows (I_{mp}/I_{sc}, V_{mp}/V_{oc}, P_{mp}/P_{mp_stc}, and FF) and the results are displayed in Fig. 6.3.

Notations: F1: short-circuit fault, F2: open-circuit fault, F3: line-to-line fault, and F4: mismatch fault (these will be used in the next Sections 6.7 and 6.8). In the experimental tests, these faults F1, F2, etc. take values 1, 2, and so on. To improve the quality of data (features), outliers should be removed from the dataset. Fig. 6.4 shows the distribution of experimental features and faults.

6.5.2 Dataset 6.2

The second dataset consists of 2700 I-V curves (normal and faulty) collected at the University of Setif, Algeria, by using a real-time PV emulator. The advantage of using a real-time PV emulator is that most faults can be emulated thus avoiding the risk of PV array damage.

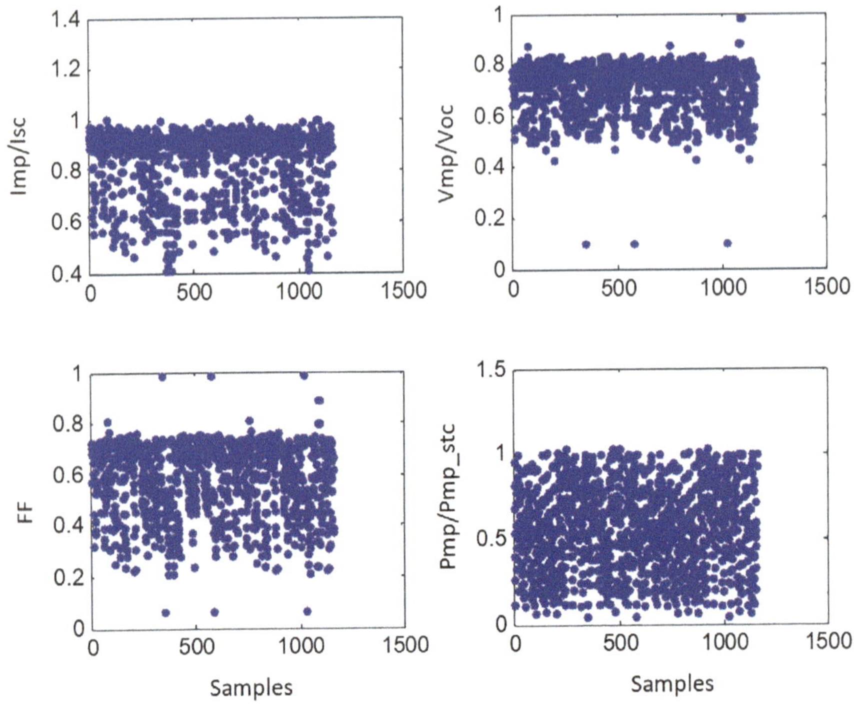

Fig. 6.3 Dataset 6.1: key features: normalized current, normalized voltage, power, and FF of the PV array.

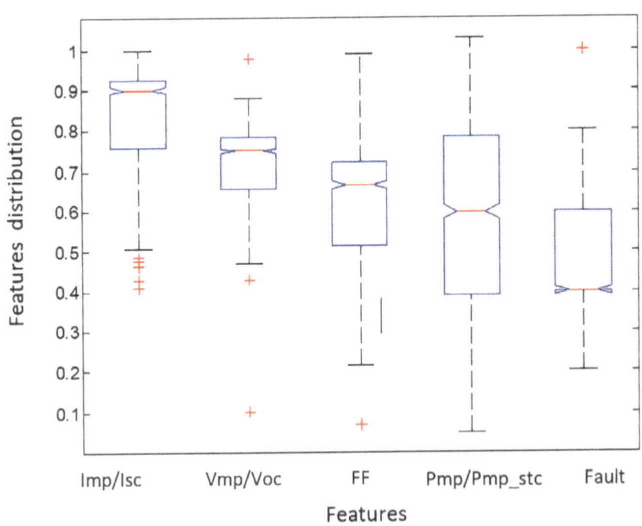

Fig. 6.4 Boxplot of the normalized features and the labeled faults.

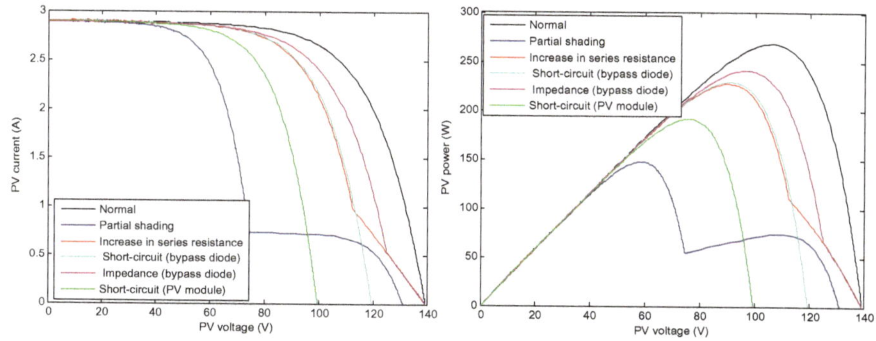

Fig. 6.5 Example of I-V and P-V curves under normal and faulty conditions (partial shading, increase in series resistance, short-circuit (bypass diode), impedance (bypass diode), and short-circuit PV module).

To generate the I-V curves Matlab®/Simulink, Control Desk and DS1104 platform are used which is connected to a PC via the PCI slot, and to the APS-1102A programmable DC/AC power source via PLC1104 module [24].

The PV string contains seven PV modules connected in series. The PV module used is JW-50P. Fig. 6.5 shows the I-V curves emulated under normal and faulty conditions. The types of faults are partial shading, increase in series resistance, short-circuit (PV module and bypass diode), and impedance (bypass diode).

Table 6.3 lists the extract main features from the I-V curves shown in Fig. 6.5.

In addition to the main features reported in Table 6.3, other features can be extracted from the I-V curves as follows [24]: dI/dV(md1), dI/dV(md2), $dI/dV(I_{sc})$, $dI/dV(V_{oc})$, dI/dV(iP), and area of the I-V curve.

Where md1 is the I-V curve slope at the midpoint between MPP and open-circuit voltage point, md2 is the I-V curve slope at the midpoint between short-circuit current

Table 6.3 The main features extracted for normal and faulty I-V curves (I_{sc}, V_{oc}, I_{mp}, V_{mp}, P_{mp}, and FF).

I-V curves with fault	Parameters					
	I_{sc}	V_{oc}	I_{mp}	V_{mp}	P_{mp}	FF
Normal I-V curve	2.9019	139.3385	2.5338	105.9745	268.5223	0.6641
Increase in series resistance	2.9128	139.4299	2.5222	90.5372	228.3562	0.5623
Short circuit (bypass diode)	2.9018	119.4726	2.5338	90.8410	230.1689	0.6639
Impedance	2.9000	139.4300	2.5194	95.9742	241.7989	0.5975
Short circuit (PV module)	2.9003	99.5419	2.5397	75.5277	191.8193	0.6644
Partial shading	2.9062	130.9940	2.5223	58.6145	147.8434	0.3884

point and MPP and iP is I-V curve slope at the MPP. In total we have 12 features normalized by using the following expression:

$$F_{\text{norm}} = \frac{Fr_{\text{n}} - Fr_{\text{a}}}{Fr_{\text{n}}} \tag{6.1}$$

where F_{norm}, Fr_{n}, and Fr_{a} are the normalized feature, normal feature, and abnormal feature, respectively.

As an example, Fig. 6.6 shows the 12 normalized features of the second dataset. To get an idea about the distribution of features, in Fig. 6.7, the distribution for the extracted 12 features is plotted.

6.5.3 *Dataset 6.3*

The third dataset consists of thermal images of a PV module at normal and faulty conditions (e.g., partial shading effect, dust accumulation, and short-circuited bypass diode). It was built at the Unit of Development of Solar Equipment, Bousmail, (Algeria) by using a thermal camera-type FLIR T640. Various tests have been considered under different climatic conditions. As an example, Fig. 6.8 shows a few pictures of thermal PV images registered on April 24, 2021 for different faults and normal PV modules.

In this dataset, hot spot phenomena are considered by investigating three causes partial shading, dust accumulation (dirty PV module), and short-circuited diode.

6.6 Feature selection and extraction

A large number of features in the dataset is one of the factors that affect both the computation time and the performance of ML models. Thus, feature selections and dimensionality reduction are important steps in ML algorithms for good performance, reducing the model complexity as well reducing the learning or training time needed by the model (computation time). In this, basically, only the most relevant input features are selected and the others that have low correlation are eliminated. There are two techniques to make dimensionality reduction: (a) feature selection and (b) feature extraction. From a PV I-V curve many features, which are summarized in Table 6.4 can be extracted. The main features are the common parameters extracted from the I-V characteristic curve, while secondary features are additional features that can be extracted from the I-V characteristic curve.

Many extraction methods are available in the literature including sequential feature selection, recursive feature selection, heuristic search (based on the application of genetic algorithms), and ML methods which include unsupervised and semi-supervised algorithms and other approaches [25].

In the next section, two examples of how to reduce the features are presented: (1) recursive feature selection (RFS) and (2) principal component analysis (PCA). To do this, the second dataset is used and by n the total number of features, and by

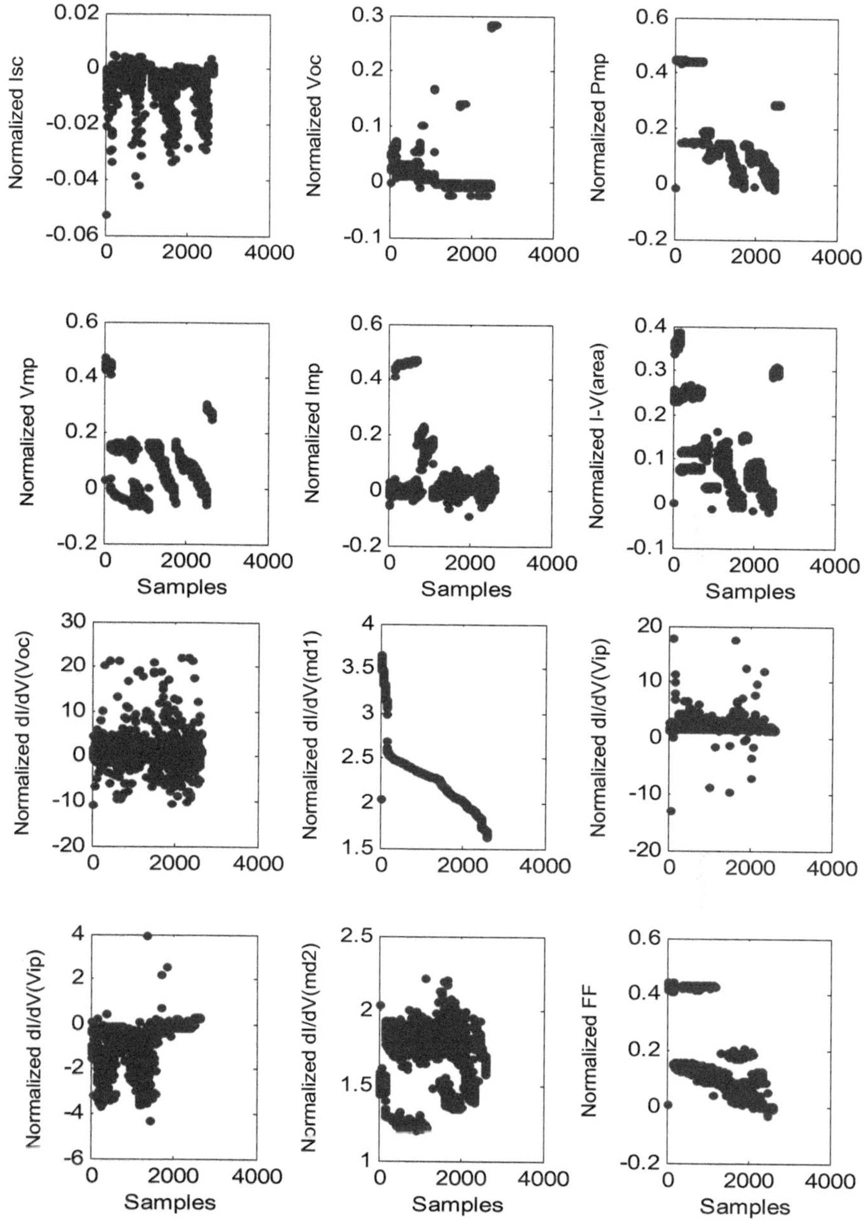

Fig. 6.6 Dataset 6.2: extracted normalized features: (feature 1 Area of I-V curve, feature 2 I_{sc}, feature 3 V_{oc}, feature 4 P_{mp}, feature 5 I_{mp}, feature 6 V_{mp}, feature 7: $dI/dV(V_{oc})$, feature 8: dI/dV (md1), feature 9: dI/dV(Vip), feature 10: $dI/dV(I_{sc})$, feature 11:dI/dV(md2), feature 12: FF)

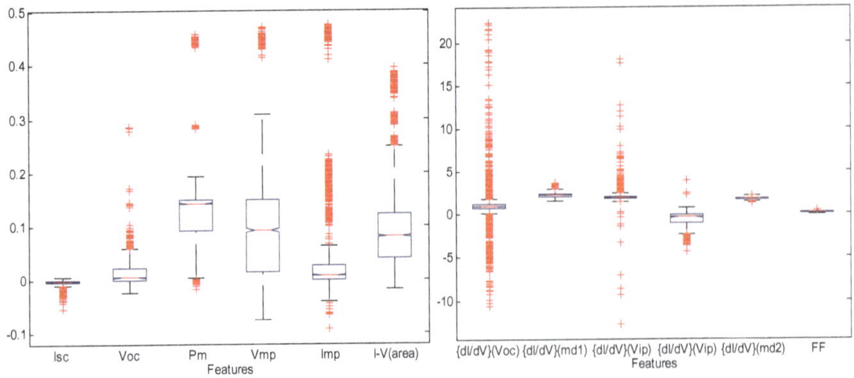

Fig. 6.7 Boxplot of the 12 normalized features.

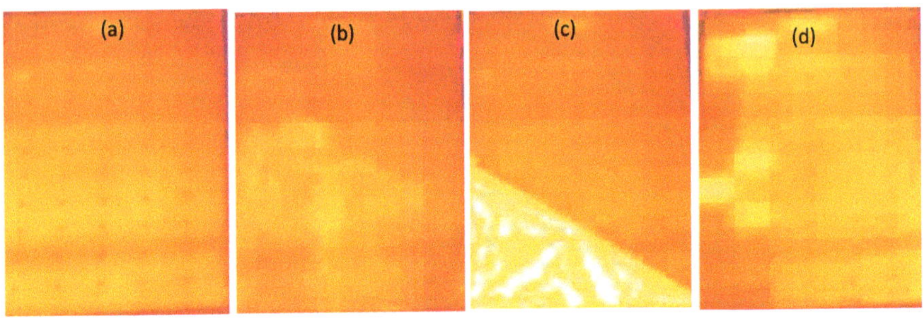

Fig. 6.8 Thermal images of a PV monocrystalline module: (A) normal, (B) dust accumulation on PV module surface, (C) partial shading, and (D) shunted bypass diode.

Table 6.4 Example of features extracted form an I-V curve (PV module, string, or array).

Features	Expression
Main futures	
I_{sc} (A)	Short-circuit current (calculated at $V=0$)
V_{oc} (V)	Open-circuit voltage (calculated at $I=0$)
I_{mp} (A)	Current at maximum power (calculated at $I(y)$)
V_{mp} (V)	Voltage at maximum power (calculated at $V(x)$)
P_{mp} (W)	Power at maximum (max(P))
FF	Fill factor ($P_{mp}/(I_{sc}*V_{oc})$)
Eff(%)	Efficiency (P_{mp}/(effective area*G_{stc}))*100%
Additional features	
SF1	$dI/dV(k1)$, $k1$ is the I-V curve slope at the midpoint between MPP and open-circuit voltage point
SF2	$dI/dV(k2)$ $k2$ is the I-V curve slope at the midpoint between short-circuit current point and MPP
SF3	Area of I-V curve
SF4	$dI/dV(I_{sc})$
SF5	$(I_{mp}-I_{sc})/V_{mp}$
SF6	$(I_{sc}/2-I_{mp})/(I_{sc}/2-V_{mp})$
..	..

m the number of features selected are denoted. Both methods can help to avoid overfitting and improve classification performance.

6.6.1 Recursive feature selection

Recursive feature selection works by searching for a subset of features by starting with all features in the training dataset and successfully removing features until the desired number remains. RFS method is one of the most frequently used for small samples [26].

Example 6.1: RFS for features reduction

In this example, the RFS algorithm is applied (named RFE) [27] available in sklearn library for feature reduction dimensionally to *m* features of the second dataset (this dataset has initially $n = 12$ features). Random forest function (`RandomForestClassifier`) is used for model development. The following listing is a basic code that permits the feature reduction using the RFS algorithm. In this example, $m = 5$ is chosen, so from 12 features only 5 relevant features are selected. The code is given in **Listing 6.1**.

#Listing 6.1 Feature selection with RFS algorithm

```
#Import libraries
import pandas as pd
from sklearn.ensemble import RandomForestClassifier
from sklearn.feature_selection import RFE
#Loading data
dataset2=pd.read_csv('e:/database_emulator.csv')
data=dataset2.values
n=12
X1=data[:,0:n]
Y1=data[:,n]
# feature selection
m=5
#model development and fit
model=RadomForestClassifier(n_estimators=200)
FRS=RFE(model, m)
Fit1=FRS.fit(X1, Y1)
print("Number of features: %d" % fit1.n_features_)
print("Selected features: %s" % fit1.support_)
```

The results are shown below:

```
Number of features=5
Selected features: [False False True True True False False False
False True False True]
```

The corresponding selected five features corresponding to "*True*" are: Feature 3 = normalized V_{oc}, Feature 4 = normalized P_{mp}, Feature 5 = I_{mp}, Feature 10 = dI/dV(vip), and Feature 12 = FF.

6.6.2 Principal component analysis

Principal component analysis (PCA) [28] is one of the most popular ML algorithms used to convert high-dimensional data to low-dimensional data by selecting the most important features. A property of PCA is that you can choose the number of dimensions or principal component in the transformed result. PCA can provide good precision results when applied to correlated attributes [29]. In **Listing 6.2**, the principal component is set to $m = 5$.

Example 6.2: PCA for features reduction

PCA function available in sklearn library is used.

 # Listing 6.2 Feature selection with PAC

```
#Import libraries
import pandas as pd
from sklearn.decomposition import PCA
#Loading data
dataset2=pd.read_csv('e:/database_emulatorcsv.csv')
data=dataset2.values
n=12
X1=data[:,0:n]
Y1=data[:,n]
m=5
#model development and fit
pca_model=PCA(n_components=m)
fit1=pca_model.fit(X)
print("Explained Variance: %s" % fit1.explained_variance_ratio_)
print(fit1.components_)
```

The explained variance results of the five principal components are as follows:

```
Explained Variance: [0.63087064 0.2388275 0.09455418 0.0222568
0.00847083]
```

It can be seen that the transformed dataset hardly resembles the source data. The importance of features can be calculated by using the following command lines:

```
model=ExtraTreesClassifier(n_estimators=100)
model.fit(X, Y)
print(model.feature_importances_)
```

In this dataset the importance five parameters are Feature 3 = normalized V_{oc}, Feature 4 = normalized P_{mp}, Feature 5 = normalized I_{mp}, Feature 2 = Normalized I_{sc}, and Feature 12 = FF.

6.7 Fault detection in photovoltaic arrays

Early fault detection in PV systems is considered as the most important step in which the fault should be immediately isolated or removed, to avoid power losses and risk of fire. The faulty PV array can be detected by different strategies:

(1) The first method aims to compare the measured and estimated data (e.g., power, current, or voltage). An accurate model is necessary for making this comparison. In addition, the thresholds should be also carefully estimated in order to avoid false alarm indications [1]. The thresholds can be calculated based on the maximum error introduced by the model uncertainty and the noise measurement. The sensors employed should be within the specifications required by the IEC 61724 standard [30] that indicates a relative error of 1%, 1%, and 2% while measuring current, voltage, and power, respectively.

(2) The second method is based on data-driven approaches, a dataset is always indispensable with a computing algorithm. In this approach, the detection precision depends mainly on the data quality (removing noisy data, outliers, etc.) as well as the algorithm efficacy. In such methods, ML algorithms are the most appropriate, especially for multiple fault detection.

(3) The third method is based on thermal images processing to detect the major defect on PV modules (e.g., hotspot, snail trails, microcrack cells, and cells aging). Also in this category, images quality (good resolution) and efficient image processing algorithms are indispensable. ML and DL algorithms are the most suitable.

In fact, both second and third methods perform a binary classification to detect faulty PV modules or arrays.

6.7.1 Fault detection based on analytical models

In this approach, the well-known one diode, two diodes, or other models [31–33] could be used to estimate the maximum power (e.g., from the plotted P-V curve), which represents the output PV power. Calculated PV power by the model is compared with the one measured. The normal operation limits are defined as follows [34]:

$$\text{Upper} = \mu + 3\sigma \quad (6.2)$$

$$\text{Lower} = \mu - 3\sigma \quad (6.3)$$

where μ and σ are the mean and standard deviation, respectively.

The upper and lower values are calculated using the ratio between the measured and predicted power at normal operating conditions. Thus, points outside the normal operation limits are considered as faults. To calculate the upper and lower values the measured power from a PV array is used with the ones estimated by using the one diode model, which accepts as input the electrical parameters from the datasheet of

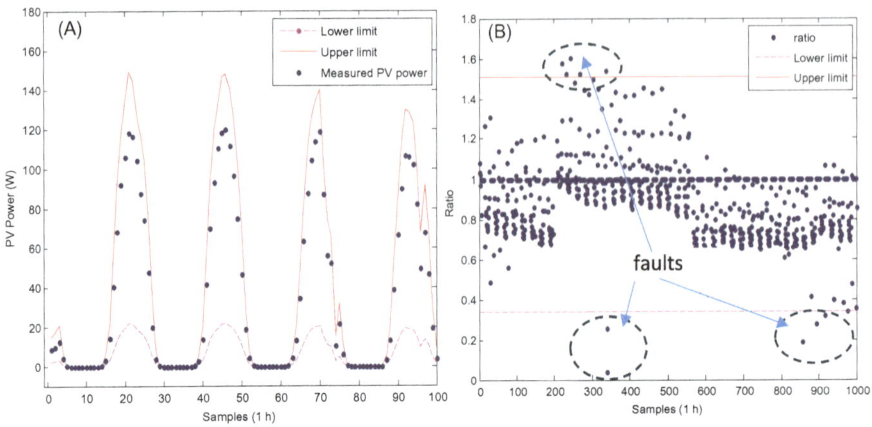

Fig. 6.9 (A) Measured power and normal operation limits are expressed in W and (B) fault detection is based on operation limits.

the PV module as well the values of solar irradiance and module temperature. Then the ratio value, i.e., the measured power/predicted power is calculated. In this case, the calculated lower and upper limits are 0.34 and 1.51, respectively.

Fig. 6.9A shows the power measured and normal operation limits, in Fig. 6.9B an example of fault detection based on the lower and upper limits is illustrated.

The accuracy of this method depends on the model precision, the defined upper and lower limits, as well as the uncertainty of the instruments used.

6.7.2 Fault detection using machine learning algorithms

PV fault detection using ML algorithms can be considered as a binary classification problem, in which the output label could be healthy ('0') or faulty ('1'). The main steps to implement a ML algorithm, under Python language, for binary classification (fault detection) problem are summarized below:

Step #1: Import all necessary libraries and functions.
Step #2: Load the dataset and define the input features and the output labels of the model ['0', '1'].
Step #3: Split the dataset into two parts: training and testing portions.
Step #4: Call the binary classifier method (e.g., DT, SVM, …).
Step #5: Select model parameters and fit the model to train set, k time to obtain the optimal performance.
Step #6: Predict the class and evaluate the model by using error metrics and matrix confusion.

Two ML algorithms are applied for fault detection of a PV array based on historical data (I-V curves). The first dataset is used to check the ability of SVM and DT machine learning algorithms in fault detection. This dataset is divided into two parts: 80% for the training of both ML algorithms and 20% for testing the models' accuracy.

Example 6.3: Support vector machine (SVM) for fault detection

The idea is to train a ML algorithm to predict the power output accurately, and then compare the predicted with the one measured by calculating the following average error (Er):

$$\mathrm{Er} = \frac{\sum_{t=1}^{T} (\hat{P}_t - P_t)}{\sum_{t=1}^{T} (\hat{P}_t)} \tag{6.4}$$

$\hat{P}$ and P are the predicted and measured power at instant t, respectively, and T is a period of time.

If the Er exceeds the defined threshold (Th) the system will report a fault. The threshold is defined based on several experiments. In this example, a model to predict the output PV power using the SVM algorithm is developed [28]. The SVM model accepts as input, measured values of G, Tc, and Ta, and as output the power delivered (P). A dataset was collected from a PV array of 5×2 modules. The code is illustrated in **Listing 6.3**.

 # Listing 6.3 Fault detection using SVM

```
import matplotlib.pyplot as plt
import pandas as pd
from sklearn.model_selection import train_test_split
from sklearn import svm
#Import libraries
import pandas as pd
from sklearn.model_selection import train_test_split
from sklearn import svm
from sklearn import metrics
%matplotlib inline

# loading dataset
col_names=['Tc','Ta', 'G', 'P']
dataset=pd.read_csv('e:/PV_Array_Detection.csv',header=None,
  names=col_names)
input_var=['Tc', 'Ta','G']
X=dataset[input_var].values
Y=dataset.P.values

#Train and fit the model
X_train, X_test, y_train, y_test=train_test_split(X, Y, test_size
  =0.20,random_state=0)
model=svm.SVR(kernel='rbf', C=200)
model.fit(X_train, y_train)
```

```
#Test the model
y_pred=model.predict(X_test)
print('mae=', metrics.mean_absolute_error(y_test, y_pred))
mse=metrics.mean_squared_error(y_test, y_pred)
print('rmse=', np.sqrt(mse)) #mse**(0.5)
r=metrics.r2_score(y_test, y_pred)
print('r=',np.sqrt(r))
Th=0.001;

#Detection step
Pm=np.array(Vect1)
Pp=np.array(Vect2)
Er=np.sum(Pp-Pm)/np.sum(Pp)
IfEr>=Th:
Print('fault')
Else print ('no fault')
```

Fig. 6.10A shows scatted curve between the predicted and the measured curves, and as can be seen from Fig. 6.10B a good correlation is obtained. To check the performance of the SVM model the errors MAE, RMSE, and *r* are calculated and listed in Table 6.5.

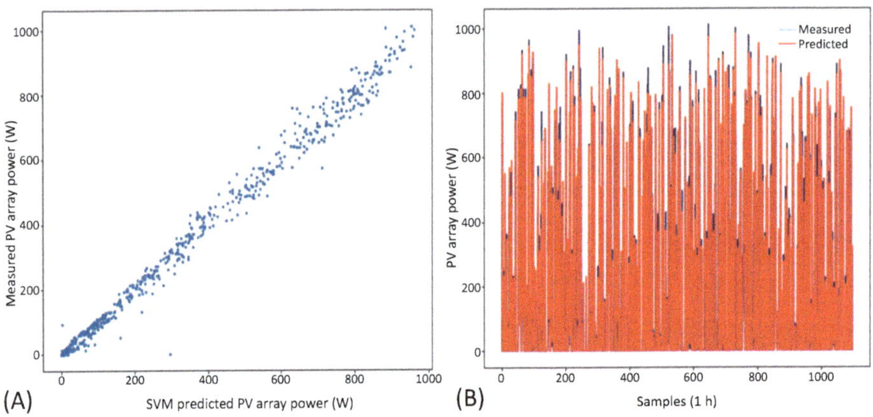

Fig. 6.10 Comparison between measured and predicted powers by SVM model: (A) scatter plot and (B) superposition curves.

Table 6.5 Error metrics: MAE, RMSE, and *r*.

SVM parameters	MAE (W)	RMSE (W)	*r* (%)
C=200, cache_size=200, coef0=0.0, degree=3, epsilon=0.1, gamma='scale', kernel='rbf', max_iter=-1, shrinking=True, tol=0.001, verbose=False	11	22.25	98

From Table 6.5 it can be concluded that the SVM model performs good, however, to improve the fault detection accuracy and avoid false alarms, the dataset should be periodically updated due to the aging of the PV modules over time, tuning the model parameters, or using more accurate ML algorithms.

Example 6.4: Decision tree (DT) algorithm for fault detection

In the second example, a C4.5 decision tree (DT) algorithm [28] and historical I-V curves from the first dataset are used. As input features of the DT model, P_{mp}, I_{mp}, V_{mp}, and FF are used, and as output, two classes are defined: healthy = '0' or faulty = '1.' The code is given in **Listing 6.4**.

#Listing 6.4 Fault detection using C4.5 DT algorithm

```
#Import libraries
import matplotlib.pyplot as plt
import pandas as pd
from sklearn.model_selection import train_test_split
from sklearn.tree import DecisionTreeClassifier
from sklearn import metrics, tree

#loading data
col_names=['Pmp','Imp','Vmp' ,'FF','F']
data=pd.read_csv('e:/database1.csv',header=None, names=col_names)
features=['Pmp','Imp','Vmp' ,'FF']
X=data[features].values
y=data.F.values
#Train and fit the model

X_train, X_test, y_train, y_test=train_test_split(X, y,
  test_size=0.20, random_state=0)
model=tree.DecisionTreeClassifier(criterion="entropy",
  max_depth=3)
model_=model.fit(X, y)

#Model Accuracy,
y_pred=model_.predict(X_test)
print("Accuracy:",metrics.accuracy_score(y_test, y_pred))
text_representation=tree.export_text(model)
print(text_representation)
fig=plt.figure(figsize=(30,30))
_=tree.plot_tree(clf, feature_names=['Pm','Imp','Vmp','FF'], class_names
  =['Normal','Fault'],filled=True)
```

The DT model parameters and the calculated accuracy are listed in Table 6.6. As can be seen, the DT model can detect the fault in the PV array with a very accurate result (98%).

Fig. 6.11 illustrates the obtained fault detection diagram based on the DT using *tree.export_test() and plot.tree()* functions.

Table 6.6 DT model parameters and accuracy.

DT model parameters	Accuracy (%)
ccp_alpha=0.0, class_weight=None, criterion='entropy', max_depth=3, max_features=None, max_leaf_nodes=None, min_impurity_decrease=0.0, min_impurity_split=None, min_samples_leaf=1, min_samples_split=2, min_weight_fraction_leaf=0.0, presort='deprecated', random_state=None, splitter='best'	98

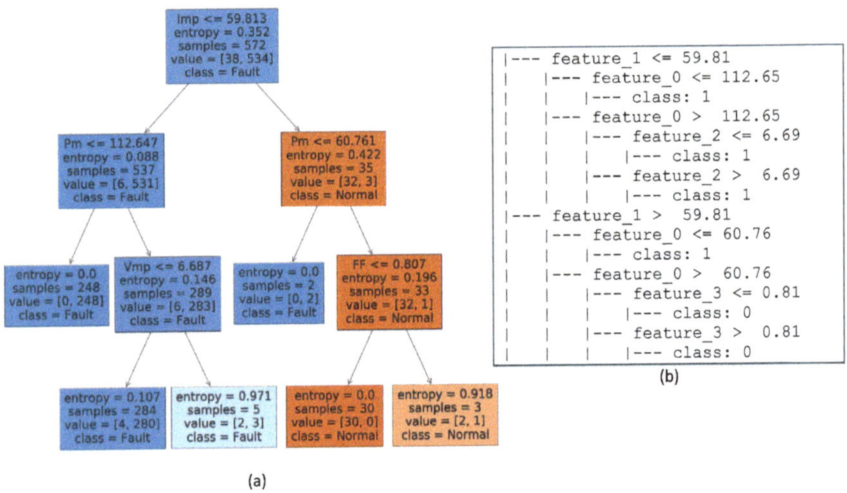

Fig. 6.11 DT-based fault detection approach: (A) DT diagram with max_depth parameter = 3 and (B) DT model (text representation).

The results demonstrate the capability of the presented C4.5 DT algorithm in the detection of faults in a PV array based on historical data (accuracy = 98%).

6.7.3 Fault detection using deep learning algorithms

In this section a deep learning algorithm like deep convolutional neural network (DCNN) to detect the defect in PV modules based on thermal images (of PV modules) is used. To do this the third dataset is employed.

Example 6.5: DCNN for fault detection-based thermography images

The main code of the DCNN model is given in **Listing 6.5**.

#Listing 6.5 Fault detection using deep CNN

```
modelCNN=keras.Sequential([
    keras.layers.AveragePooling2D(6,3, input_shape=(640,480,3)),
    keras.layers.Conv2D(64, 1, activation='relu'),
    keras.layers.MaxPool2D(2,2),
    keras.layers.Conv2D(32, 1, activation='relu'),
    keras.layers.MaxPool2D(2,2),
    keras.layers.Dropout(0.5),
    keras.layers.Flatten(),
    keras.layers.Dense(128, activation='relu'),
    keras.layers.Dense(64, activation='relu'),
    keras.layers.Dense(2, activation='softmax')
])
modelCNN.compile(optimizer='Adamax',
      loss=keras.losses.CategoricalCrossentropy(),
      metrics=['accuracy'])
results=modelCNN.fit(trainX, trainY, epochs=20, batch_size=32,
  validation_data=(testX, testY))
```

The DCNN designed model consists of 10 layers: 1 AveragePooling2D layer, 2 Conv2D layers, 2 MaxPool2D layers, 1 droop layer, 1 flatten layer, and 3 dense layers (Table 6.7).

As shown in Fig. 6.12A, the training accuracy reached 1, and the loss is 0.03 which is acceptable. With reference to the confusion matrix shown in Fig. 6.12B, the classification rate is 8/8, which means the model detects a fault with good precision.

6.8 Fault classification of photovoltaic arrays

When a fault is detected, the next step is to classify and identify the fault. This step is very crucial, and it can be considered a multiclass classification problem. Some faults can be identified or classified by developing a simple algorithm based on (*if-then-else*) rules. In this, a comparison is made between the measured and modeled I-V curves and then it is used to classify the possible faults. A simple case of a PV string with few modules connected in series is considered here. The faults considered are: open-circuit, short-circuit, shading, and line-to-line. Two thresholds are defined, Th1 and Th2, which are calculated based on the maximum error introduced by the model

Table 6.7 DCNN model parameters and accuracy.

DT model parameters	Accuracy (%)
`batch_size =32, optimizer='Adamax',loss=keras.losses. CategoricalCrossentropy, ctivation fucntions=relu and softmax`	100

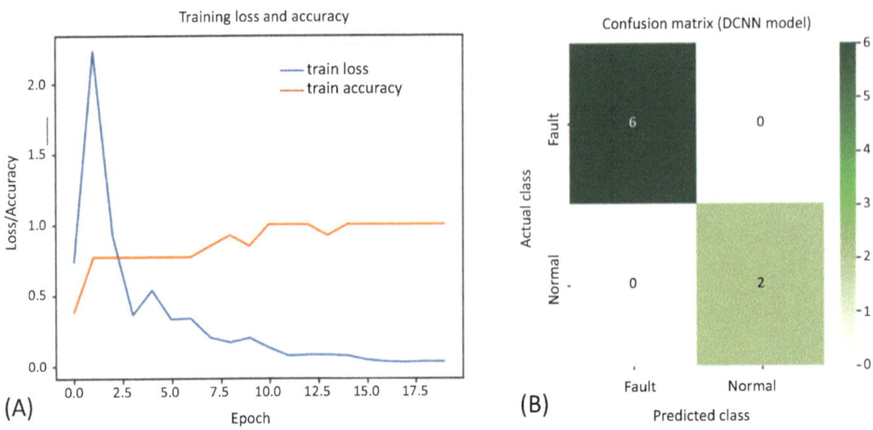

Fig. 6.12 (A) Training and loss accuracy of the DCNN model and (B) confusion matrix.

uncertainty and the measurement noise. The measured parameters are denoted by I_{sc}, V_{oc}, I_{mp}, V_{mp}, and P_{mp}, while I^*_{sc}, V^*_{oc}, V^*_{mp}, I^*_{mp}, and P^*_{mp} are the estimated parameters. N represents the number of peaks on the I-V curve. A basic fault defection algorithm is shown below:

```
If N<=1 then
        If Isc=Isc* then open circuit fault
                   Else if Isc>Th1 then multiples faults (shadow effect
with faulty bypass diode)
                                 Else if Imp>Th1 then
                                         If Voc> Th2 then a group of
faults (short-circuit, open-circuit, diode fault, line-to-line and mis-
match fault)
                                         Else line-to-line fault
Else Imp>Th1 then
                   If Vmp>Th2 then two possible faults (shading or
line-to-line)
Else shading fault.
```

Despite the fact that this approach can be easily implemented and provide acceptable classification results [11], it has some limits particularly in the case of multiple faults or faults that have the same symptoms (i.e., the same faulty I-V curves, but the source of the fault is different). In this situation, the application of ML or DL algorithms is much appropriate [1]. Fig. 6.13 shows a 3D plot of FF, I, and V without normalization. As can be seen, it is hard to apply a simple algorithm to classify these faults (F1, F2, F3, and F4).

This section deals with the presentation and comparison of some multiclass ML algorithms for fault classification in PV arrays. To evaluate the performance of the

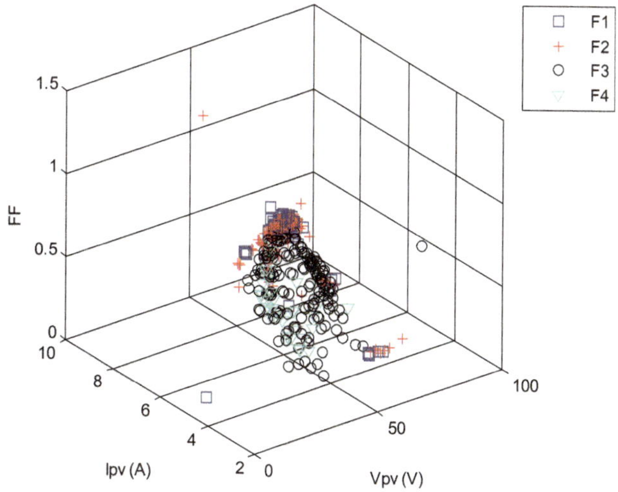

Fig. 6.13 3D plot of the selected features under various faulty conditions.

ML classifiers, the confusion matrix (CM) or error matrix is plotted [35,36]. The common performance metrics that can be calculated from the CM are accuracy, precision recall, and F1-score (see Appendix D).

6.8.1 Fault classification using machine learning algorithms

The first dataset is used to develop multiclass ML models for PV fault classification. The fault classes are denoted by F1 = '1', F2 = '2', F3 = '3', and F4 = '4', thus four faulty classes exist. To classify faults, six ML algorithms are considered (K-NN, DT, RF, XGBoost, CatBoost, and LGBM). The main steps to implement a ML classifier, for the multiclass classification problem, for PV array are described below:

Step #1: Import all necessary libraries and functions.
Step #2: Load the dataset and define in the input features and the output labels of the model.
Step #3: Split the dataset into two parts: training and testing the model.
Step #4: Call the classifier method (e.g., K-NN, RF, DT, …).
Step #5: Select model parameters and fit the model to train set, k time to get the optimal performance.
Step #6: Predict the classes and evaluate the model by using the confusion matrix and accuracy score.

Example 6.6: K-Nearest Neighbors (K-NN) classifier

In this example, the K-nearest neighbors (K-NN) algorithm [28] is used to develop the classifier model. The corresponding code is given in **Listing 6.6**: This code is the main for other ML algorithms examined in this section.

#Listing 6.6: Fault classification using k-NN

```
import numpy as np
import pandas as pd
from sklearn.model_selection import train_test_split
from sklearn.neighbors import KNeighborsClassifier
from sklearn import metrics
from sklearn.metrics import accuracy_score, f1_score,
  precision_score, recall_score, classification_report,
  confusion_matrix
from sklearn.datasets import make_classification
from sklearn.metrics import plot_confusion_matrix
import matplotlib.pyplot as plt
%matplotlib inline

col_names=['irra', 'tem', 'pm', 'c_i', 'c_v', 'ff', 'fault']
dataset1=pd.read_csv('e:/dataFDnew.csv',header=None,
  names=col_names)
feature_cols=[ 'pm', 'c_i', 'c_v', 'ff']
X=dataset1[feature_cols]
y=dataset1.fault
X_train, X_test, y_train, y_test=train_test_split(X, y,
  test_size=0.20,random_state=40)
#call the k-NN model
model=KNeighborsClassifier(n_neighbors=40)
#fit the k-NN model
model.fit(X_train, y_train)
#model prediction
y_pred=model.predict(X_test)
#Print confusion matrix
print(confusion_matrix(y_test,y_pred))
print(classification_report(y_test,y_pred))
print(accuracy_score(y_test, y_pred))
print("Accuracy:",metrics.accuracy_score(y_test, y_pred))

#Plot confusion matrix
plot_confusion_matrix(cmodl X_test, y_test)
plt.show()
```

First, import the KNeighborsClassifier module and create K-NN classifier object by passing the argument number of neighbors in *KNeighborsClassifier()* function available in the Sklearn library. The most important parameter to optimize the K-NN algorithm is the value of K, which represents the number of nearest neighbors. To do this the code is run for K between 1 and 100.

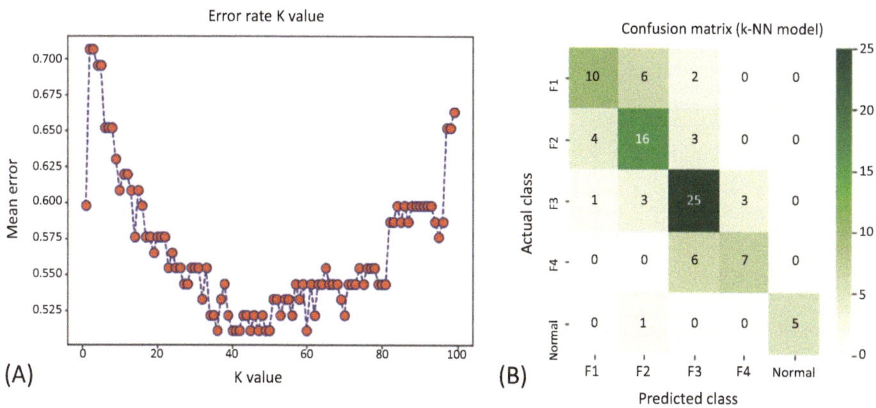

Fig. 6.14 (A) Optimal K value of K-NN model and (B) confusion matrix of K-NN model.

```
for k in range(1, 100):
     k_NN=KNeighborsClassifier(n_neighbors=k)
     k_NN.fit(X_train, y_train)
     pred_k=knn.predict(X_test)
     error.append(np.mean(pred_i !=y_test))
```

From the plotted curve of mean error versus K values, shown in Fig. 6.14A, it can be seen that the values of K of 36, 39, 40, 41, 44, 45, 46, and 47 are optimal.

To find the most influenced input features, all possible combinations are made based on the following expression:

$$PC = q!(n! - q)! \tag{6.5}$$

where PC is the possible combination, n is the total number of features, and q is the reduced selected number. To plot the CM, the following code is used. The obtained CM is shown in Fig. 6.14B.

```
CM=confusion_matrix(y_test, y_pred)
class=pd.DataFrame(CM, index=['F1','F2','F3','F4','Normal'],
columns=['F1','F2','F3','F4','Normal'])
plt.figure(figsize=(5, 5))
sns.heatmap(class, annot=True, cmap='Greens', fmt='g')
plt.xlabel('Predictedclass',fontsize=14)
plt.ylabel('Actualclass',fontsize=14)
plt.title('Confusionmatrix',fontsize=14)
plt.show()
```

Accuracy can be computed by comparing actual test set values and predicted values.

From the first line in the CM (Fig. 6.14B) of the 18 F1 classes, the model classifies 6 and 2 fall in F2 and F3 classes, respectively. In the second line of the 23 F2 classes,

Table 6.8 Error metrics: precision, recall, F1-score and accuracy.

K-NN model parameters	Features	Precision	Recall	F1-Score	Accuracy (%)

`algorithm='auto',`
`leaf_size=30,`
`metric='minkowski',`
`metric_params=None,`
`n_jobs=None,`
`n_neighbors=13,`
`p=2, Vmp/`
`weights='uniform'`
	I_{mp}/I_{sc}, V_{mp}/V_{oc}, and FF				

Class	F1	0.67	0.56	0.61	68.5
	F2	0.62	0.70	0.65	
	F3	0.69	0.78	0.74	
	F4	0.70	0.54	0.61	
	F5:Normal	1.00	0.83	0.91	

the model classifies 4 and 3 fall in F1 and F3 classes, respectively. In the third line, 25 F3 classes are correctly classified, while 1, 3, and 3 fall in F1, F2, and F4, respectively. In the fourth line, 7 F4 classes are classified correctly, while 6 fall in F3. In the last line, 5 normal classes F5 are correctly classified from the total six normal classes, one is incorrectly classified in F2 as a fault class.

Error metrics are reported in Table 6.8. The classification accuracy is 68.5% which is acceptable. The best configuration is obtained with three features (V_{mp}/V_{oc}, I_{mp}/I_{sc}, and FF) as shown in Table 6.8. K-NN performs better with a lower number of features than a large number of features.

Example 6.7: Decision tree (DT) classifier

To develop a DT [28] classifier, the same code as the one shown in **Listing 6.6** is used. A little modification should be conducted in **Listing 6.6**. So, the DT function is imported *(DecisionTreeClassifier)* from *sklearn.tree* library, and adding the following command line:

```
model=DecisionTreeClassifier(criterion='entropy', random_state=10)
```

Table 6.9 shows the results of the classification stage of the DT classifier. The classification accuracy is 78% which is acceptable, and an improvement in the accuracy compared with K-NN classifier (68.5%) can be observed. In this case, the best configuration is obtained with four features as shown in Table 6.9.

The decision tree model is illustrated in Fig. 6.15A and the calculated CM is shown in Fig. 6.15B.

Table 6.9 Error metrics: precision, recall, F1-score and accuracy.

DT model parameters	Features	Precision	Recall	F1-Score	Accuracy (%)
(ccp_alpha=0.0, class_weight=None, criterion='entropy', max_depth=None, max_features=None, max_leaf_nodes=None, min_impurity_decrease=0.0, min_i min_samples_leaf=1, min_samples_split=2, min_weight_fraction_leaf=0.0, presort='deprecated', random_state=10, splitter='best')	I_{mp}/I_{sc}, V_{mp}/V_{oc}, FF, and P_{mp}/P_{mp_stc}				
Class	F1	0.71	0.67	0.69	78
	F2	0.69	0.78	0.73	
	F3	0.84	0.84	0.84	
	F4	0.83	0.77	0.80	
	Normal	1.00	0.83	0.91	

```
|--- feature_2 <= 6.27
|   |--- feature_1 <= 52.25
|   |   |--- feature_0 <= 149.68
|   |   |   |--- class: 4
|   |   |--- feature_0 > 149.68
|   |   |   |--- class: 3
|   |--- feature_1 > 52.25
|   |   |--- feature_0 <= 160.87
|   |   |   |--- class: 2
|   |   |--- feature_0 > 160.87
|   |   |   |--- class: 3
|--- feature_2 > 6.27
|   |--- feature_1 <= 59.28
|   |   |--- feature_1 <= 52.25
|   |   |   |--- class: 2
|   |   |--- feature_1 > 52.25
|   |   |   |--- class: 1
|   |--- feature_1 > 59.28
|   |   |--- feature_2 <= 6.55
|   |   |   |--- class: 2
|   |   |--- feature_2 > 6.55
|   |   |   |--- class: Normal
```
(A)

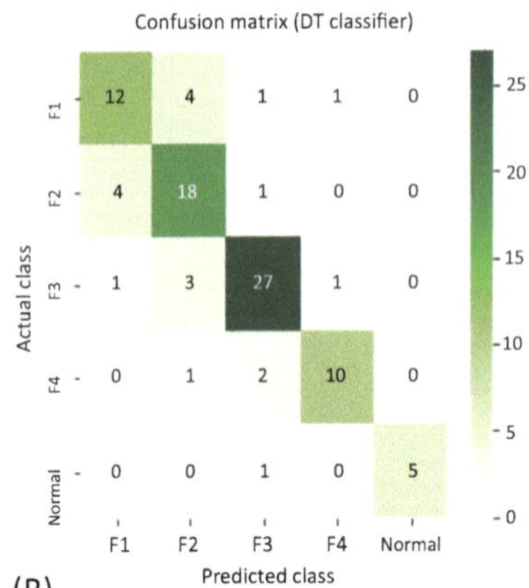

(B)

Fig. 6.15 DT classifier: (A) DT model and (B) confusion matrix of DT classifier.

By comparing the confusion matrices of K-NN and DT models, it can be observed that the value of each diagonal element of $CM(i, i)_{DT}$ is greater than $CM(i, i)_{K\text{-}NN}$ which confirms the superiority of the DT model in the fault classification.

Example 6.8: Random Forest (RF) classifier

To develop the RF [28] classifier for fault classification, the *RandomForestClassifier* function is used. The following command line should be inserted in **Listing 6.6**.

```
from sklearn.ensemble import RandomForestClassifier
model=RandomForestClassifier(n_estimators=100, random_state=20)
```

After several experiments, a good result is obtained with all features in the input. Error metrics are reported in Table 6.10 with a good classification accuracy of 88%.

The following command lines allow the visualization of the feature importance.

```
feature_imp=pd.Series(cl.feature_importances_, index=X_train.col-
umns).sort_values(ascending=False)
#feature_imp
# Creating a bar plot
sns.barplot(x=feature_imp, y=feature_imp.index)
# Add labels to your graph
plt.xlabel('Feature Importance Score', fontsize=14)
plt.ylabel('Features',fontsize=14)
plt.title("Visualizing Important Features", fontsize=14)
plt.legend()
plt.show()
```

Table 6.10 Error metrics: precision, recall, F1-score and accuracy.

RF model parameters	Features	Precision	Recall	F1-Score	Accuracy (%)
bootstrap=True, ccp_alpha=0.0, class_weight=None, criterion='gini', max_depth=None, max_features='auto', max_leaf_nodes=None, max_samples=None, min_impurity_decrease=0.0, min_impurity_split=None,min_samples_leaf=1, min_samples_split=2,min weight_fraction_leaf=0.0, n_estimators=100, n_jobs=None, oob_score=False, random_state=0, verbose=0, warm_start=False	I_{mp}/I_{sc} V_{mp}/V_{oc}, FF, G, T, and P_{mp/Pmp_stc}				
Class	F1	1.00	0.67	0.80	88
	F2	0.77	1.00	0.87	
	F3	0.91	0.91	0.91	
	F4	0.92	0.85	0.88	
	Normal	1.00	1.00	1.00	

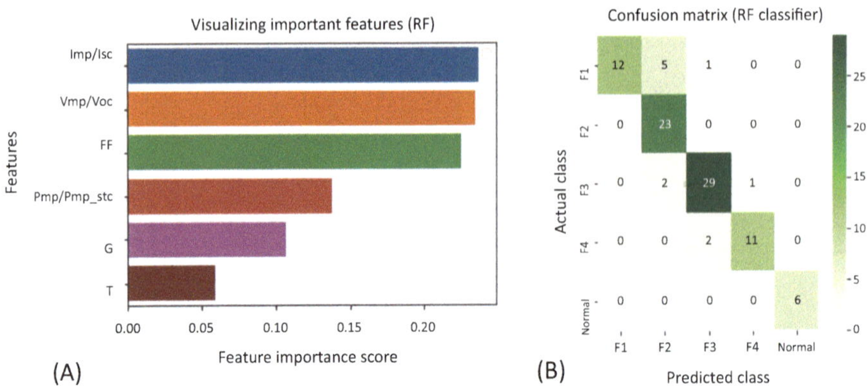

Fig. 6.16 (A) RF classifier: important factures and (B) confusion matrix of the RF classifier.

Features importance and confusion matrix results are illustrated in Fig. 6.16. From Fig. 6.16A it can be seen that *G* and *T* have low importance compared to other features. As can be observed from the confusion matrix Fig. 6.16B, the normal case is correctly classified. Based on the accuracy, the RF classifier outperforms both K-NN and DT classifiers.

Regarding the CM shown in Fig. 6.16B, in the first row, 12 samples of class F1 are correctly classified, while 6 are incorrectly classified. In the second and the last rows, no error is given and the model classified the fault F2 and normal case correctly. In the third row, 29 samples are correctly classified, while 3 are misclassified. In the fourth row, 11 samples of class F4 are correctly classified, and only 2 samples are misclassified.

Example 6.9: Boosting tree (XGBoost) classifier

In this example, a type of boosted tree (XGboost) algorithm is used [28,29]. Boosting falls under the category of the ML algorithms. *XGBClassifier* function is an advanced version of the gradient boosting function. To check its capability in fault classification, the same code illustrated in **Listing 6.6** is used by inserting the following command lines:

```
import xgboost as XGBClassifier
model=XGBClassifier(objective="multi:softprob", random_state=40)
```

Classification performance of this model is listed in Table 6.11. As reported in Table 6.11, this model presents several parameters which can be tuned. The classification accuracy is 85% and the normal class is estimated correctly. Other faults are estimated with variable recall, score, and precision. For more details, in Fig. 6.17 the feature importance and the confusion matrix are plotted.

The following function allows the user to get an idea about the feature importance.

```
xgb.plot_importance(model)
```

Table 6.11 Error metrics: precision, recall, F1-score and accuracy.

BT model parameters	Features	Precision	Recall	F1-score	Accuracy (%)
base_score=0.5, booster='gbtree', colsample_bylevel=1,colsample_bynode=1, colsample_bytree=1, gamma=0, gpu_id=-1, importance_type='gain', interaction_constraints='', learning_rate=0.300000012, max_delta_step=0, max_depth=6, min_child_weight=1, missing=nan, monotone_constraints='()',n_estimators=100, n_jobs=4, num_parallel_tree=1,objective='multi:softprob', random_state=40, reg_alpha=0, reg_lambda=1, scale_pos_weight=None, subsample=1, tree_method='exact', use_label_encoder=True,validate_parameters=1, verbosity=None)	I_{mp}/I_{sc} V_{mp}/V_{oc}, FF, G, T, and P_{mp/Pmp_stc}				
Class	F1	0.91	0.56	0.69	85
	F2	0.73	0.96	0.83	
	F3	0.88	0.94	0.91	
	F4	0.91	0.77	0.83	
	Normal	1.00	1.00	1.00	

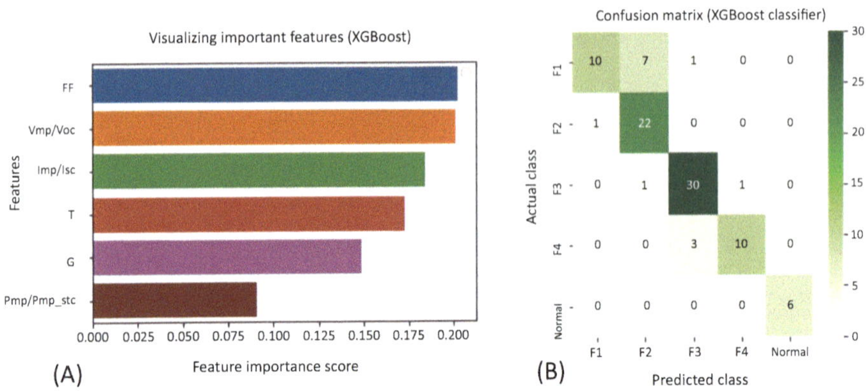

Fig. 6.17 (A) XGBoost classifier: feature importance and (B) confusion matrix of the XGBoost classifier.

Fig. 6.17A shows the most important feature to others, and as can be seen, the normalized P_{mp} and V_{mp} are the most influenced features.

Regarding the CM shown in Fig. 6.17B, in the first row, 10 samples of class F1 are correctly classified, while 8 are incorrectly classified. In the second row, 22 are correctly classified from 23. In the third row, 30 samples are correctly classified, while 2 are misclassified. In the fourth row, 10 samples of class F4 are correctly classified, and only 3 samples are misclassified. In the last row, all samples are correctly classified, i.e., six samples from six.

Example 6.10: CatBoost classifier

In this example, a type of boosted tree (BT) algorithm is used, which is CatBoost [29]. CatBoost is an open-source algorithm based on gradient boosted decision trees.

CatBoostClassifier function is an advanced version of the gradient boosting function. To check its capability in fault classification, the same code illustrated in **Listing 6.6** is used by inserting the following command lines:

```
from catboost import CatBoostClassifier
model=CatBoostClassifier()
```

Classification performances of this model are listed in Table 6.12. As reported in Table 6.12 this model presents a few parameters which can be tuned. The classification accuracy is 90% and the normal class is estimated correctly. Other faults are estimated with variable recall, score, and precision. RF and CatBoost classifiers have close results, and both belong to the boosting algorithms. For more details, in Fig. 6.18 the feature importance and the confusion matrix are plotted.

Fig. 6.18A shows the most important feature to others, and as can be seen, the normalized V_{mp} is the most influenced feature, and the second one is the irradiance.

From the CM shown in Fig. 6.18B, in the first row 14 samples of class F1 are correctly classified from 18 samples. In the second row, 22 samples class F2 are correctly classified from 23. In the third row, 30 samples are correctly classified, while 2 are

Table 6.12 Error metrics: precision, recall, F1-score and accuracy.

CatBoost model parameters	Features	Precision	Recall	F1-Score	Accuracy (%)
iterations=2000, learning_rate=0.1	I_{mp}/I_{sc} V_{mp}/V_{oc}, FF, G, T, and P_{mp}/P_{mp_stc}				
Class	F1	0.93	0.78	0.85	90
	F2	0.81	0.96	0.8	
	F3	0.91	0.94	0.92	
	F4	0.91	0.77	0.83	
	Normal	1.00	1.00	1.00	

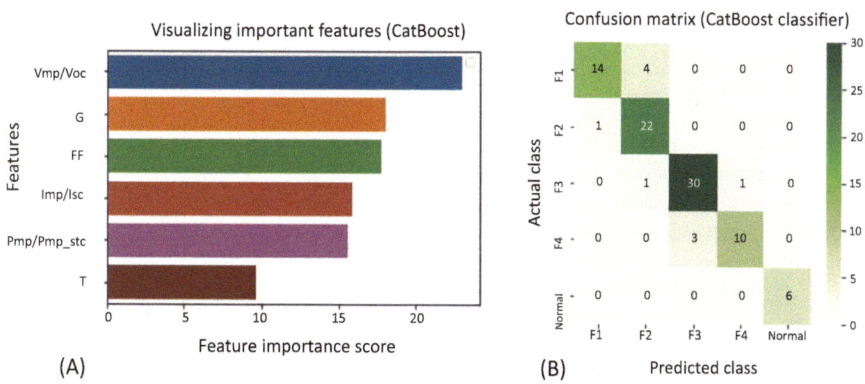

Fig. 6.18 (A) CatBoost classifier: feature importance and (B) confusion matrix of the CatBoost classifier.

misclassified. In the fourth row, 10 samples of class F4 are correctly classified, and only 3 samples are misclassified. In the last row, all samples are correctly classified, i.e., six samples from six.

Example 6.12: LGBM classifier

In this example, a lightGBM (LGBM) algorithm is used [29], which is a fast, distributed, high-performance gradient boosting framework based on decision tree algorithms. LGBMClassifier function is a gradient boosting framework that uses tree-based learning algorithms. To check its capability in fault classification, the same code illustrated in **Listing 6.6** is used by inserting the following command lines:

```
from lightgbm import LGBMClassifier
model_LGBMClassifier()
```

Classification performances of this model are listed in Table 6.13. As reported in Table 6.13 this model presents several parameters which can be tuned. The

Table 6.13 Error metrics: precision, recall, F1-score and accuracy.

BT model parameters	Features	Precision	Recall	F1-Score	Accuracy (%)
boosting_type='gbdt', class_weight=None, colsample_bytree=1.0, importance_type='split', learning_rate=0.1, max_depth=-1, min_child_samples=20, min_child_weight=0.001, min_split_gain=0.0, n_estimators=100, n_jobs=-1, num_leaves=31, objective=None, random_state=None, reg_alpha=0.0, reg_lambda=0.0, silent=True, subsample=1.0, subsample_for_bin=200000, subsample_freq=0	I_{mp}/I_{sc} V_{mp}/V_{oc}, FF, G, T, and P_{mp/Pmp_stc}				
Class	F1	0.75	0.67	0.71	85
	F2	0.73	0.83	0.78	
	F3	0.94	0.94	0.94	
	F4	0.92	0.85	0.88	
	Normal	1.00	1.00	1.00	

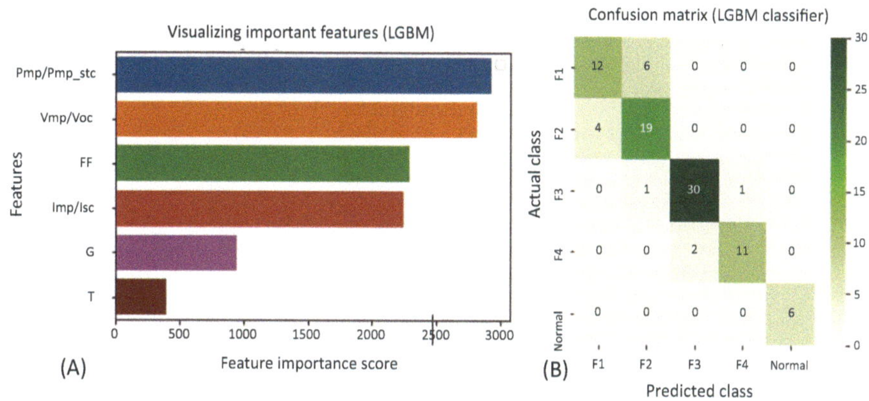

Fig. 6.19 (A) LGBM classifier: feature importance and (B) confusion matrix of the LGBM classifier.

classification accuracy is 85% and the normal class is estimated correctly. Other faults are estimated with variable recall, score, and precision. XGBoost and LGBM classifiers have very close results, and both belong to the boosting algorithms. For more details, in Fig. 6.19 the feature importance and the confusion matrix are plotted.

Fig. 6.19A shows the most important feature to others, and as can be seen, the normalized P_{mp} and V_{mp} are the most influenced features.

Example 6.13: Boosting classifier using k-fold cross validation

In this example, we use the three boosting algorithms (CatBoost, XGBoost, and LightGBM) with k-fold cross-validation method [30] to classify PV faults. The main code is shown in **Listing 6.7**. In this validation method, the most k value used is 10, but it could be modified; more details about the k-fold cross validation is shown in Appendix F.

First, all functions like *cross_val_score* and *make_pipeline* available in the sklearn library should be imported.

#Listing 6.7 K-fold cross validation

```
from sklearn.model_selection import cross_val_score
from sklearn.pipeline import make_pipeline
from catboost import CatBoostClassifier
from lightgbm import LGBMClassifier
from xgboost import XGBClassifier

classifier=make_pipeline(StandardScaler(), LGBMClassifier())
#classifier=make_pipeline(StandardScaler(),XGBClassifier())
#classifier=make_pipeline(StandardScaler(),CatBoostClassifier
  (iterations=1000, learning_rate=0.1))
scores=cross_val_score(classifier, X=X_train, y=y_train, cv=10,
  n_jobs=1)
print('Cross Validation accuracy scores: %s'% scores)
```

```
print('Cross Validation accuracy: %.3f+/-%.3f'%(np.mean(scores),np.
  std(scores)))
classifier.fit(X_train, y_train)
#Predicting the Test set results
y_pred=classifier.predict(X_test)
#Print confusion matrix
print(confusion_matrix(y_test,y_pred))
print(classification_report(y_test,y_pred))
print(accuracy_score(y_test, y_pred))
print("Accuracy:",metrics.accuracy_score(y_test, y_pred))
```

The obtained accuracy is 85.2%, 88%, and 85% for XGboost, CatBoost, and LightGMB, respectively. There is no remarkable difference between these results and those obtained in the last examples without using k-fold cross validation. This could be due to the size of the dataset.

Remark. From the above examples, each classifier requires a specific number of features to obtain good accuracy for fault classification. To optimize the classifiers some input parameters have to be adjusted carefully (such as K, number of estimators, criterion, max_depth, etc.). It has been demonstrated that CatBoost and RF classifiers outperform XGboost, LGBM, DT, and K-NN, K-NN is the worst classifier.

For other ML algorithms that are not investigated in this chapter, such multiLayer perceptron (MLP) neural networks, Naïve Bayes (NB), support vector machine (SVM), and logistic regression (LR), the same code (**Listing 6.6**) can be used. The following functions are available in sklearn library and can be easily used to implement the classifiers: MLP: *MLPClassifier()*, LR: *LogisticRegression()*, SVM: *SVC()*, and NB: *GaussianNB()*.

6.8.2 Fault classification using ensemble learning algorithms

Ensemble learning (EL) is the process by which multiple models (such as classifiers), are strategically generated and combined to solve aproblem [28,37].

Example 6.14: Ensemble learning (EL) classifier

In this example, an EL algorithm is used to classify PV array faults. The second dataset is used, and the most relevant features are estimated by the PCA algorithm (see Section 6.6.2). Five features are extracted, i.e., Feature 3 = V_{oc}, Feature 5 = P_{mp}, Feature 4 = I_{mp}, Feature 10 = dI/dV(vip), and Feature 12 = FF, have been selected.

To develop an EL classifier, boosting and bagging algorithms [28] are considered like Bagging classifier, Extra tree classifier, AdaBoost classifier, histogram-based gradient boosting, and gradient boost classifiers. Voting ensemble is one of the simplest methods to combine the prediction from multiple ML algorithms [38,39]. The EL procedure for fault classification is summarized as follows:

Step #1: Import the available boosting and bagging algorithms from sklean. ensemble library.
Step #2: Loading the second dataset and dividing it into two subsets (80% for training and 20% for testing the models).
Step #3: Execute each algorithm separately and then call Voting classifier function.
Step #4: Evaluate the model accuracy using the confusion matrix and calculate f1-score, recall, precession, and accuracy.

The main code of EL classifier is illustrated in **Listing 6.8**.

#Listing 6.8 Ensemble learning for PV array fault classification

```
#import libraries and functions
import pandas as pd
import sklearn as sk
from sklearn import metrics
from sklearn.ensemble import VotingClassifier
from sklearn import model_selection
from sklearn.ensemble import BaggingClassifier
from sklearn.ensemble import ExtraTreesClassifier
from sklearn.ensemble import AdaBoostClassifier
from sklearn.ensemble import GradientBoostingClassifier
from sklearn.ensemble import HistGradientBoostingClassifier
from sklearn.experimental import enable_hist_gradient_boosting
import seaborn as sns
import matplotlib.pyplot as plt
%matplotlib inline

#Loading data
col_names=['s1', 's2', 's3', 's4', 's5', 's6', 's7', 's8',
  's9', 's10','s11','s12','fault']
dataset2=pd.read_csv('e:/database_emulatorcsv4.csv',header=None,
  names=col_names)

feature_cols=['s3', 's4', 's5', 's10', 's12']
X=dataset2[feature_cols]
y=dataset2.fault
X_train, X_test, y_train, y_test=train_test_split(X, y, test_size=
  0.20,random_state=10)

#Voting Ensemble learning methods for faults classification
#Training classifiers
HGBC_model=HistGradientBoostingClassifier()
ETC_model=ExtraTreesClassifier(n_estimators=100)
ABC_model=AdaBoostClassifier(n_estimators=100)
```

```
GBC_model=GradientBoostingClassifier(n_estimators=100,
  random_state=0)
RFC_model=RandomForestClassifier(n_estimators=100)
Vot_model=VotingClassifier(estimators=[('HGBC', HGBC_model),
  ('ETC', ETC_model),('ABC', ABC_model),('GBC', GBC_model),('RFC',
  RFC_model)], voting='soft', weights=[2, 1, 2, 1,2])
HGBC_model.fit(X_train, y_train)
ETC_model.fit(X_train, y_train)
GBC_model.fit(X_train, y_train)
ABC_model.fit(X_train, y_train)
RFC_model.fit(X_train, y_train)
Vot_model.fit(X_train, y_train)
y_pred=vot_model.predict(X_test)
#Plot confusion matrix
CM=confusion_matrix(y_test, y_pred)
data_form=pd.DataFrame(CM, index=['F1','F2','F3','F4','F5'], columns=
  ['F1','F2','F3','F4','F5'])
plt.figure(figsize=(5, 5))
sns.heatmap(data_form , annot=True, cmap='Greens', fmt='g')
plt.xlabel('Predictedclass',fontsize=14)
plt.ylabel('Actualclass',fontsize=14)
plt.title('Confusionmatrix',fontsize=14)
plt.show()
#Print confusion matrix
print(confusion_matrix(y_test,y_pred))
print(classification_report(y_test,y_pred))
print(accuracy_score(y_test, y_pred))
print("Accuracy:",metrics.accuracy_score(y_test, y_pred))
```

Table 6.14 lists the calculated error metrics. Very good accuracy is obtained (95%) by the EL algorithm. Faults F1, F3, and F5 are accurately classified from the plotted

Table 6.14 Errors metrics: precision, recall, F1-score, and accuracy.

EL model	Features	Precision	Recall	F1-Score	Accuracy (%)
	s3, s4, s5, s10, s12				95
Class	F1: shading	1.00	0.99	1.00	
	F2:Inrease in Rs	0.81	0.91	0.86	
	F3:Bypass short-circuit	1.00	1.00	1.00	
	F4:Bypass impedance	0.94	0.87	0.90	
	F5: PV module short circuit	1.00	1.00	1.00	

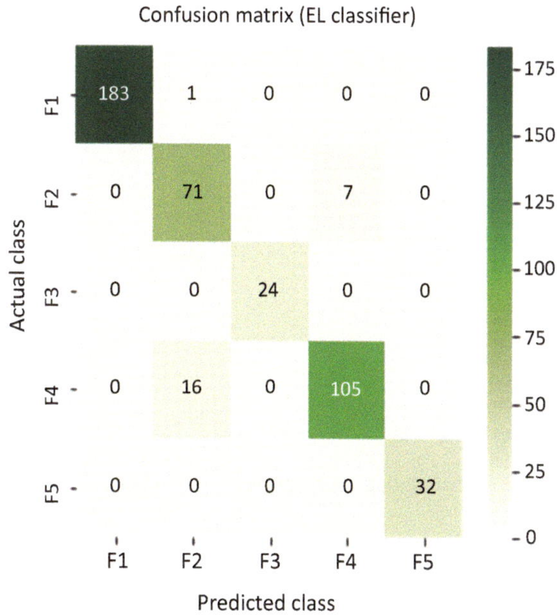

Fig. 6.20 Confusion matrix of the EL classifier.

confusion matrix shown in Fig. 6.20. For case F2, 71 are correctly classified, while 7 are incorrectly classified. In case F4, 105 points are correctly classified and 16 are incorrectly classified which fall in the F2 class.

Although the fault classification results provided by the PV emulator (second dataset) are more accurate than those measured (first dataset), but the experimental data are more closer for the real-world applications.

It is worth noting that most ML algorithms investigated in this chapter converge very quickly, except EL algorithm which requires more computation time. Additionally, the computing time depends on other parameters such as the size and the number of input features and target classes. Tuning the parameters' values for ML algorithms will effectively improve classifier performance.

6.8.3 Fault classification using deep learning algorithms

This section aims to apply a DCNN for fault classification of a PV array based on thermography images. The investigated faults are dust accumulation, partial shading, and shunted diode. All these faults have a negative impact on the PV output, and clearly show the PV module as a hot spot problem (see Fig. 6.8, dataset 6.3).

Example 6.15: DCNN for fault classification based on thermography images

The code is shown in **Listing 6.9**, we need just to label the three classes in the dataset as F1: Dust accumulation, F2: Partial shading, and F3: Short-circuited diode. The epoch is set to 50, the training and validation accuracy and loss are shown in Fig. 6.21A and B. Both training and validation accuracy reach 1 after a few iterations, and the loss is very close to 0.0001.

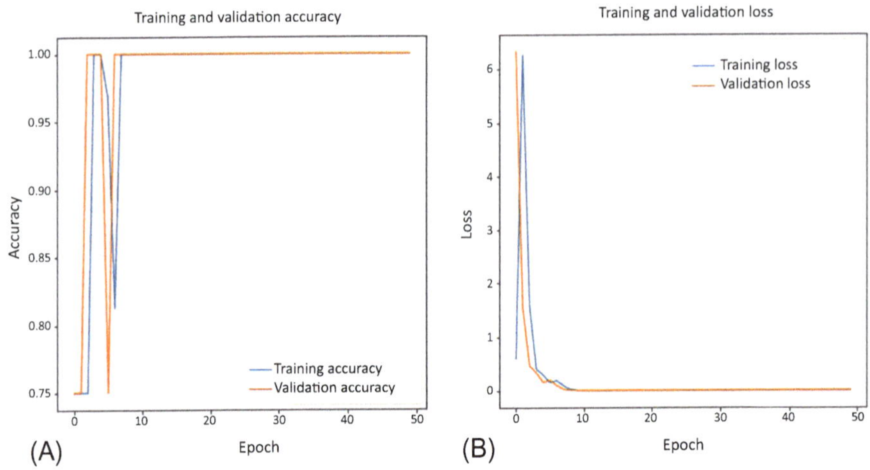

Fig. 6.21 (A) Training and validation accuracy and (B) training and validation loss.

#Listing 6.9: Fault classification using deep CNN

```
modelCNN=keras.Sequential([
    keras.layers.AveragePooling2D(6,3, input_shape=(640,480,3)),
    keras.layers.Conv2D(64, 1, activation='relu'),
    keras.layers.MaxPool2D(2,2),
    keras.layers.Conv2D(32, 1, activation='relu'),
    keras.layers.MaxPool2D(2,2),
    keras.layers.Conv2D(32, 1, activation='relu'),
    keras.layers.MaxPool2D(2,2),
    keras.layers.Conv2D(64, 1, activation='relu'),
    keras.layers.MaxPool2D(2,2),
    keras.layers.Dropout(0.5),
    keras.layers.Flatten(),
    keras.layers.Dense(128, activation='relu'),
    keras.layers.Dense(64, activation='relu'),
    keras.layers.Dense(32, activation='relu'),
    keras.layers.Dense(2, activation='softmax')
])
model.compile(optimizer='Adamax',
     loss=keras.losses.CategoricalCrossentropy(),
     metrics=['accuracy'])
results=modelCNN.fit(trainX, trainY, epochs=20, batch_size=32,-
  validation_data=(testX, testY))
```

The calculated precision, recall, F1-score, and accuracy are reported in Table 6.15. The accuracy is 73%, which is acceptable regarding the limited dataset (479 images).

With reference to the confusion matrix shown in Fig. 6.22, the misclassified 100-accuracy = 100–73 = 27%. From the first line of the CM, the recall of F1 is 7/16 = 43.75%, eight samples are classified incorrectly in F2, and one sample in

Table 6.15 Errors metrics: precision, recall, F1-score, and accuracy.

Model DCNN Structure	Class	Precision	Recall	F1-Score	Accuracy (%)
one average pooling layer, 4 Max pooling layers, one dropout layer, one flatten layer, 4 dense layers, optimizer=Adamax loss=keras.losses. CategoricalCrossentropy(), metrics=accuracy	F1: Dust accumulation F2: Partial shading F3: Short circuited diode	0.54 0.68 0.82	0.44 0.62 0.93	0.48 0.65 0.87	73

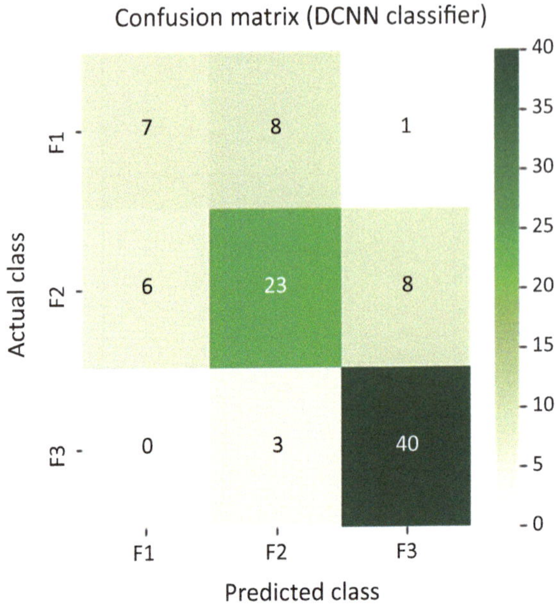

Fig. 6.22 Confusion matrix: DCNN for faulty thermal images classification.

F3. From the second line of the CM, the recall rate of F2 is 23/37 = 62.16%, six samples are classified incorrectly in F1, and eight samples in F3. Form the third line of the CM, the recall rate of F3 = 40/43 = 93.02%, only three samples are incorrectly classified in F2.

6.9 Summary

This chapter begins with a brief introduction to the type of faults in PV arrays, protection of PV systems, and then the datasets employed in this chapter are presented in detail. Subsequently, in Section 6.6 feature selection and extraction procedures with two explained examples are presented. The Sections 6.7 and 6.8 cover the big part of the chapter, which is the application of ML, EL, and DCNN to detect and classify possible faults on PV systems. Only methods based on the I-V curves and thermal images have been investigated in this chapter. The accuracy of the methods presented depends on some factors, like the dataset quality, dataset size, and tuning model parameters.

In this chapter, readers can understand when and how to apply ML, EL, and DL algorithms in this topic based on the presented 15 examples. It should be pointed out that the implementation of such techniques for real-time application is still an open challenge. The more challenging issue is the automatic fault localization for large-scale PV plants, of course, few attempts have been done to apply aerial thermal images, but this technique is relatively expensive and cannot detect all defects of photovoltaic systems.

References

[1] A. Mellit, S. Kalogirou, Artificial intelligence and internet of things to improve efficacy of diagnosis and remote sensing of solar photovoltaic systems: challenges, recommendations and future directions, Renew. Sust. Energ. Rev. 143 (2021), 110889.
[2] A. Mellit, G.M. Tina, S.A. Kalogirou, Fault detection and diagnosis methods for photovoltaic systems: a review, Renew. Sust. Energ. Rev. 91 (2018) 1–17.
[3] S. Silvestre, A. Chouder, E. Karatepe, Automatic fault detection in grid connected PV systems, Sol. Energy 94 (2013) 119–127.
[4] W. Chine, A. Mellit, A.M. Pavan, S.A. Kalogirou, Fault detection method for grid-connected photovoltaic plants, Renew. Energy 66 (2014) 99–110.
[5] C. Kapucu, M. Cubukcu, A supervised ensemble learning method for fault diagnosis in photovoltaic strings, Energy 227 (2021), 120463.
[6] A. Mellit, Recent applications of artificial intelligence in fault diagnosis of photovoltaic systems, in: A Practical Guide for Advanced Methods in Solar Photovoltaic Systems, Springer, Cham, 2020, pp. 257–271.
[7] A.R. Espinosa, M. Bressan, L.F. Giraldo, Failure signature classification in solar photovoltaic plants using RGB images and convolutional neural networks, Renew. Energy 162 (2020) 249–256.
[8] B. Li, C. Delpha, D. Diallo, A. Migan-Dubois, Application of Artificial Neural Networks to photovoltaic fault detection and diagnosis: a review, Renew. Sust. Energ. Rev. (2020) 110512.
[9] R.H.F. Alves, G.A. de Deus Júnior, E.G. Marra, R.P. Lemos, Automatic fault classification in photovoltaic modules using convolutional neural networks, Renew. Energy 179 (2021) 502–516.
[10] M.K. Alam, F. Khan, J. Johnson, J. Flicker, A comprehensive review of catastrophic faults in PV arrays: types, detection, and mitigation techniques, IEEE J. Photovoltaics 5 (3) (2015) 982–997.
[11] W. Chine, A. Mellit, V. Lughi, A. Malek, G. Sulligoi, A.M. Pavan, A novel fault diagnosis technique for photovoltaic systems based on artificial neural networks, Renew. Energy 90 (2016) 501–512.
[12] M. Hussain, M. Dhimish, S. Titarenko, P. Mather, Artificial neural network based photovoltaic fault detection algorithm integrating two bi-directional input parameters, Renew. Energy 155 (2020) 1272–1292.
[13] A. Haque, K.V.S. Bharath, M.A. Khan, I. Khan, Z.A. Jaffery, Fault diagnosis of photovoltaic modules, Energy Sci. Eng. 7 (3) (2019) 622–644.
[14] UL1741 Standard, Inverters, Converters, Controllers and Interconnection System Equipment for Use with Distributed Energy Resources, 2010.
[15] IEC 61140, Protection Against Electric Shock—Common Aspects for Installation and Equipment, 2016.
[16] IEC 62548, Low-voltage electrical installations—Part 7-712: Requirements for special installations or locations—Photovoltaic (PV) systems, 2017.
[17] A. Niccolai, A. Gandelli, F. Grimaccia, R. Zich, S. Leva, Overview on photovoltaic inspections procedure by means of unmanned aerial vehicles, in: 2019 IEEE Milan PowerTech, IEEE, 2019, pp. 1–6.
[18] N. Wang, Z.L. Sun, Z. Zeng, K.M. Lam, Effective segmentation approach for solar photovoltaic panels in uneven illuminated color infrared images, IEEE J. Photovoltaics 11 (2) (2020) 478–484.
[19] T. Takashima, J. Yamaguchi, K. Otani, K. Kato, M. Ishida, Experimental studies of failure detection methods in PV module strings, in: 2006 IEEE 4th World Conference on Photovoltaic Energy Conference, vol. 2, IEEE, 2006, pp. 2227–2230.

[20] T. Takashima, J. Yamaguchi, M. Ishida, Disconnection detection using earth capacitance measurement in photovoltaic module string, Prog. Photovolt. Res. Appl. 16 (8) (2008) 669–677.
[21] M.U. Saleh, C. Deline, E. Benoit, S. Kingston, A.S. Edun, N.K.T. Jayakumar, J.B. Harley, C. Furse, M. Scarpulla, An overview of spread spectrum time domain reflectometry responses to photovoltaic faults, IEEE J. Photovoltaics 10 (3) (2020) 844–851.
[22] L. Chen, S. Li, X. Wang, Quickest fault detection in photovoltaic systems, IEEE Trans. Smart Grid 9 (3) (2016) 1835–1847.
[23] H. Zhang, X. Hong, S. Zhou, Q. Wang, Infrared image segmentation for photovoltaic panels based on Res-UNet, in: Chinese Conference on Pattern Recognition and Computer Vision (PRCV), Springer, Cham, 2019, pp. 611–622.
[24] A. Belaout, F. Krim, A. Mellit, B. Talbi, A. Arabi, Multiclass adaptive neuro-fuzzy classifier and feature selection techniques for photovoltaic array fault detection and classification, Renew. Energy 127 (2018) 548–558.
[25] G. Chandrashekar, F. Sahin, A survey on feature selection methods, Comput. Electr. Eng. 40 (1) (2014) 16–28.
[26] X.W. Chen, J.C. Jeong, Enhanced recursive feature elimination, in: Sixth International Conference on Machine Learning and Applications (ICMLA 2007), IEEE, 2007, pp. 429–435.
[27] J. Brownlee, Machine Learning Mastery With Python, vol. 527, Machine Learning Mastery Pty Ltd, 2016, pp. 100–120.
[28] A. Géron, Hands-on Machine Learning With Scikit-Learn, Keras, and TensorFlow: Concepts, Tools, and Techniques to Build Intelligent Systems, O'Reilly Media, 2019.
[29] J. Brownlee, Gradient Boosting With Scikit-Learn, XGBoost, LightGBM, and CatBoost, Machine learning, 2020. *https://machinelearningmastery.com/gradient-boosting-with-scikit-learn-xgboost-lightgbm-and-catboost/*.
[30] R. Kohavi, A study of cross-validation and bootstrap for accuracy estimation and model selection, in: Proceedings of the Fourteenth International Joint Conference on Artificial Intelligence, vol. 14(2), Morgan Kaufmann, San Mateo, CA, 1995, pp. 1137–1145.
[31] I.T. Jolliffe, Principal components in regression analysis, in: Principal Component Analysis, Springer, New York, NY, 1986, pp. 129–155.
[32] International standard IEC 61724. *https://webstore.iec.ch/preview/info_iec61724%7Bed1.0%7Den.pdf*.
[33] F. Ghani, G. Rosengarten, M. Duke, J.K. Carson, The numerical calculation of single-diode solar-cell modelling parameters, Renew. Energy 72 (2014) 105–112.
[34] A.M. Pavan, A. Mellit, V. Lughi, Explicit empirical model for general photovoltaic devices: experimental validation at maximum power point, Sol. Energy 101 (2014) 105–116.
[35] D.L. King, J.A. Kratochvil, W.E. Boyson, Photovoltaic Array Performance Model, Department of Energy, United States, 2004, pp. 1–43.
[36] R. Platon, J. Martel, N. Woodruff, T.Y. Chau, Online fault detection in PV systems, IEEE Trans. Sustainable Energy 6 (4) (2015) 1200–1207.
[37] A. Tharwat, Classification assessment methods, Appl. Comput. Inform. 17 (2020) 168–192.
[38] R. Polikar, Ensemble learning, in: Ensemble Machine Learning, Springer, Boston, MA, 2012, pp. 1–34.
[39] S. Saha, A. Ekbal, Combining multiple classifiers using vote based classifier ensemble technique for named entity recognition, Data Knowl. Eng. 85 (2013) 15–39.

Control and optimal management of grid-connected photovoltaic systems and micro-grids using artificial intelligence and metaheuristic techniques

7

Chapter outline

7.1 Introduction

In the last decade, the integration of distributed generation (DG) resources like photovoltaic (PV), wind energy (WE) generators, and other sources of energy into distribution networks has received considerable attention [1]. The PV and WE sources present the advantages of cleanness, durability, and zero emissions, thus are widely developed for various applications of DG systems. The growth of local renewable energy sources and heavy loads in power distribution networks cause several issues with a direct impact on the stability of the electrical grid [2].

The strong penetration of smaller sources such as wind, PV, hydropower, and fuel cells in the electricity market is seen as an alternative to green energy and a flexible extension of the power grid. Therefore, this scenario distinguishes the complementary infrastructure of the micro-grid (MG), which offers several advantages of the power system, like reduction of peak load, fault reliability, and high quality of power through a flexible control system [3]. A MG is a power system that combines different types of

Handbook of Artificial Intelligence Techniques in Photovoltaic Systems. https://doi.org/10.1016/B978-0-12-820641-6.00007-7
Copyright © 2022 Elsevier Ltd. All rights reserved.

energy generators like PV, wind turbines, fuel cell plants, and others; energy storage units like batteries, ultra-capacitors, flywheels, and others; and AC-DC loads. A MG can be stand-alone or interconnected to the distribution network.

Power sharing in MGs plays an essential role to in maintaining the stability of the grid, i.e., limit voltage and frequency deviation. Therefore, the problem is how to share power in the load with inverters working in parallel. Many attempts have been made to solve this issue by using different control strategies; droop control, adaptive virtual impedance, and other methods combine PID (proportional, integral, and derivative), and metaheuristic approaches to optimize the power sharing [4,5].

Energy management system (EMS) is another interesting issue of DG systems. For example, in MGs, one of the objectives of the EMS is to provide operational reference signals to MG units, state monitoring, and device communication to build a bi-directional interaction between the power production and the consumption [6]. Thus, the EMS is able to autonomously make decisions and determine hour by hour the correct dispatch of generators with the final goal of minimizing the global energy costs [7–9]. In addition, it schedules and coordinates the user's energy consumption, distributed generation, and storage to minimize energy costs, while providing high-power quality [10].

From the above, the DC-AC power inverter is the key technology that transfers and converts the available power at the DC side of this kind of systems into AC power at the utility side. Thus, the control scheme associated with the grid side converter is responsible to [2]:

- manage of the active and reactive power flow
- stabilize of the DC link voltage regardless of the operating conditions variation
- enhance the quality of the power supplied to local loads and/or utility grid
- reject the distribution network perturbations (like voltage dips and harmonics)
- power sharing

A considerable number of control strategies have been published in literature to address the above issues [10–13]. This chapter covers three essential sections: (1) control of grid-connected PV system, (2) energy management system of MGs, and (3) power sharing in MGs.

With reference to the literature [14–16], it has been shown that controllers based on artificial intelligence (AI) techniques outperform classical controllers to solve the above topics, and currently with the advance in reconfigurable circuits and embedded technology, these techniques are the most appropriate to be implemented in reconfigurable circuits like field-programmable gate arrays (FPGA) for real-time applications [17–20]. All examples presented in this chapter have been developed and simulated under Matlab®/Simulink and Python languages.

7.2 Control of grid-connected and hybrid photovoltaic systems

A photovoltaic (PV) array converts photon energy into electricity by means of the photovoltaic effect, where the maximum power is generated at a unique point at given climatic conditions. Thus, the incorporation of a maximum power point tracking

(MPPT) controller in a PV array, commonly associated with DC-DC or DC-AC converter, is of the main interest such that it ensures the continuous extraction of the maximum power for varying climatic conditions [21]. Currently, the development of PV plants integrated to the public network is growing worldwide. In this type of application, the key technology in the power electronic interface between the PV array and utility grid is the inverter that allows the efficient DC-AC conversion and transfer of electric power. An inverter should be carefully controlled in order to accomplish important tasks like the mitigation of the DC-link voltage oscillations, the control of active and reactive powers, the generation of sinusoidal current with low harmonic distortion (THD < 5%), and other less important operations. Hence, different control schemes for DC-AC power converters have been proposed in the literature, where the voltage oriented control (VOC) and the direct power control (DPC) are the most widely used [22,23]. In the next two examples, fuzzy logic (FL) and particle swarm optimization (PSO) algorithm is applied for the direct power control of grid-connected PV systems.

Example 7.1: Fuzzy-based voltage oriented control of a grid-connected PV system

The fuzzy logic (FL) is widely applied for the optimization of classical control algorithms due to the features of auto-adaptation, robustness against perturbations, and ability to handle uncertainties [24]. The configuration of the two-stage grid-connected PV system considered in this example is illustrated in Fig. 7.1A. The

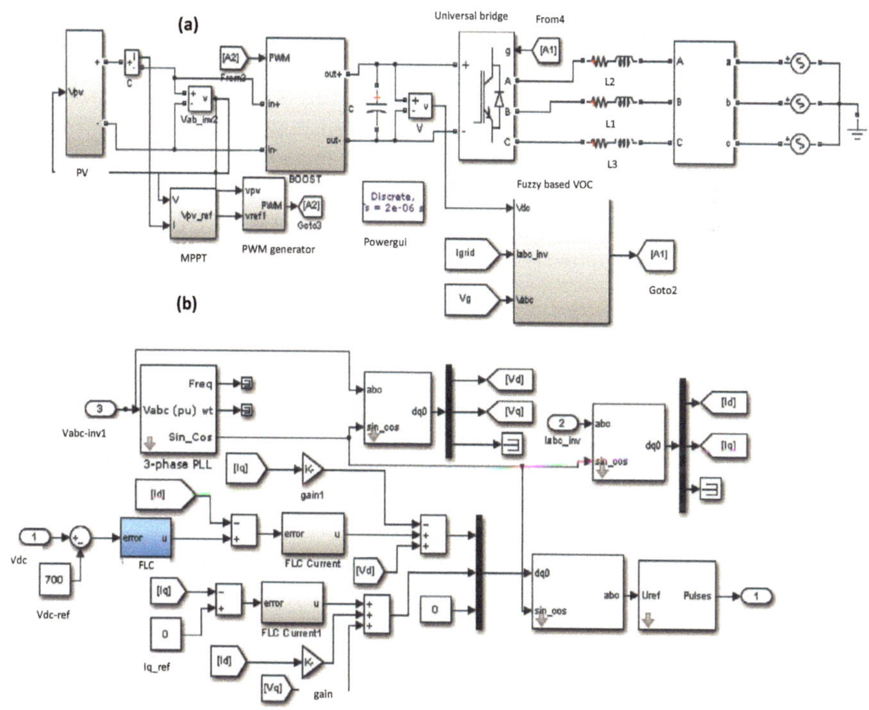

Fig. 7.1 (A) Grid-connected PV system in Matlab/Simulink environment and (B) VOC based on FLCs.

DC-DC boost converter is controlled through the IncCond algorithm [25], used for tracking the maximum power from PV modules/arrays that determine the reference voltage (V_{DC}ref) corresponding to maximum power point (MPP) of the PV array. A modified version of the VOC scheme is applied for the control of the three-phase inverter.

As shown in Fig. 7.1B, FL controllers have been used instead of PI controllers in traditional VOC scheme. Thus, the regulation of the DC-link voltage as well as the d and q-axis grid current is performed through fuzzy logic controllers (FLCs).

The goal is to enhance the power quality and the control performance for time-varying operation conditions. The zero-order Sugeno-fuzzy inference system (FIS) is applied with two inputs (the error and the change of error) and single output. The membership functions of the inputs and output are depicted in Fig. 7.2. The fuzzy rules of the FLCs employed are reported in Table 7.1.

The results of the PV system simulated at standard test conditions (STC) are presented in Fig. 7.3. It can be seen that the FLC improved the tracking performance. The DC-link voltage converges quickly to the set point after a response time of less than 0.04 s. The three-phase current at the inverter output has a sinusoidal waveform with minimal harmonic distortion (less than 4%).

To check the dynamic response of the fuzzy-based VOC scheme, the PV system is simulated for varying insolation intensity (it changes from 1000 to 600 W/m^2 at the instant 0.45 s then, rises to 800 W/m^2 at 0.8 s, $T_a = 25°C$). It is obvious from the obtained results in Fig. 7.4 that the FLC outperforms the PI controllers and offers more accurate results in terms of transient and static responses.

Example 7.2: PSO-based direct power control of a grid-connected PV system

The general topology of a grid-connected PV system is shown in Fig. 7.5. It consists of a PV array, a DC-DC boost converter, and a three-phase inverter interconnected to the grid by means of an L-filter. The boost converter, controlled by the IncCond-MPPT algorithm, helps to raise the PV voltage to the desired level at the inverter input. The direct power control (DPC) scheme is adopted for the regulation of active and reactive powers at the output of the three-phase inverter [21].

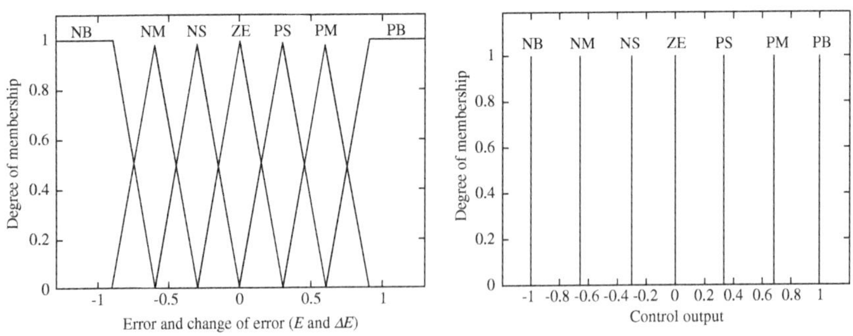

Fig. 7.2 Membership functions of inputs and output of FLC.

Table 7.1 Rule base of FLCs applied for current and voltage regulation.

ΔE / E	NB	NM	NS	ZE	PS	PM	PB
NB	NB	NB	NB	NB	NM	NS	ZE
NM	NB	NM	NM	NM	NS	ZE	PS
NS	NB	NM	NS	NS	ZE	PS	PM
ZE	NM	NM	ZE	ZE	ZE	PM	PM
PS	NM	NS	ZE	PS	PS	PM	PB
PM	NS	ZE	PS	PM	PM	PM	PB
PB	ZE	PS	PM	PB	PB	PB	PB

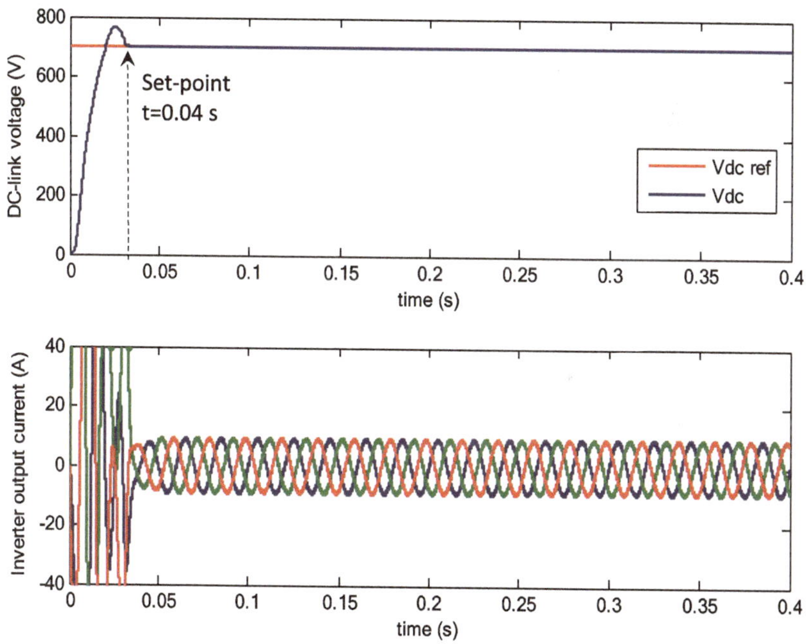

Fig. 7.3 Performance of fuzzy VOC under stable insolation.

Fig. 7.6 gives the details of the controllers' blocks developed in Matlab/Simulink environment.

The PSO algorithm is adopted for the offline tuning of the PI controller, applied for the DC-link voltage (V_{DC}) regulation. With this method, optimized values of proportional (K_P) and integral (K_I) gains of PI regulator can be obtained without a need for mathematical modeling. The objective function of the optimization problem is formulated in term of integral time absolute error (ITAE) and the maximum overshoot ($d_{\max}$):

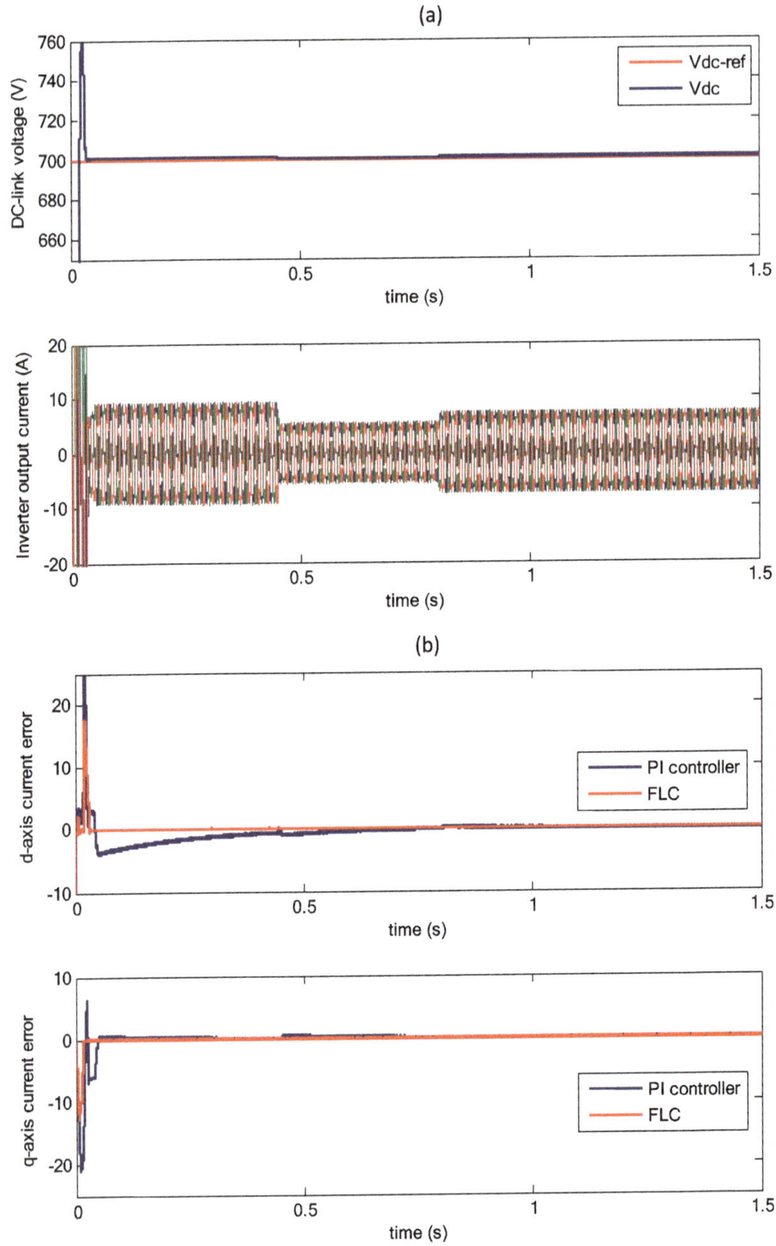

Fig. 7.4 Performance of fuzzy VOC under changing insolation.

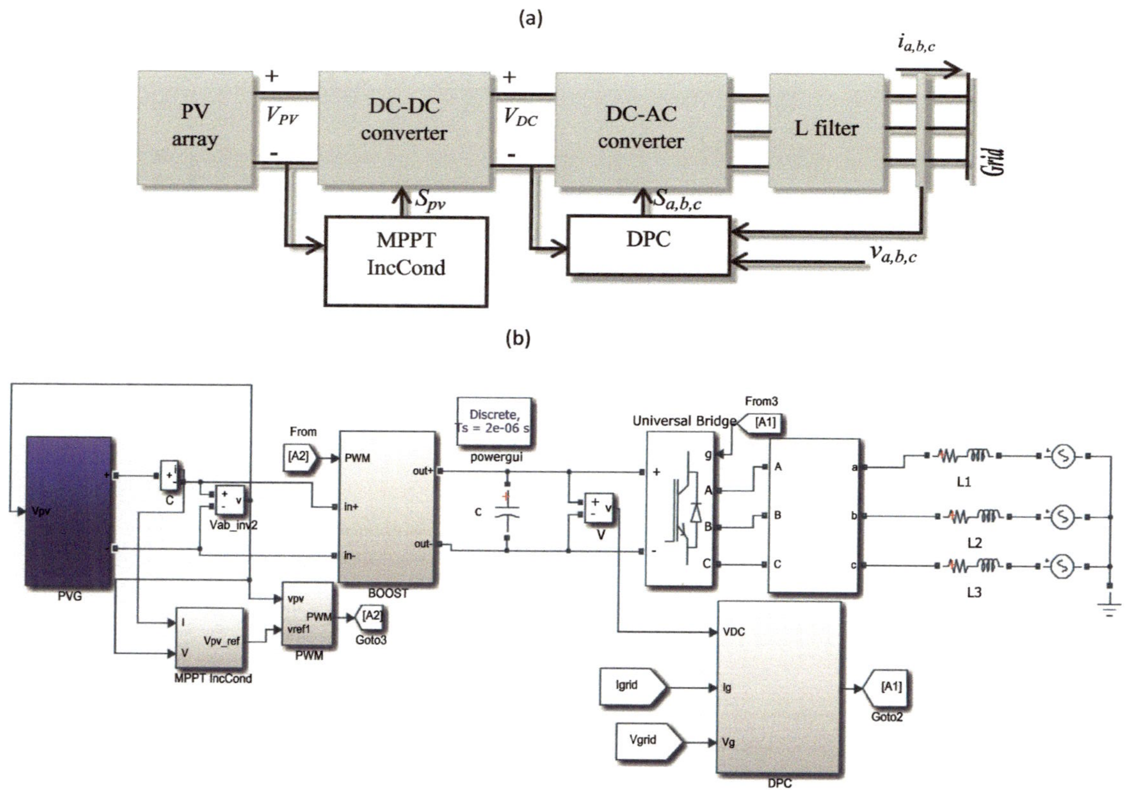

Fig. 7.5 (A) General architecture of two-stage grid-connected PV system and (B) developed PV system in Matlab/Simulink environment.

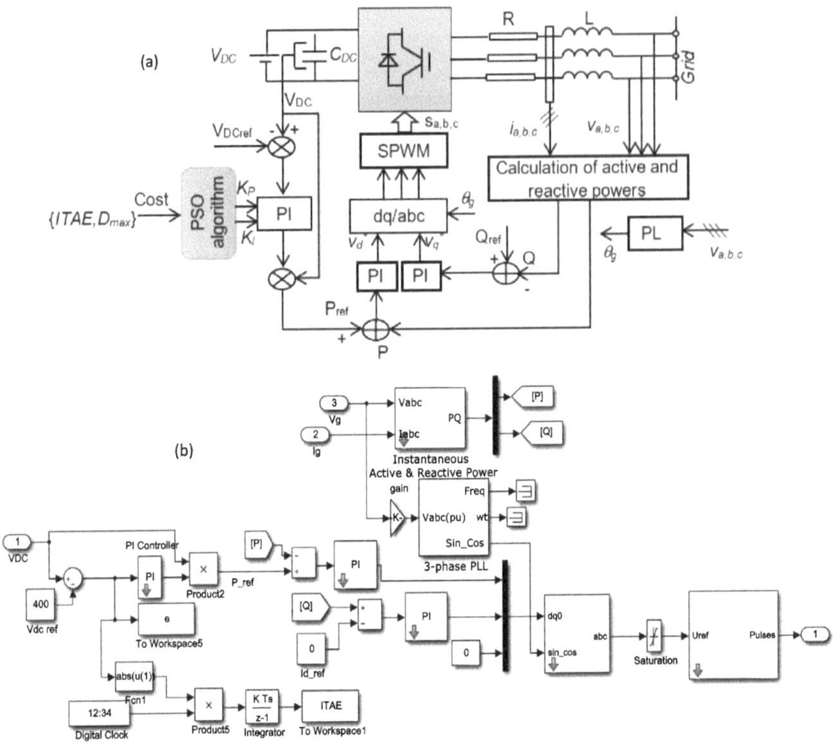

Fig. 7.6 (A) Details of DPC scheme applied for the inverter control and (B) DPC scheme developed in Matlab/Simulink environment.

$$\text{ITAE} + d_{\max} = \int_0^T t|e(t)|dt + \max(e(t)) \tag{7.1}$$

With $e(t)$ being the error between the set-point V_{DCref} and the measured voltage V_{DC}, ($e(t) = V_{\text{DCref}} - V_{\text{DC}}(t)$). The optimal parameters of K_P and K_I, obtained at the end of simulation with the PSO algorithm, will ensure the minimization of voltage error in the steady state as well as the reduction of maximum overshoot.

Each particle of population presents a candidate solution of the minimization problem. It should be noted that the gains K_P and K_I of the PI controller are defined. For each iteration, the particle updates their vectors of position and velocity according to the following expressions:

$$\begin{aligned} v_{i,l}(k+1) = {} & w * v_{i,l}(k) + c_1 r_{1i,l}(k)\left(P_{bi,l}(k) - x_{i,l}(k)\right) \\ & + c_2 r_{2i,l}(k)\left(G_{bl}(k) - x_{i,l}(k)\right) \end{aligned} \tag{7.2}$$

$$x_{i,l}(k+1) = x_{i,l}(k) + v_{i,l}(k+1), l \in \{1, \ldots, d\}, i \in \{1, \ldots, N\} \tag{7.3}$$

$x_{i,l}(k)$ presents the lth component of the position vector at the iteration k. $v_{i,l}(k)$ is the lth component of the velocity vector at k. N is the number of population particles, d is the dimension of the position vector, w is the inertial coefficient, and c_1 and c_2 are the social and cognitive parameters. $r_{1i,l}(k)$ and $r_{2i,l}(k)$ are the random numbers with uniform distribution in the interval [0,1]. Each particle of swarm i should remember its local best position $P_{bi} = (P_{bi,1}, \ldots, P_{bi,d})$. The global best position attained by the swarm is defined as $G_b = (G_{b_1}, \ldots, G_{b_d})$. In order to converge rapidly to the optimal solution, a dynamic inertia weight is applied:

$$w(k) = w_{\max} - (w_{\max} - w_{\min})\left(\frac{k}{K_{\max}}\right) \tag{7.4}$$

where $K_{\max}$ is the maximum number of iterations. The parameters $w_{\max}$ and $w_{\min}$ are the maximal and minimal values of the inertia factor, respectively.

The stopping criteria of the PSO algorithm are based on the ability to reach the desired value of the cost function or achieve the maximum number of iterations. The main steps of the search of optimal parameters of the PI regulator tuned offline by PSO have been summarized in the following algorithm:

```
Begin
Define the cost function (Equ.7.1) and problem parameters (k=1; d=2;
kmax=20...);
Initialize randomly the positions (xi) and velocities (vi) of N parti-
cles of population.
Evaluate the objective function for each particle.
For each particleset: local best cost = Current cost and local best
position = Current position
Set global best cost = min (local best costs)
Set Gb = arg min (local best costs)
While (k<= kmax) do
 Loop for i=1:N do
  For l=1:d do (% d=2)
   Update the velocity and position of each particle according to
   Equ.7.2 and Equ.7.3.
   Check boundary violations.
   End for
 Evaluate the fitness of each individual of population (Simulate the
Simulink file using "sim" function of Matlab).
End loop for
For i=1:N do
Find the local best position Pbi:
 If Current cost function < local best cost then
 Set local best cost = Current cost and Pbi(k+1)=xi(k+1)
```

```
      Else
        Keep the previous value of local best cost (Pbi(k+1)=Pbi(k))
      End if
    Update the global best cost and position:
        Set global best cost = min (local best costs)
        Set Gb(k+1)=argmincost(Pbi(k+1))
  End for
 Update the while counter k= k+1
 End while
 Give the optimal values of Kp and Ki
```

Some simulation parameters of PSO-based DPC scheme simulated in Matlab/Simulink environment are given in Table 7.2.

7.3 Application of AI techniques for management and control of hybrid micro-grids

In a hybrid MG, the power management (PM) is a crucial task that should ensure the optimal supervision of the power flow and the coordination among power production units and the consumer' loads. The PM goals are to guarantee an economic use of electrical energy as well as an efficient energy supply for the end users. Besides, the time-varying dynamics of MG components should be carefully controlled through local controllers, where the power references are imposed by the PM system. The framework commonly used for the management/control of hybrid MG is illustrated in Fig. 7.7.

The nonlinear characteristic of renewable energy sources, the random variations of loads demand, and the grid perturbations makes the MG management/control very challenging. The use of AI techniques is a promising alternative in the field without a need for mathematical development. Many researchers have applied FL, artificial neural networks (ANNs), and PSO algorithms for the control of MGs that help to

Table 7.2 Parameters of PSO based DPC scheme in Matlab/Simulink.

Parameter	Value
Maximum number of iteration	$K_{max}=20$
Number of swarm particles	$N=6$
Dimension of the particle	$d=2$
The cognitif et social coefficients	$C_1=C_2=2$
Final and initial inertia weight	$w_{max}=0.9$, $w_{min}=0.4$
DC-link voltage reference	$V_{DCref}=400\,V$
Reactive power reference	$Q_{ref}=0\,VAr$

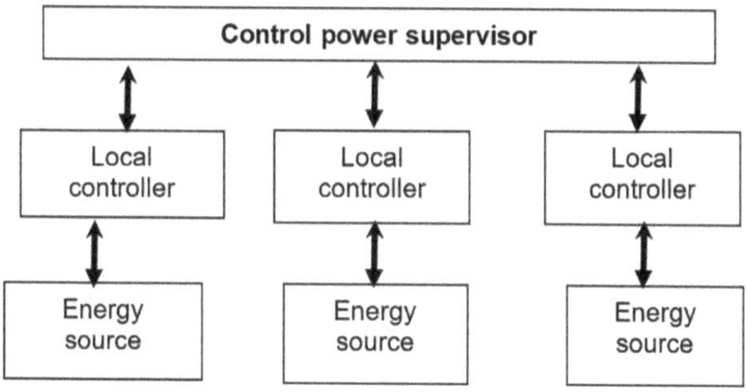

Fig. 7.7 Framework of the centralized power supervision in hybrid system.

improve the power quality and the control performance under time-varying conditions. In the next example, the intelligent control for a grid-connected MG using FL and an adaptive neuro-fuzzy inference system (ANFIS) is described.

Example 7.3: Intelligent control of MG using FL, ANN, and ANFIS

The MG considered consists of a PV array, a solid oxide fuel cell (SOFC), a battery energy storage system (BESS), and AC loads. An intelligent control strategy proposed in Ref. [2] is used to control the MG.

An online-trained ANN is used for the tracking of the optimal operating point of the PV array. Fuzzy logic and ANFIS approaches are applied for the control of the SOFC, BESS, and the three-level neutral-point clamped (3L-NPC) converter. The bloc diagram of the hybrid system, developed in Matlab/Simulink, is given in Fig. 7.8.

The main idea is to use FL and ANFIS for the online tuning of the PI controllers employed in local controllers of power converters. The main advantage of this approach is to make the linear PI regulators more adaptive for dynamic change of power supply and loads demand. Furthermore, more robustness against uncertainties and external perturbations can be achieved. For this purpose, Sugeno-fuzzy gain tuners are proposed for the adaptation of proportional and integral gains of PI controllers, used for the regulation of output current of SOFC and BESS. The BESS controller implemented in Matlab/Simulink is illustrated in Fig. 7.9. The parameters of PI controller are online tuned using the information of the current error and its derivative received at the input of FIS. The fuzzy rules of the BESS controller are summarized in Table 7.3 and its implementation in Matlab/Simulink is shown in Fig. 7.10.

On the other side, the virtual flux-oriented control (VFOC) scheme modified as shown in Ref. [2] by using neuro-fuzzy gain tuners for the *dq*-axis current controllers is used as shown in Fig. 7.11.

The neuro-fuzzy gain-scheduling scheme, also developed in Ref. [2], is based on the first-order Sugeno FIS with two inputs and one output. Each adaptive PI controller uses two ANFIS tuners for both proportional and integral gains. The current set point and the integral of current error are defined as the inputs of each ANFIS gain tuner. The optimal values of K_P and K_I are adjusted at an appropriate manner at the output of gain tuners. The parameters of ANFIS regulators like those of the membership functions are determined offline using hybrid learning techniques. The training data are

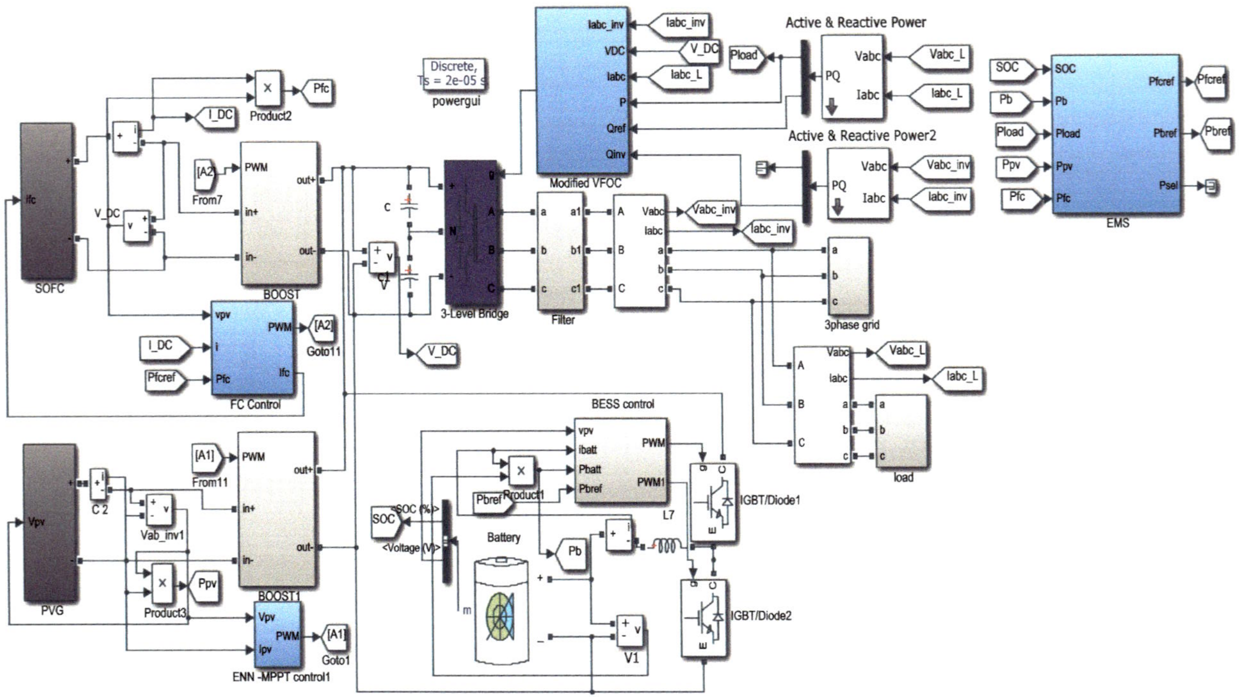

Fig. 7.8 Block diagram of the grid-connected hybrid system.

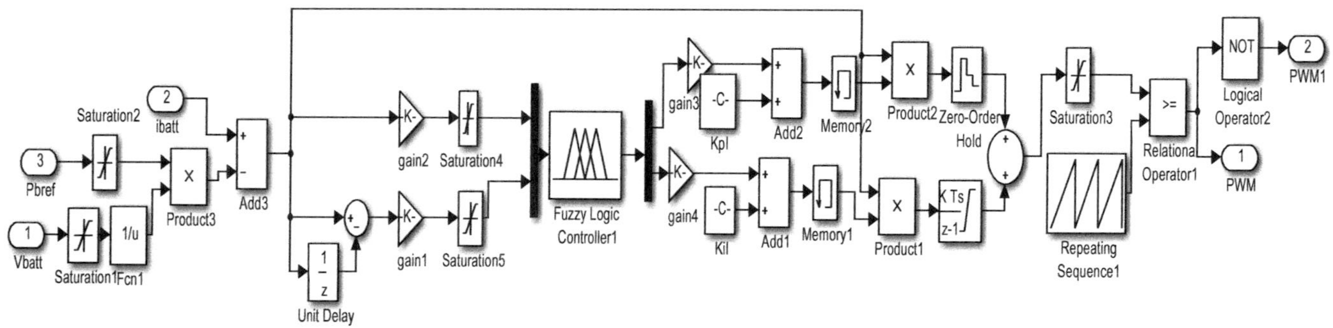

Fig. 7.9 BESS controller implemented in Matlab/Simulink.

Table 7.3 Rules base for regulation of BESS controller gains.

ΔE_{BESS}	NB	NM	NS	ZE	PS	PM	PB
E_{BESS}							
K_P							
NB	B	B	B	B	B	B	B
NM	MB	MB	MB	MS	MB	MB	MB
NS	MB	MS	MS	MS	MS	MS	MB
ZE	MS	MS	S	S	S	MS	MS
PS	MB	MS	MS	MS	MS	MS	MB
PM	MB	MB	MB	MS	MB	MB	MB
PB	B	B	B	B	B	B	B
K_I							
NB	S	MB	MB	MB	MB	MB	S
NM	MB	MS	MS	MB	MS	MS	MB
NS	MB	MB	B	B	B	MB	MB
ZE	B	B	B	B	B	B	B
PS	MB	MB	B	B	B	MB	MB
PM	MB	MS	MS	MB	MS	MS	MB
PB	B	MB	MB	MB	MB	MB	B

collected by using different values of PI gains under different conditions of operation. An example of the result for the variable operating conditions is given in Fig. 7.12.

Example 7.4: Adaptive neural networks based control of MG

Fig. 7.13 shows a grid-connected MG, which consists of a PV array, a wind turbine generator (WTG), SOFC, BESS DC, and AC loads. The key idea is to substitute classical controllers by intelligent regulators that help to enhance the static and dynamic performance. Hence, a single neuron radial basis function network (SN-RBFN) is employed for the MPPT of the PV array in order to enhance the tracking performance of the IncCond MPPT method. Moreover, adaptive ADALINE regulators have been developed to replace the classical PI controllers commonly used for the control of WT speed, SOFC output power, and the BESS charge-discharge current. An example of Matlab/Simulink implementation of ADALINE-based local controller of BESS is illustrated in Fig. 7.14. A control structure of a grid-connected MG based on artificial techniques is employed [26,27].

Besides, a modified control scheme for the three-phase inverter is developed based on online-trained neural networks. The active and reactive powers exchanged with loads and/or grid are indirectly controlled by acting on the *dq*-axis components of the inverter output current. As shown in Fig. 7.14, recurrent Elman neural networks (ENN) are employed for the control of direct and quadrature currents. The control of the DC-link voltage is performed through an adaptive feed-forward neural network (FFNN). The developed ENN and FFNN have the ability to learn online by using the adaptive interaction algorithm. This feature makes the control scheme more

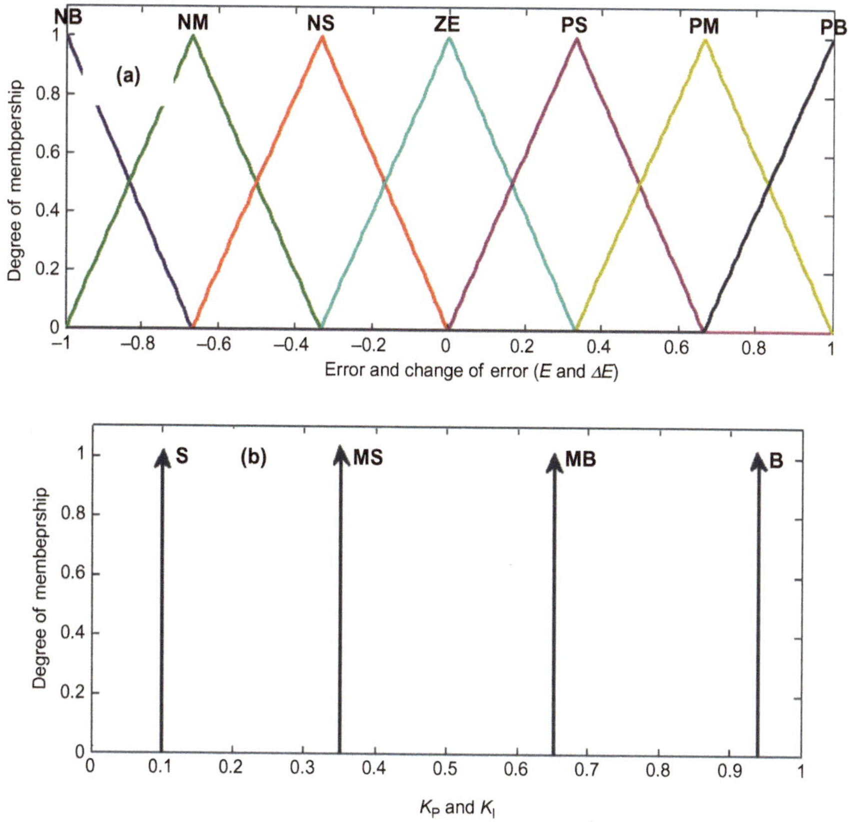

Fig. 7.10 Membership functions assigned to: (A) input variables of fuzzy gain tuner and (B) PI gains.

adaptive to the continuous change of operating conditions and time-varying dynamics of energy sources. It should be noted that the synchronization with the grid is ensured by means of a VF estimator of grid voltage. The Simulink block diagram of the DPC control scheme of three-phase inverter is represented in Fig. 7.15.

The supervision of the power flow in the MG is performed through a Mamdani-fuzzy inference system. The management goal is to minimize the power purchased from the utility grid, to coordinate the MG components, and to keep the efficient operation of the system. For different generation conditions and power demand, the fuzzy PM system decides the values of power references for both the SOFC and BESS. To prolong the lifetime of BESS, the state of charge (SoC) should be kept in secure range between minimal and maximal values SoC_{min} and SoC_{max}, respectively. The net power (P_{net}) is defined as the difference between the power generated by renewables resources and the loads demand:

$$\begin{cases} P_{net} = P_{RES} - P_L \\ \text{where } P_{RES} = P_{PV} + P_{WT}, P_L = P_{Lac} + P_{Ldc} \end{cases} \tag{7.5}$$

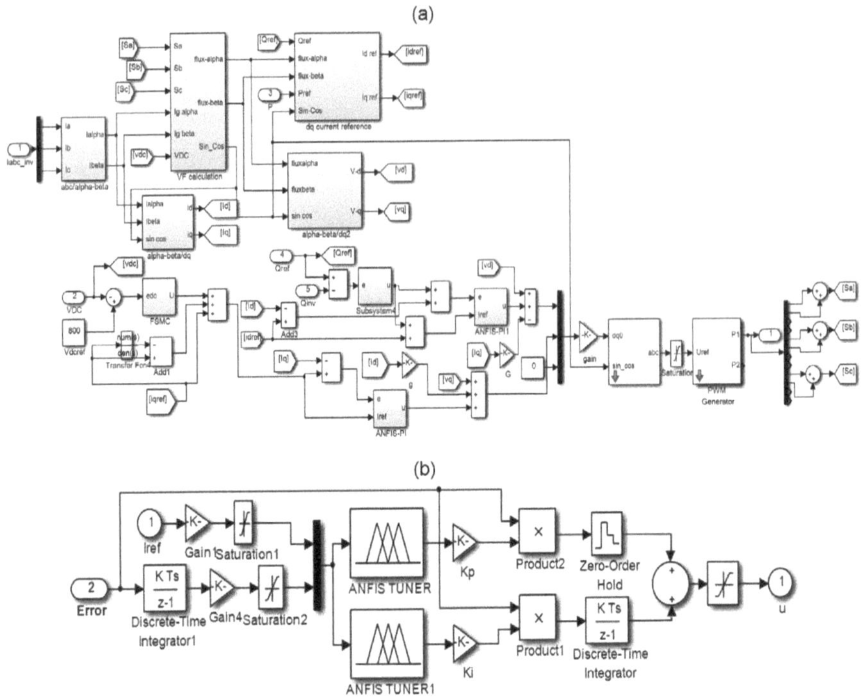

Fig. 7.11 (A) Matlab/Simulink implementation of modified VFOC scheme and (B) ANFIS-based gain tuner of PI current controllers.

where P_{RES} is the power available from RES, P_{PV}, P_{WT} are the power produced by the PVG and the WTG, respectively. P_L is the total power demand and P_{Ldc} and P_{Lac} are the DC and AC loads power, respectively. The Mamdani FIS receives at the input the net power P_{net} and the value of SOC of BESS to determine the power set points for the SOFC (P_{FC}^{*}) and BESS (P_b^{*}). The main rules of the fuzzy-based PM system are summarized in Table 7.4, where the labels NB, NM, NS, ZE, PS, PM, PB, PB+, L, M, and H mean, respectively, negative big, negative medium, negative small, zero, positive small, positive medium, positive big, positive very big, low, medium, and high. The distribution of membership functions, defined for the inputs and outputs of the PM system, is illustrated in Fig. 7.16. The main rules of fuzzy PM system can be given as follows:

- When the produced renewable energy surpasses the power demand (of DC and AC loads) and the BESS is full charged ($\text{SoC} > \text{SoC}_{\text{max}}$), the energy excess has to be injected to the utility grid.
- In the case of deficit of power generation from RES ($P_{\text{net}} < 0$), the SOFC and the BESS have to generate the power to respond to the loads request.
- When $\text{SoC} < \text{SoC}_{\text{min}}$ during the energy lack period ($P_{\text{net}} < P_{fc-\text{max}}$), the SOFC should ensure the charge of the BESS and the satisfaction of the energy demand.

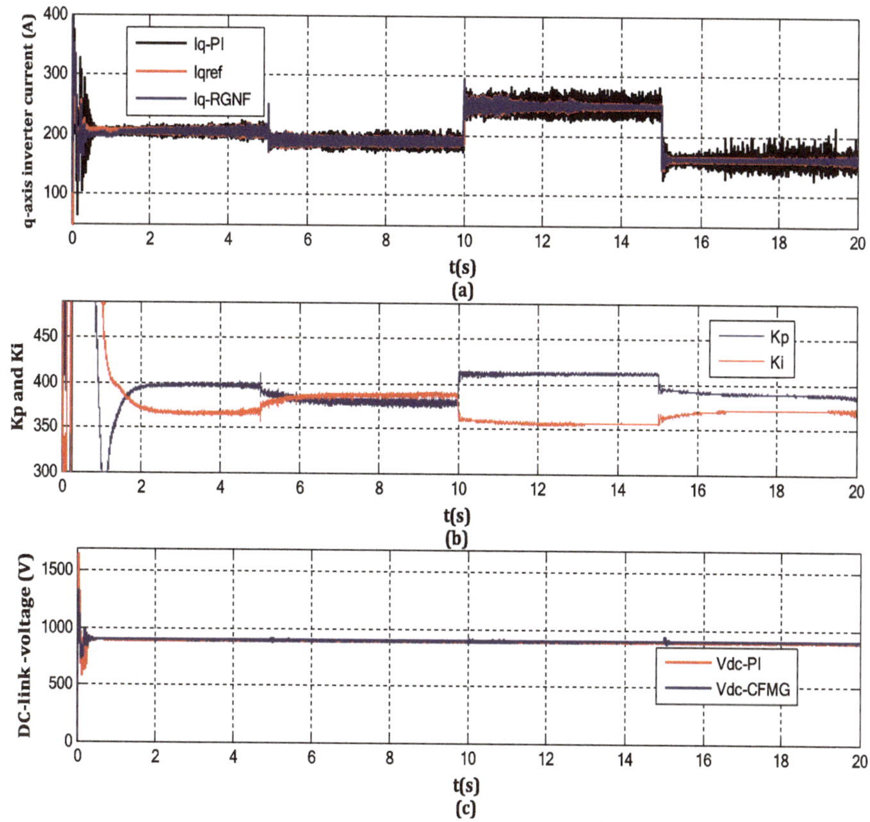

Fig. 7.12 Variation of: (A) q-axis inverter current, (B) gains of PI controller of the q-axis current, and (C) DC-link voltage.

Table 7.4 Fuzzy rule base of PM system used to determine: (a) P_{fc}^{*}, (b) P_{b}^{*}.

P_{net}	NM	NS	PS	PM	PB	PB+
SOC						
a						
H	–	–	–	PM	PB	PB
M	–	–	PS	PB	PB	PB
L	–	–	PM	PB	PB	PB
b						
H	ZE	ZE	PB	PS	PB	PB
M	NB	NS	ZE	ZE	PS	PB
L	NB	NS	NB	NS	NS	NS

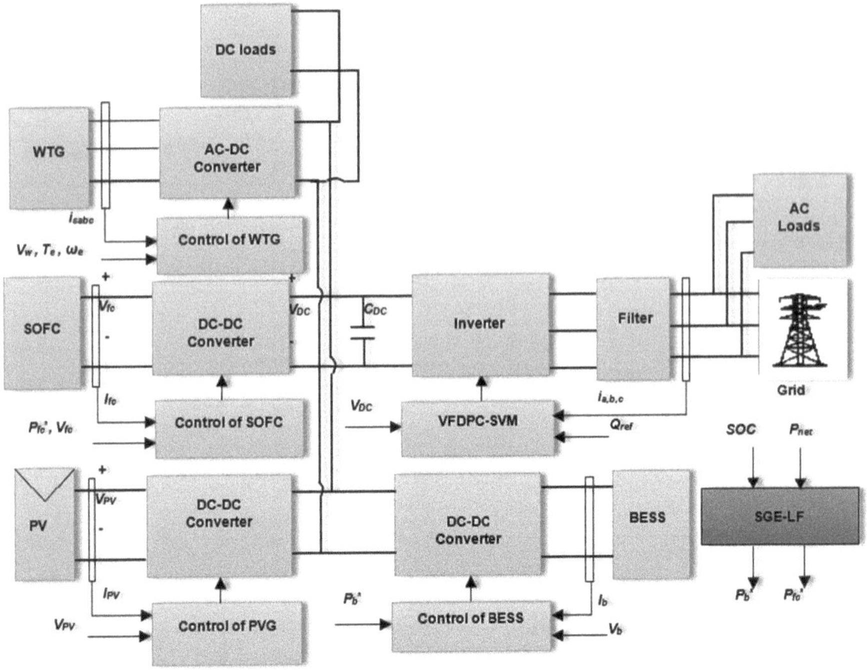

Fig. 7.13 General structure of the grid-connected MG considered in this example.

- Whenever the $SoC < SoC_{min}$, the BESS should operate in the charge mode despite the conditions of generation and demand.
- When the net power is greater than the nominal power of SOFC ($P_{net} > P_{fc-nom}$), the BESS should operate in discharge mode even if $SoC > SoC_{min}$.
- In the case where the power available from MG is lower than the power demand, the public grid covers the energy shortage.

Fig. 7.17 gives an example of simulation results obtained for variable operating conditions of MG.

7.4 Energy management of micro-grids

A good number of energy management strategies have been developed in literature [28] and the main objective of these is to manage the power optimally [29]. In these, the authors used linear programming methods, heuristic methods [30], and recently artificial intelligence techniques [31].

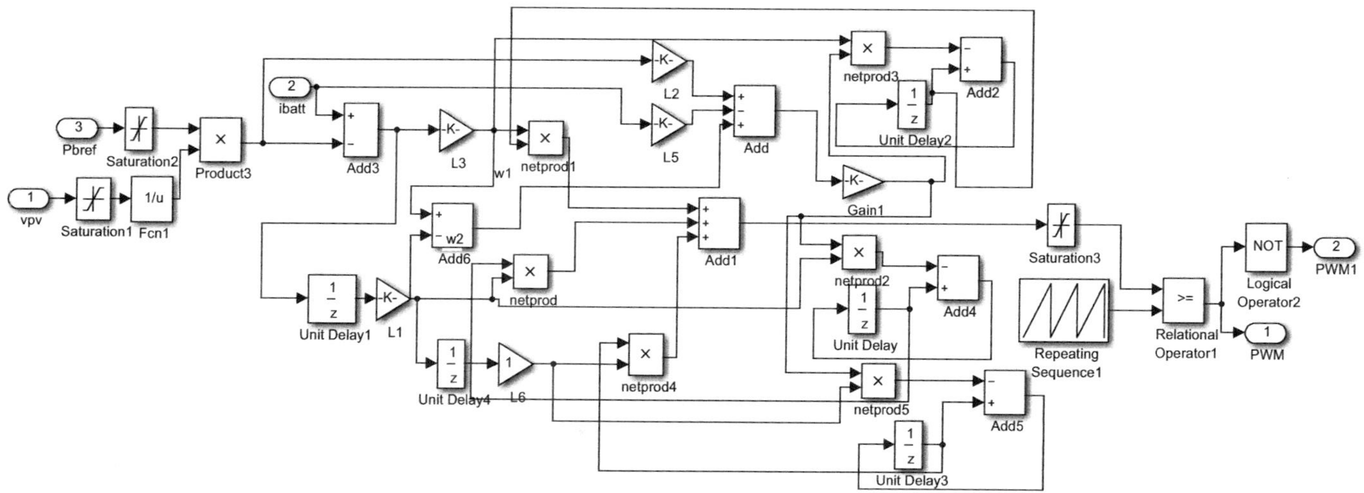

Fig. 7.14 ADALINE-based local controller of BESS.

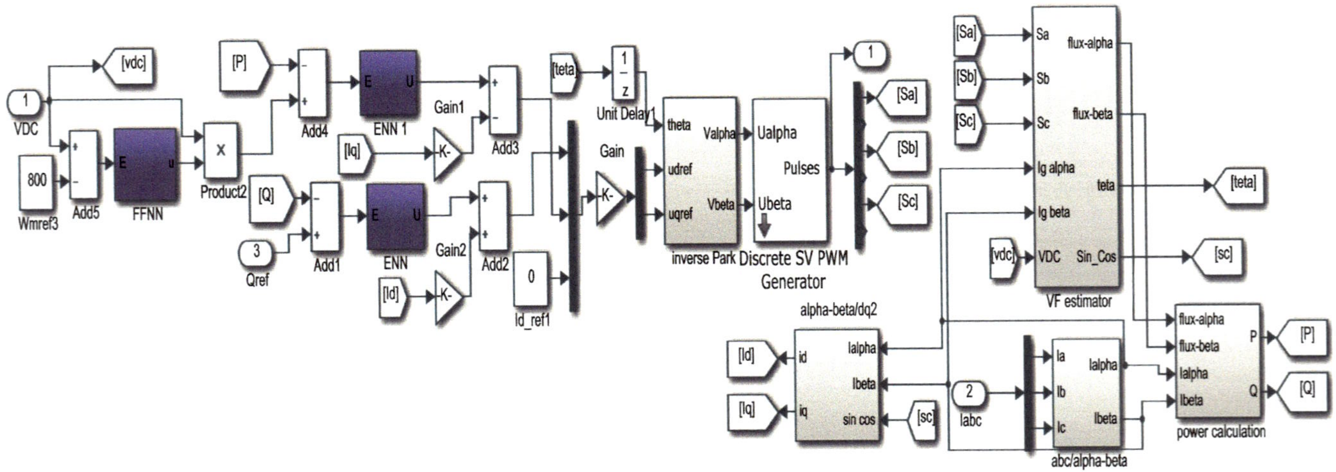

Fig. 7.15 DPC scheme of grid-connected inverter.

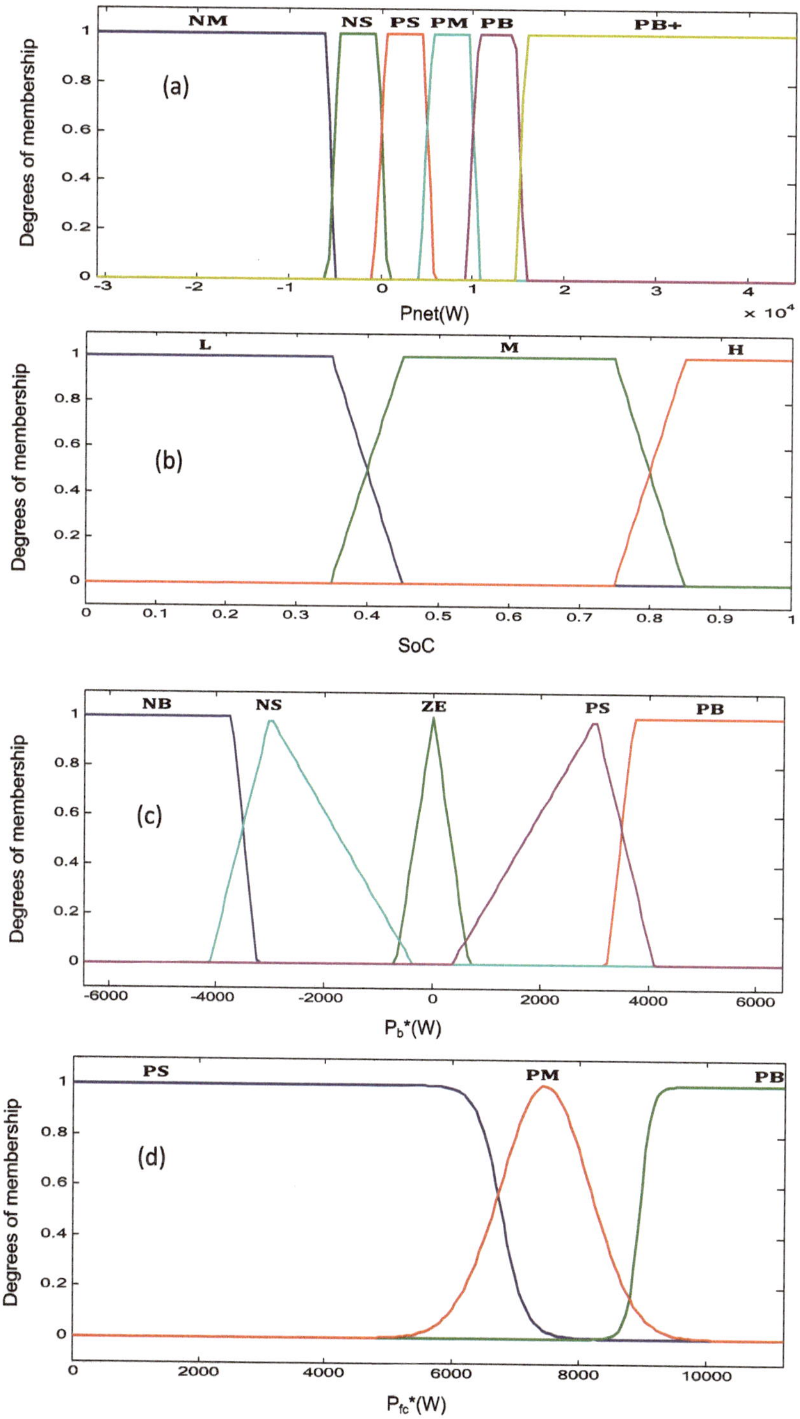

Fig. 7.16 Membership functions of: (A) P_{net}, (B) SoC, (C) P_b^*, and (D) P_{fc}^*.

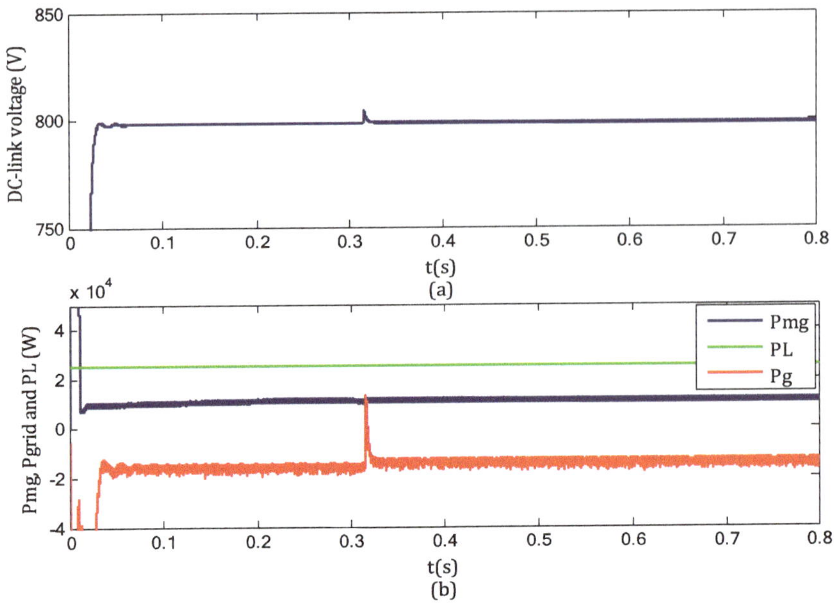

Fig. 7.17 Control performance of DPC scheme for varying conditions.

A simple EMS of a MG for supplying of an electrical vehicle charging station can be solved by using a conventional *if-then-else* algorithm [32]:

```
Take measurement EV (presence=1, absent=0), SoC_EV, V_EV, P_pv, SoC_B,
SoC_B_min, SoC_B_max, P_min, SoC_EV, SoC_EV_min
Loop If car=1 then
       If SoC_EV<SoC_EV_min then
             If P_pv>P_min then
                    Start charging the EV from PV (PV2EV)
                              else
                    If SoC_B>SoC_B_min and V_b>V_EV then
                          Start discharging battery to EV, B2EV
                    else
                                                     start  charging
from the grid to EV, G2B
       else
       if P_pv>P_min then
                    if SoC_B>SoC_B_max then
                          connect to the grid from PV, PV2G
                    else
                          Charging battery from PV, PV2B
             else loop, to test if the car is present
End
PV2B: Photovoltaic to battery
PV2G: Photovoltaic to grid
B2EV: Battery to electrical vehicle
PV2EV: Photovoltaic to electrical vehicle
G2B: Grid to battery
```

This section aims to present an intelligent energy management strategy, to manage the power of a MG, using artificial neural networks.

7.4.1 Micro-grid with an electrical vehicle station

The structure of the MG investigated is shown in Fig. 7.18, and consists of 3.9 kWp PV array, 4.6 kVA inverter with a 10 kWh lithium iron phosphate batteries, an interface board performing the connection with the low voltage (LV) 230 V grid, and a 22 plus 22 kW charging station. An EMS optimizes the energy flows between the different parts of the MG, where the data are exchanged through the Campus LAN [33,34].

7.4.2 Dataset 7.1

Fig. 7.19 shows an example of the collected data for 1 day. A dataset of 150 days at 5 min steps has been used to develop an EMS based on artificial neural networks. It consists of energy demand by the electrical vehicle (EdCar), produced power by a PV array of 4 kW (Ppv), state of charge of batteries (SoC), the exchanged power (EEx) with the grid, and the active power (P_{act}) [35].

7.4.3 Energy management system based on artificial neural networks

Fig. 7.20 shows the structure of EMS using artificial neural networks. This configuration aims to predict the active power based on the forecasted state of charge, PV power produced, energy demand by the car, and power exchanged. The first block, recurrent neural network, is developed to predict in 1-h ahead the different values of SoC, Ppv, EdCar, and EEx. The type of RNN used here is long-short term memory (LSTM) [35]. In a second block, a feed-forward neural network is used to estimate the reactive power in order to make a decision about the charging/discharging of batteries.

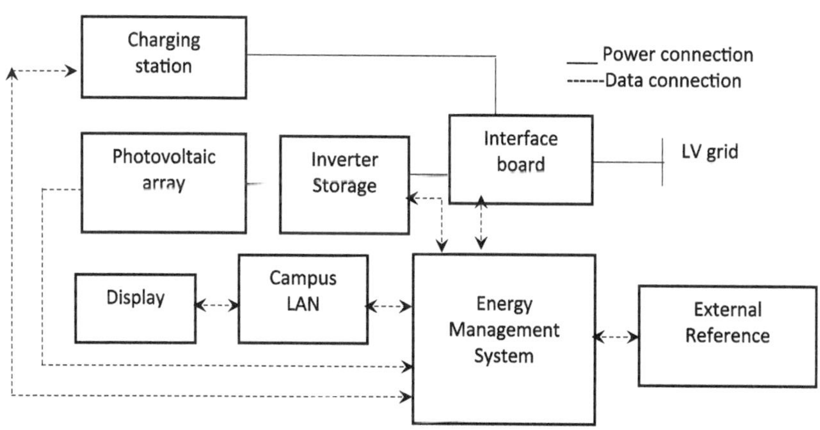

Fig. 7.18 The structure of the MG considered.

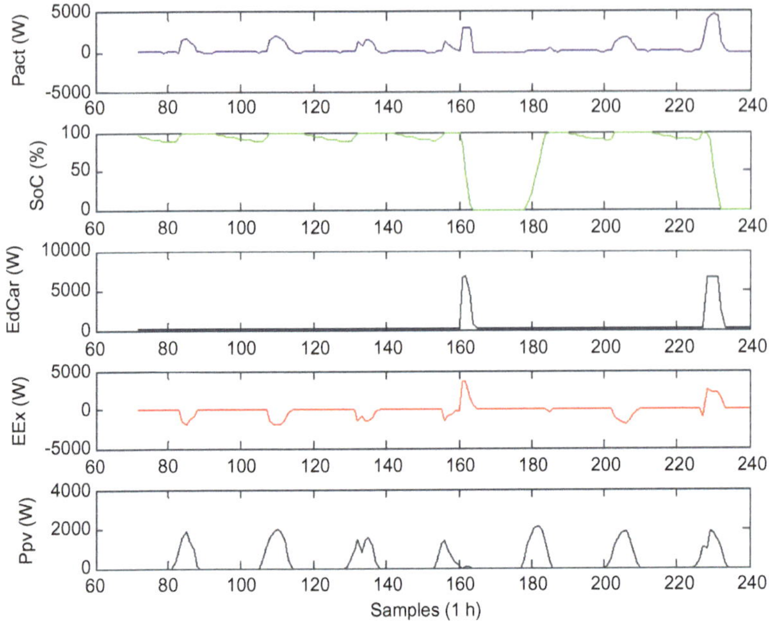

Fig. 7.19 Dataset 7.1: Variation of the monitored data: Ppv, EEx, PdCar, SoC, and Pact (from January 3 to 10, 2020 with time step of 1 h).

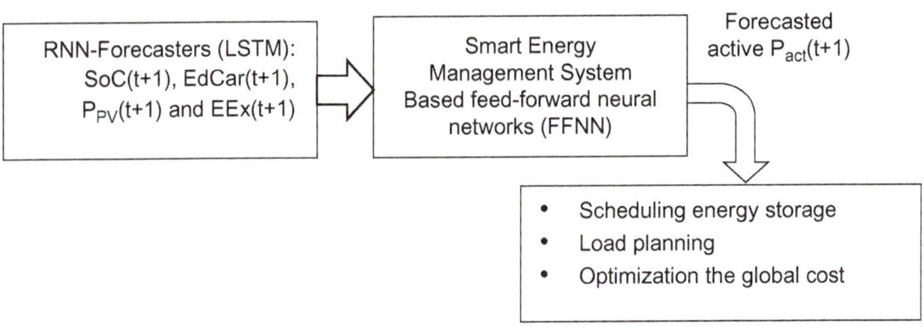

Fig. 7.20 The general structure of the EMS-based ANNs (LSTM and FFNN).

The EM algorithm is designed by using a simple FFNN. Fig. 7.21 shows the structure of the EMS-based FFNN, which consists of one input layer of four inputs (SoC, Ppv, EdCar, and EEx), two hidden layers, and one output layer corresponding to the active power (P_{act}). The EMS model can be formulated as follows:

$$P_{act} = \text{ffnn}(\text{SoC}, \text{Ppv}, \text{EdCar}, \text{EEx}) \quad (7.6)$$

Example 7.5: EMS based on artificial neural networks

The dataset of 2091 samples was divided into two parts: training (80%) and testing (20%). In practice with shallow NNs, two hidden layers are enough. The codes of the

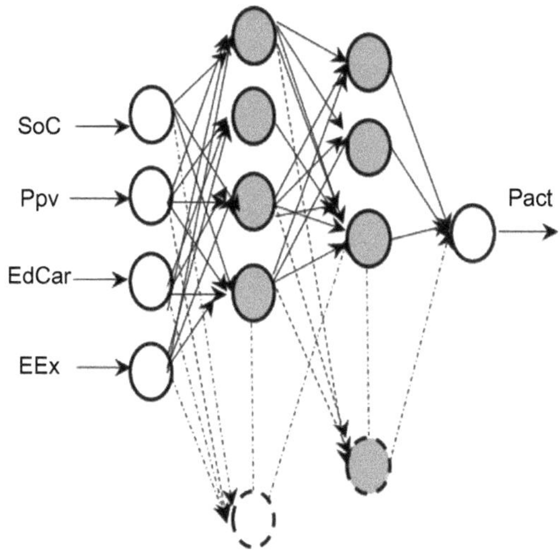

Fig. 7.21 The structure of the employed FFNN to design the EMS.

```
#Listing7.1 LSTM forecasting model
model=Sequential()
model.add(LSTM(units=100, activation='relu',input_shape=(X_train.
  shape[1], 1)))
model.add(Dense(units=1))
model.compile(loss='mean_squared_error', optimizer='adam')
Resu=model.fit( X_train,  y_train,  epochs=100, batch_size=32,
  shuffle=False,  validation_split=0.1)
model.evaluate(X_test, y_test)
```

```
#Listing7.2 FFNN estimator model
net=newff(minmax(p),[5,1],{'logsig','purelin'},'trainbr');
net.trainParam.show=500;
net.trainParam.epochs=1000;
net.trainParam.goal=1e-8;
net.trainParam.lr=0.08;
[net,tr]=train(net,p,t);
```

first and the second blocks are depicted in the following listings. The first **Listing 7.1** is written under Python, while the second **Listing 7.2** has been developed under Matlab.

To evaluate the accuracy of the forecasters as well as the EM system, the well-known error metrics like correlation coefficient (r), root mean squared error (RMSE), and mean absolute percent error (MAE) have been employed.

Fig. 7.22 shows the scatter curves between measured and forecasted (SoC, Ppv, EEx, and EdCar). From Fig. 7.22, an acceptable agreement between the measured and forecasted values is observed. To quantify the results, the three error metrics

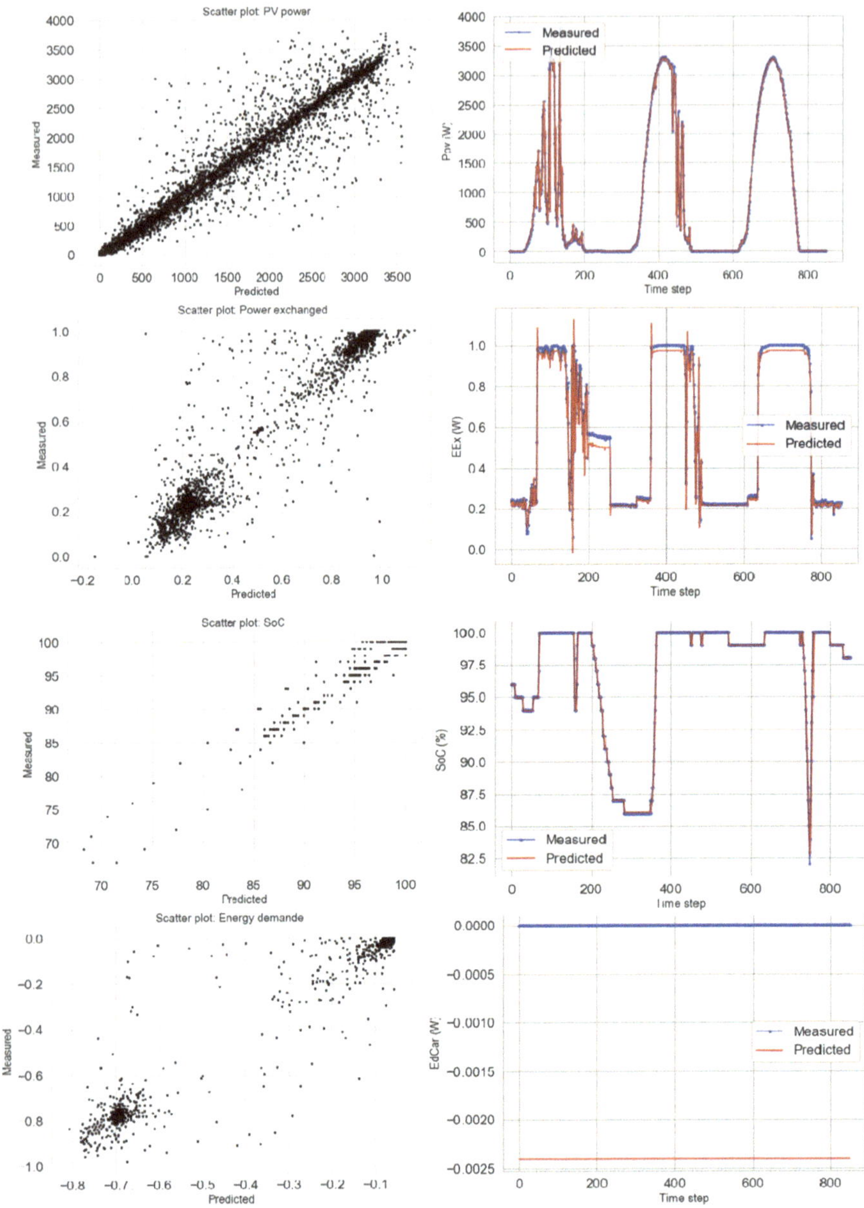

Fig. 7.22 Measured versus forecasted parameters (SoC, Ppv, EEx, and EdCar).

are calculated and the results are reported in Table 7.5. The LSTM model forecasts the three parameters with good accuracy, however, energy demanded by the car (EdCar) presents the lower accuracy according to the metrics obtained $r=91\%$, RMSE $=0.53$ kW, and MAE $=0.142$ kW. This means that it is relatively difficult to estimate energy demanded by the car.

Table 7.5 Error metrics: *r*, RMSE, and MAE.

LSTM model Unit = 150, epoch = 100, optimizer = Adam, batch size = 32	***r* (%)**	**RMSE (kW)**	**MAE (kW)**
PV power (Ppv)	98.00	0.057	0.023
Power exchanged (EEx)	99.00	0.058	0.030
State of charge (SoC)	98.64	0.0041	0.001
Energy demand (EdCar)	91.00	0.531	0.142

The data forecasted have been used to feed the second block (FFNN) in order to predict the value of the reactive power for the next hour, which can be used for (1) scheduling the energy, (2) storage and load planning, and (3) optimization of the global cost. Fig. 7.23 shows the obtained results and as can be seen a very good accuracy is obtained by a simple FFNN structure with one hidden layer (five neurons).

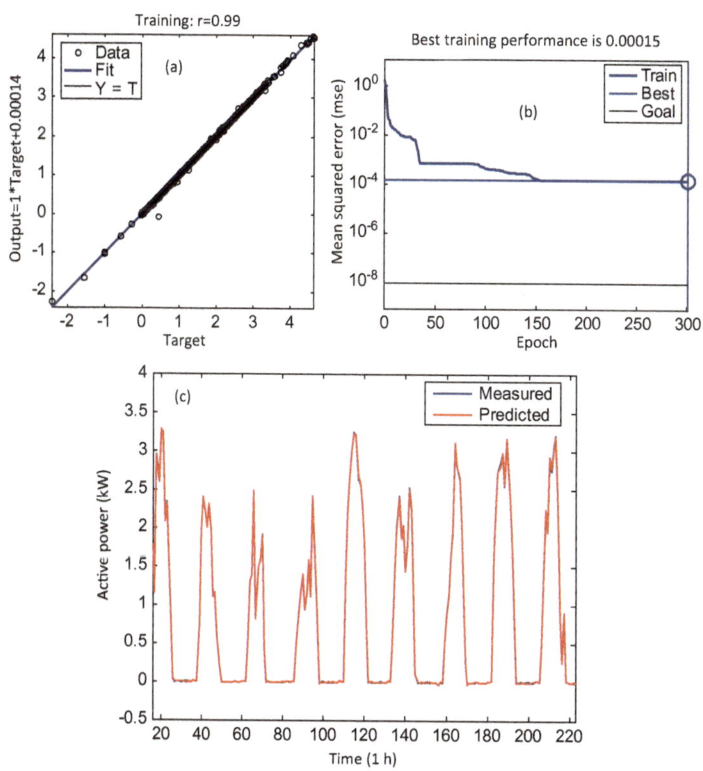

Fig. 7.23 (A) Scatter curve between measured and predicted values, (B) loss (MSE) function evolution, and (C) superposed curves: measured versus predicted active power using FFNN.

Table 7.6 Error metrics: r, RMSE, and MAE.

FFNN model **Total dataset = 127,016, Training dataset (80%): 116854, Testing (20%): 10162. FFNN, activation function = trainlm, Epoch = 300**	r **(%)**	**RMSE (kW)**	**MAE (kW)**
FFNN ($4 \times 5 \times 1$)	99.99	0.007	0.4

With reference to the results reported in Table 7.6, very good accuracy is obtained, so the FFNN model can perfectly predict the reactive power.

7.5 Power sharing in micro-grids

MG can work in two modes: network mode and island status modes. In such a system, the MG operates according to the control schemes of the inverter, so that the control scheme can play a key role in ensuring high-performance operation of the distributed power units. However, problems such as energy quality, stabilities [36], and distortion of harmonics [10] must be taken into account to improve the quality of the power supply. Conventional droop control is the most used technique for DGs power sharing in islanded MG, where real power and reactive power can be independently controlled by adjusting the frequency and the output voltage of the inverters.

The P-W droop equations are given as follows [37]:

$$W_1 = W_0 - n.P_1 \tag{7.7}$$

$$W_2 = W_0 - n.P_2 \tag{7.8}$$

The Q-V droop equations are defined as follows:

$$E_1 = E_0 - m.Q_1 \tag{7.9}$$

$$E_2 = E_0 - m.Q_1 \tag{7.10}$$

where W_1, W_2, E_1, and E_2 are the output voltage frequency and amplitude at no load and n and m are the droop frequency and amplitude coefficients, respectively. P_1, P_2, Q_1, and Q_2 are DG1 and DG2 active powers and reactive powers, respectively.

7.5.1 Virtual impedance for reactive power sharing

Accurate reactive power sharing can be achieved by inserting a big virtual impedance into the voltage regulation loop, despite the fact that the sharing of reactive power is improved; the sharing error cannot be eliminated. To eliminate the error in sharing

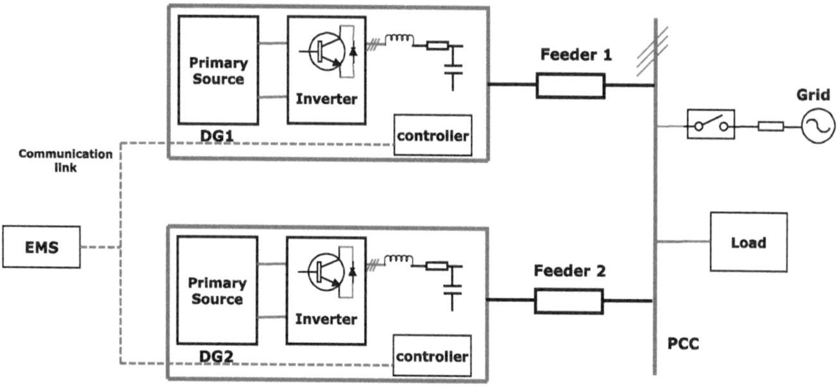

Fig. 7.24 Classical MG with two DG units.

reactive energy in a MG, as can be seen in Fig. 7.24, for an adaptive virtual impedance can be used as follows [38]:

$$z_v = c \int Q^* - Q \tag{7.11}$$

where z_v is the virtual impedance, c is the adaptation coefficient, Q^* is the reactive power reference, and Q is the measured reactive power.

From Eq. (7.11), the adaptive virtual impedance (AVI) is proportional to the integral of the error, between the reference of the reactive power and the measured reactive power. The integral term is used to eliminate the error of reactive power sharing. The reference of each DG is calculated by micro-energy management system (mEMS) according to its operating power and the total load and sends it back to each unit, after each DG shares its reactive power real load with the (EMS) via the communication link.

Fig. 7.25 shows a basic configuration of a MG-mashed structure. In the meshed structure, there are four loads (local load 1, local load 2, common load 1, and common load 2) connected directly to two DG units output or to common load points. Generally, in this structure, the number of loads is different (two by one DG unit) compared to classical structure, where all loads can be considered as a unique load. Therefore, a load change occurs often, during the adaptation of virtual impedance.

7.5.2 Reactive power sharing analysis for meshed microgrid

Considering the meshed MG shown in Fig. 7.25, output voltage magnitude for each DG is given by Eqs. (7.12), (7.13). Generally, the resistive behavior of the supply impedance is negligible compared to the inductive behavior. To this end, feeder impedances are considered to be purely inductive [10]. The drop in voltage caused by the power delivered across the impedance between the DG and the point of the load can be written as follows:

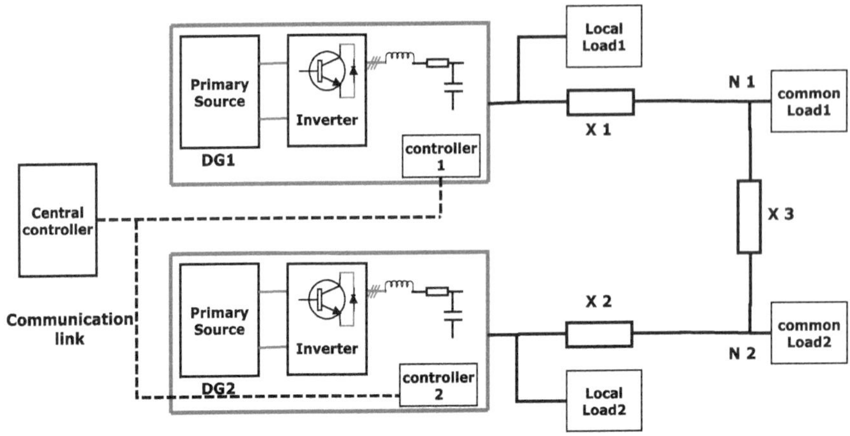

Fig. 7.25 Meshed MG structure configuration for two DG units.

$$E_1 - V_1 = \frac{X_1(Q_1 - Q_{LL1})}{E_0} \tag{7.12}$$

$$E_2 - V_2 = \frac{X_2(Q_2 - Q_{LL2})}{E_0} \tag{7.13}$$

In addition, the voltage drop caused by the power circulating between the point of load 1 and load 2 can be given by the following relation:

$$V_1 - V_2 = \frac{X_3 Q_{N1}}{E_0} = \frac{-X_3 Q_{N2}}{E_0} \tag{7.14}$$

where V_1 and V_2 are the voltage magnitude at points of load 1 and load 2, respectively, X_1, X_2, and X_3 are feeders' impedances of DG_1, DG_2, and feeder impedance between points of loads respectively, and Q_{LL1}, Q_{LL2} are DG_1, DG_2, local loads, and reactive powers, respectively.

The reactive power in the load points can be expressed as follows:

$$Q_{N1} = Q_1 - Q_{LL1} - Q_{L1} \tag{7.15}$$

$$Q_{N2} = Q_2 - Q_{LL2} - Q_{L2} = -Q_{N1} \tag{7.16}$$

where Q_{N1}, Q_{N2} are the loads points 1 and 2 reactive power, respectively.

By using Eqs. (7.3), (7.4), (7.6)–(7.10), reactive power accuracy error can be expressed as

$$\Delta Q = (Q_1 - Q_2) = \frac{(X_3 + X_2 + X_1)(Q_{LL1} + Q_{L1} - Q_{LL2} - Q_{L2}) - 2X_2Q_{L2} + 2X_1Q_{L1}}{2E_0m + (X_3 + X_2 + X_1)} \tag{7.17}$$

From Eq. (7.17), it can be concluded that power sharing error depends on the value of the difference between the supply impedances, but also depends on the characteristics of the local load and the values of the impedances between the load points.

Thus, to eliminate this reactive power sharing inaccuracy, virtual impedances X_{V1}, X_{V2} are added to the power loop of the DGs local controllers. To eliminate feeders' impedances mismatch completely $X_{V1} = -X_{V2}$. Adaptive virtual impedance variation ΔX_{V1} and ΔX_{V2} of X_{V1} and X_{V2} from time $t = 0$ to time $t = T1$ can be expressed as follows:

$$\Delta X_{V1} = -\Delta X_{V2} = \Delta X_V \tag{7.18}$$

From Eqs. (7.11), (7.12), reactive power sharing accuracy variation $\frac{\partial \Delta Q}{\partial t}$ can be expressed as follows:

$$\frac{\partial \Delta Q}{\partial t} == \frac{4\Delta X_V (Q_{L2} + Q_{L1})}{2E_0m + (X_3 + X_2 + X_1)} \tag{7.19}$$

As can be seen from Eq. (7.19), the variation of power sharing error depends on the virtual impedance variation value and also on the reactive power sum demanded by the common loads; which can therefore decrease reactive power sharing error elimination rapidity or perturb completely the controller. A large variation in virtual impedance ΔX_V can speed up the controller dynamics but can also affect the sharing accuracy or the system stability. To this end, the use of a controller where these coefficients are adaptable can solve the problems of rapidity and the precision of the reactive power sharing especially in cases, where the reactive loads demand varies quickly.

7.5.3 Virtual impedance adaptation using ant colony optimization algorithm

Ant colony algorithms are systems forming a class of metaheuristics proposed to solve difficult optimization problems, which are not possible to solve with conventional methods [38]. These algorithms are inspired by the collective deposition and tracking behavior observed in real ant colonies. A colony of simple agents (ants) communicates indirectly via dynamic modifications of their environment (pheromone tracks) and thus builds a solution to a problem, based on their collective experience.

An ant colony optimization (ACO) is used to adjust the virtual impedance. After the generation of the reference voltage from the power loop; this voltage is corrected subsequently by a virtual impedance, multiplied by the output current of the DG as can be seen in Fig. 7.26. The value of this virtual impedance is adjusted by the ACO adaptation algorithm (VI-ACO) [39].

ACO algorithm

Step #1 Lunch new iteration of Ants.
Step #2 Find new solution.
Step #3 Solution evolution of ITEA.
Step #4 if Criteria test is achieved **then** solution feed, go to **step #7.**
Step #5 else Pheromone deposition.
Step #6 Pheromone evaporation, retour to step #3.

The integral temporal absolute error (ITAE) is used as a cost function. The mathematical expression of the ITAE performance index is defined by the following equation:

$$\text{ITAE} = \int t \cdot e(t) \tag{7.20}$$

where t is the time and $e(t)$ is the difference between the reference set point and the controlled signal.

The error $e(t)$ is calculated by using the following equation:

$$e(t) = \sqrt{(Q* - Q)^2} \tag{7.21}$$

The pheromone was updating then by the dropping off ants' pheromone in each path. The selection of ACO parameters ($\theta = 0.08$, $\lambda = 0.8$) is carried by many experiments.

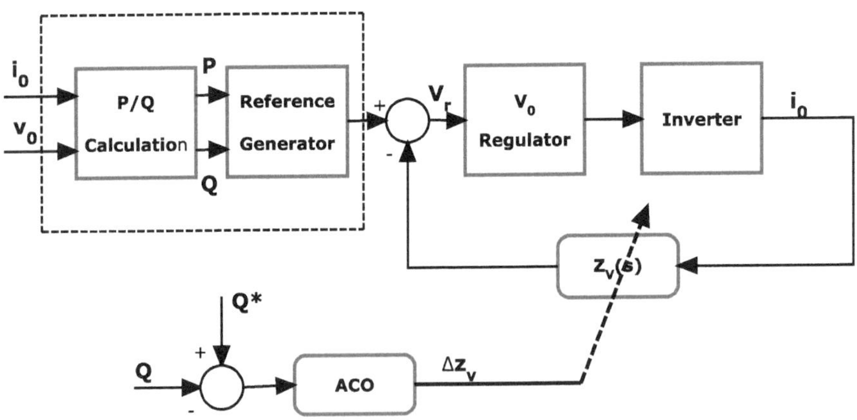

Fig. 7.26 The adaptive virtual impedance controller using ACO (VI-ACO controller).

Example 7.6: Power sharing in a MG using ACO algorithm

In this example, a control strategy is presented based on a metaheuristic method named ACO to tune virtual impedance (VI) method (VI-ACO), i.e., to improve the accuracy of reactive power sharing. The controller scheme shown in Fig. 7.27 is implemented under Matlab/Simulink.

Fig. 7.28A and B shows the proportional and the integral virtual impedances coefficients K_P and K_I variation for a DG unit. As can be observed in Fig. 7.28A, the proportional coefficient K_P takes big values at the beginning of the VI adaptation in order to accelerate the reactive power sharing error elimination. Then, its value is decreased by ACO algorithm when reactive cost function value is near to 0 to minimize the dynamic error (low overshoots) and to guarantee that large oscillations cannot be created when reactive power sharing accuracy is achieved. However, the integral coefficient variation shown in Fig. 7.28B takes a low value at the beginning of the adaptation, which assures more system stability and takes a large value when the accuracy is reached to ensure that reactive power sharing accuracy error is eliminated. Fig. 7.28C shows the variation of the cost function using virtual impedance based on ACO adaptation, as can be seen the error of the reactive sharing is quite eliminated.

The controller is tested under various operating conditions using Matlab/Simulink environment. MG and controller's parameters are reported in Table 7.7.

Case 1: *Performance evaluation of the controller (VI-ACO) for different MG operating conditions*

The performance of the controller is tested during various MG operating conditions. As can be observed from Fig. 7.29A and B, the active and reactive power sharing accuracy is reached.

Moreover, settling and rise times of reactive power sharing error elimination are improved as shown in Fig. 7.29C. Maximum dynamic reactive power error of the adaptive virtual impedance is 80 Var. Fig. 7.29D shows the virtual impedance variation when the controlled is active until $t = 1.5$ s. As can be observed from Fig. 7.29D,

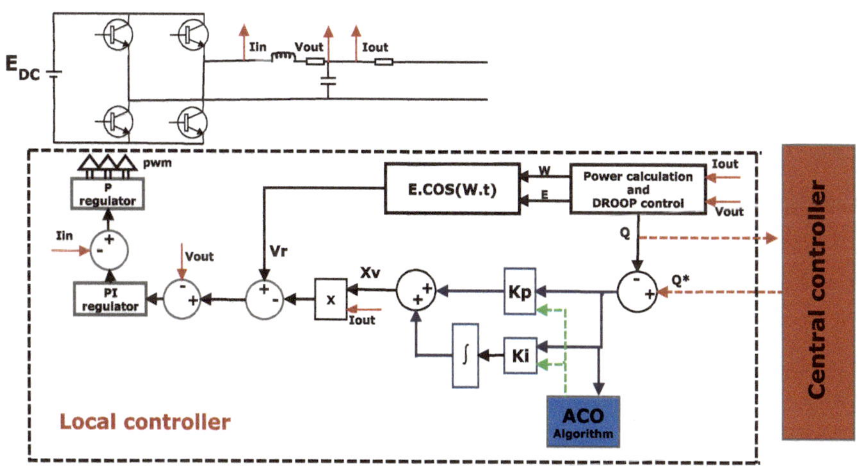

Fig. 7.27 Block diagram of controller scheme for virtual impedance using ACO algorithm (VI-ACO controller).

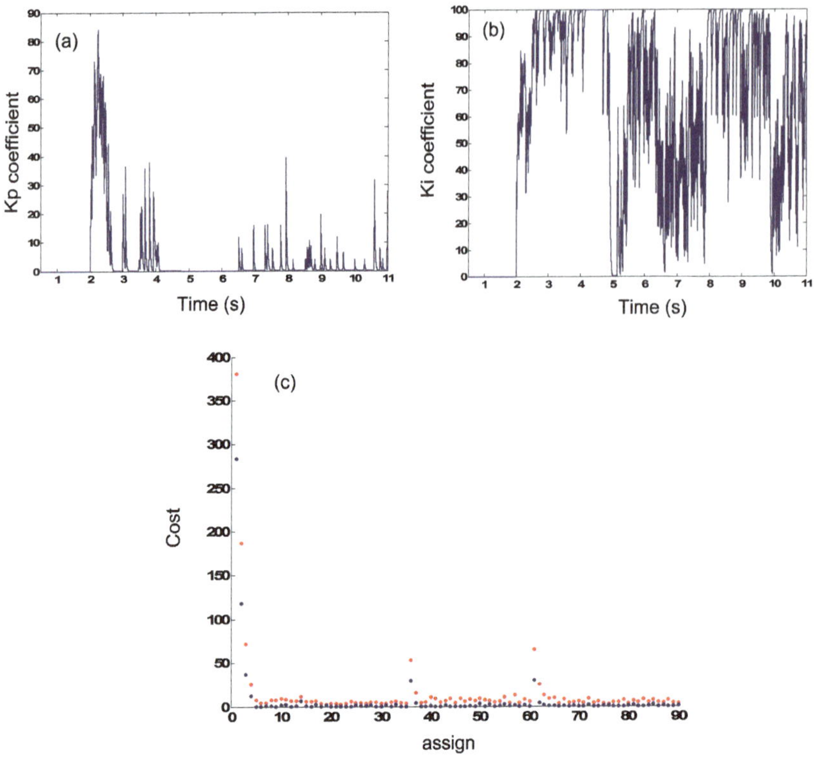

Fig. 7.28 Virtual impedance adaptation coefficient variations, (A) integral coefficient (K_P), (B) proportional coefficient (K_I), and (C) cost function variation under virtual impedance adaptation and load change.

the virtual impedance is changed adaptively to achieve reactive power sharing in different operating conditions.

Case 2: *Performance evaluation of the controller under load change during virtual impedance adaptation*

Table 7.7 Controller parameters used for Matlab/Simulink simulation.

Description	Parameter	Value
DG unit rating	S	2.5 kVA
Nominal voltage	V_0	220 V
Inverters filter	L,C	5 mH, 7 5 F
Switching frequency	fs	20 kHz
Frequency droop coefficient	m	0.002 rad/(s W)
Voltage droop coefficient	n	0.0035 V/var
Feeder 1 impedance	R_1+jX_1	$1.2+j2.3\,\Omega$
Feeder 2 impedance	R_2+jX_2	$0.9+j1.8\,\Omega$

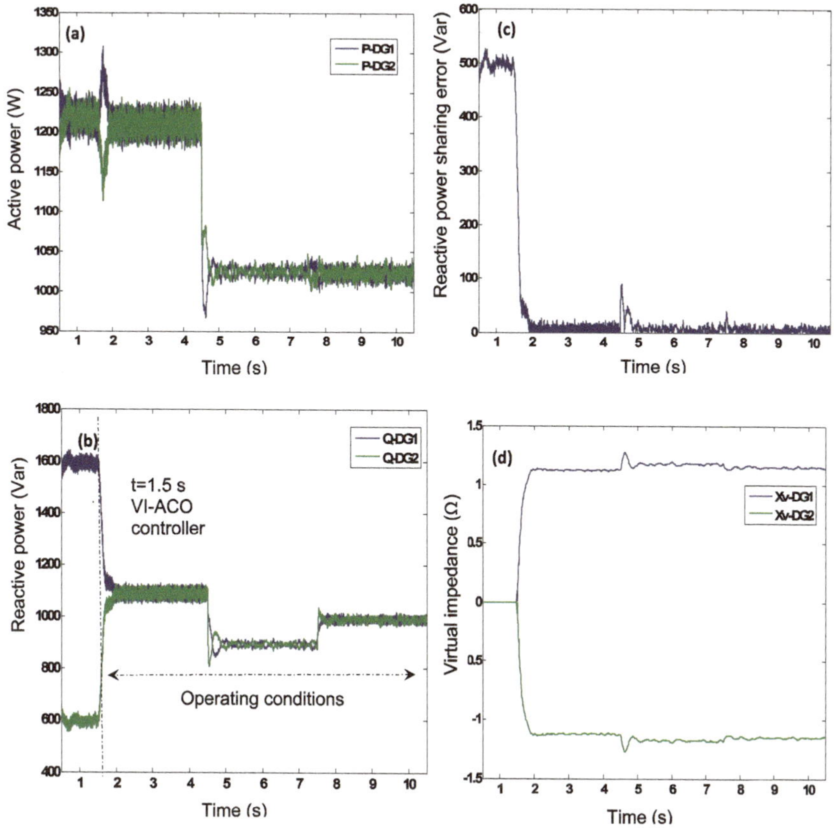

Fig. 7.29 Performance evolution of the controller (VI-ACO): Case 1: (A) active power, (B) reactive power, (C) reactive power sharing error, and (D) virtual impedance variation.

MG is considered to operate under conventional droop control during the period: $0 < t < 1.5$. At time $t = 1.5$ s the controller VI-ACO is activated, at time $t = 1.6$ s a common load 1 change occurred, and at time $t = 1.7$ s the local load 1 change occurred. Fig. 7.30 shows DG1 and DG2 reactive power and virtual impedances variation under local loads and common loads reactive power variation. As can be seen from Fig. 7.30A, the reactive power accuracy, settling time, and dynamic error performances are still acceptable until the load change, when the virtual impedance adapting is in process. In addition, the system remains stable, although the MG structure variation and load demand variation during the adaptation of the virtual impedance as shown in Fig. 7.30B.

Case 3: Performance of the controller under small load values

In this case, the MG operating point is far from its nominal operating point, active and reactive power demand are set to be one-third of the nominal DGs active and

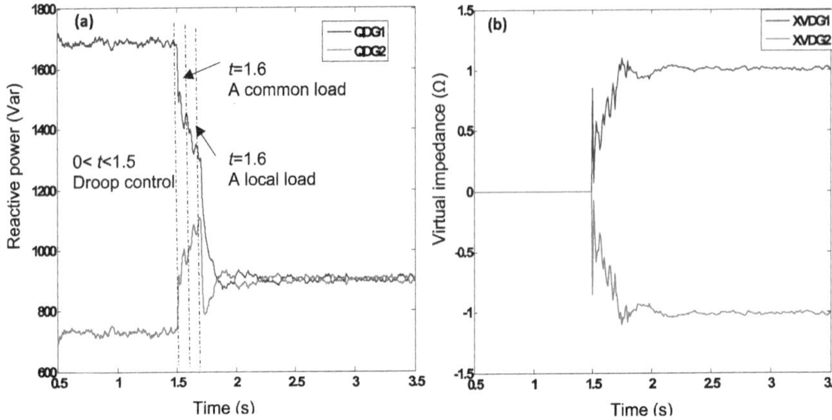

Fig. 7.30 Performance evolution of the controller (VI-ACO): Case 2 (A) reactive power and (B) virtual impedance variation.

reactive power. MG is operating under conventional droop control during period time $0<t<0.75\,\text{s}$, at time $t=0.75\,\text{s}$, the controller is activated. Results are shown in Fig. 7.31, where reactive power sharing accuracy is achieved within 1 s, and dynamic error is still acceptable. Thus, it can be concluded that the controller eliminates, in a short time ($\cong$0.25 s), the effect of load reactive power demand.

In this example, the ability of a metaheuristic ACO algorithm to improve the power sharing in MG is demonstrated; the main role of the ACO algorithm is to adjust the virtual impedance.

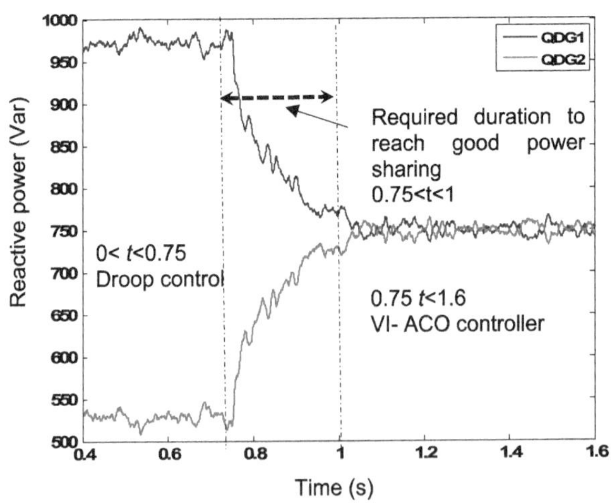

Fig. 7.31 Performance evolution of the controller (VI-ACO): Case 3 reactive power variation.

7.6 Summary

This chapter covered three main applications: (1) control of grid-connected and hybrid photovoltaic systems, (2) energy management of MGs with EV station, and (3) power sharing in MGs with parallel inverters. To solve these issues, different artificial intelligence techniques and metaheuristic algorithms are applied and examined, like long-short-term memory which is a kind of recurrent neural network, feed-forward neural networks (or multilayer perceptron), fuzzy logic, adaptive neuro-fuzzy inference system, which combines neural network with fuzzy logic, ant colony optimization, and particle swarm optimization. These algorithms have been simulated and implemented under Matlab/Simulink and Python languages.

Examples described can help the readers to learn and understand how to apply the methods investigated to develop intelligent controllers, which permit to solve key problems of distributed generation systems. Although these intelligent controllers are efficient and robust, real-time implementation of such techniques (e.g., using reconfigurable FPGA) need special skills and knowledge.

References

[1] S. Ruiz-Romero, A. Colmenar-Santos, F. Mur-Pérez, Á. López-Rey, Integration of distributed generation in the power distribution network: the need for smart grid control systems, communication and equipment for a smart city—use cases, Renew. Sust. Energ. Rev. 38 (2014) 223–234.

[2] N. Chettibi, A. Mellit, Intelligent control strategy for a grid connected PV/SOFC/BESS energy generation system, Energy 147 (2018) 239–262.

[3] M. Meiqin, L. Chang, D. Ming, Integration and intelligent control of micro-grids with multi-energy generations: a review, in: *2008 IEEE International Conference on Sustainable Energy Technologies*, IEEE, 2008, November, pp. 777–780.

[4] N.M. Dawoud, T.F. Megahed, S.S. Kaddah, Enhancing the performance of multi-microgrid with high penetration of renewable energy using modified droop control, Electr. Power Syst. Res. 201 (2021) 107538.

[5] H. Sellamna, A.M. Pavan, A. Mellit, J.M. Guerrero, An iterative adaptive virtual impedance loop for reactive power sharing in islanded meshed microgrids, Sustain. Energy Grids Netw. 24 (2020) 100395.

[6] J. Ma, X. Ma, A review of forecasting algorithms and energy management strategies for microgrids, Syst. Sci. Control. Eng. 6 (1) (2018) 237–248.

[7] U.B. Tayab, J. Lu, F. Yang, T.S. AlGarni, M. Kashif, Energy management system for microgrids using weighted salp swarm algorithm and hybrid forecasting approach, Renew. Energy 180 (2021) 467–481.

[8] X. Xu, H. Jia, D. Wang, C.Y. David, H.D. Chiang, Hierarchical energy management system for multi-source multi-product microgrids, Renew. Energy 78 (2015) 621–630.

[9] M.F. Zia, E. Elbouchikhi, M. Benbouzid, Microgrids energy management systems: a critical review on methods, solutions, and prospects, Appl. Energy 222 (2018) 1033–1055.

[10] J. Kim, J.M. Guerrero, P. Rodriguez, R. Teodorescu, K. Nam, Mode adaptive droop control with virtual output impedances for an inverter-based flexible AC microgrid, IEEE Trans. Power Electron. 26 (3) (2011) 689–701.

[11] T.B. Lopez-Garcia, A. Coronado-Mendoza, J.A. Domínguez-Navarro, Artificial neural networks in microgrids: a review, Eng. Appl. Artif. Intell. 95 (2020) 103894.
[12] H. Sellamna, A. Mellit, A. Rabhi, B. Hajji, B. Oral, Micro-grid reactive power sharing using an adaptive virtual impedance technique, in: 2019 International Conference on Wireless Technologies, Embedded and Intelligent Systems (WITS), April 3, IEEE, 2019, pp. 1–5.
[13] M.A. Zamani, T.S. Sidhu, A. Yazdani, Investigations into the control and protection of an existing distribution network to operate as a microgrid: a case study, IEEE Trans. Ind. Electron. 61 (2014) 1904–1915.
[14] C.M. Hong, C.H. Chen, Intelligent control of a grid-connected wind-photovoltaic hybrid power systems, Int. J. Electr. Power Energy Syst. 55 (2014) 554–561.
[15] A. Vinayagam, A.A. Alqumsan, K.S.V. Swarna, S.Y. Khoo, A. Stojcevski, Intelligent control strategy in the islanded network of a solar PV microgrid, Electr. Power Syst. Res. 155 (2018) 93–103.
[16] Q. Hu, R. Tang, Design and intelligent control of building integrated photovoltaic system, J. Renew. Sustain. Energy 9 (4) (2017) 043701.
[17] E. Koutroulis, K. Kalaitzakis, V. Tzitzilonis, Development of an FPGA-based system for real-time simulation of photovoltaic modules, Microelectron. J. 40 (7) (2009) 1094–1102.
[18] N. Chettibi, A. Mellit, FPGA-based real time simulation and control of grid-connected photovoltaic systems, Simul. Model. Pract. Theory 43 (2014) 34–53.
[19] R. Faraji, A. Rouholamini, H.R. Naji, R. Fadaeinedjad, M.R. Chavoshian, FPGA-based real time incremental conductance maximum power point tracking controller for photovoltaic systems, IET Power Electron. 7 (5) (2014) 1294–1304.
[20] H. Mekki, A. Mellit, S.A. Kalogirou, A. Messai, G. Furlan, FPGA-based implementation of a real time photovoltaic module simulator, Prog. Photovolt. Res. Appl. 18 (2) (2010) 115–127.
[21] N. Chettibi, A. Mellit, PSO based direct power control of grid connected photovoltaic system, in: 2015 4th International Conference on Electrical Engineering (ICEE), IEEE, 2015, December, pp. 1–6.
[22] M. Malinowski, M.P. Kazmierkowski, A. Trzynadlowski, Review and comparative study of control techniques for three-phase PWM rectifiers, Math. Comput. Simul. 63 (3–5) (2003) 349–361.
[23] J. Alonso-Martínez, J.E.G. Carrasco, S. Arnaltes, Table-based direct power control: a critical review for microgrid applications, IEEE Trans. Power Electron. 25 (12) (2010) 2949–2961.
[24] N. Chettibi, A. Mellit, G. Sulligoi, A.M. Pavan, Fuzzy-based power control for distributed generators based on solid oxide fuel cells, in: 2015 International Conference on Clean Electrical Power (ICCEP), 2015, June, pp. 580–585.
[25] G.J. Yu, Y.S. Jung, J.Y. Choi, G.S. Kim, A novel two-mode MPPT control algorithm based on comparative study of existing algorithms, Sol. Energy 76 (4) (2004) 455–463.
[26] N. Chettibi, A. Mellit, G. Sulligoi, A.M. Pavan, Adaptive neural network-based control of a hybrid AC/DC microgrid, IEEE Trans. Smart Grid 9 (3) (2016) 1667–1679.
[27] E. Unamuno, J.A. Barrena, Hybrid ac/dc microgrids—part II: review and classification of control strategies, Renew. Sust. Energ. Rev. 52 (2015) 1123–1134.
[28] A. Ghaffari, A. Askarzadeh, Design optimization of a hybrid system subject to reliability level and renewable energy penetration, Energy 193 (2020) 116754.
[29] V. Indragandhi, R. Logesh, V. Subramaniyaswamy, V. Vijayakumar, P. Siarry, L. Uden, Multi-objective optimization and energy management in renewable based AC/DC microgrid, Comput. Electr. Eng. 70 (2018) 179–198.

[30] W.M. Lin, C.S. Tu, M.T. Tsai, Energy management strategy for microgrids by using enhanced bee colony optimization, Energies 9 (1) (2016) 5.

[31] A. Rabiee, M. Sadeghi, J. Aghaeic, A. Heidari, Optimal operation of microgrids through simultaneous scheduling of electrical vehicles and responsive loads considering wind and PV units uncertainties, Renew. Sust. Energ. Rev. 57 (2016) 721–739.

[32] R. Bouhedir, A. Mellit, N. Rouibah, Simulation of a micro-grid for electric vehicles charging station, in: International Conference on Electronic Engineering and Renewable Energy, Springer, Singapore, 2020, April, pp. 565–571.

[33] A.M. Pavan, V. Lughi, M. Scorrano, Total cost of ownership of electric vehicles using energy from a renewable-based microgrid, in: 2019 IEEE Milan PowerTech, IEEE, 2019, June, pp. 1–6.

[34] M. Scorrano, R. Danielis, S. Pastore, V. Lughi, A. Massi Pavan, Modeling the total cost of ownership of an electric car using a residential photovoltaic generator and a battery storage unit—an Italian case study, Energies 13 (10) (2020) 2584.

[35] A. Mellit, A.M. Pavan, V. Lughi, Deep learning neural networks for short-term photovoltaic power forecasting, Renew. Energy 172 (2021) 276–288.

[36] M.A. Zamani, T.S. Sidhu, A. Yazdani, Investigations into the control and protection of an existing distribution network to operate as a microgrid: a case study, IEEE Trans. Ind. Electron. 61 (4) (2013) 1904–1915.

[37] J.M. Guerrero, L.G. De Vicuna, J. Matas, M. Castilla, J. Miret, Output impedance design of parallel-connected UPS inverters with wireless load-sharing control, IEEE Trans. Ind. Electron. 52 (4) (2005) 1126–1135.

[38] H. Mahmood, D. Michaelson, J. Jiang, Accurate reactive power sharing in an islanded microgrid using adaptive virtual impedances, IEEE Trans. Power Electron. 30 (3) (2014) 1605–1617.

[39] H. Sellamna, N. Rouibah, A. Mellit, G.M. Tina, H. Bekkay, Power flow control in autonomous micro-grid operation using ants colony optimization under variable load conditions, in: International Conference on Electronic Engineering and Renewable Energy, Springer, Singapore, 2018, pp. 392–398.

Internet of things (IoT) and embedded systems for photovoltaic systems

8

Chapter outline

8.1 Introduction

After verification of the capability of machine learning, ensemble, and deep learning to solve major problems of photovoltaic (PV) systems, presented in previous chapters, another important step is to check their effectiveness in real-time applications by implementing these algorithms in programmable digital devices or by co-simulations (hardware-in-loop). This step is very important before attempting to design a prototype for real-world applications, although most methods have been experimented and verified at laboratories.

In the literature, some works related to the real-time implementation in PV systems can be found, addressing some issues like PV monitoring and remote sensing systems (by designing data-acquisition and monitoring systems), tracking the maximum power (by implementing appropriate maximum power point tracking algorithms or reconfigurable array to avoid losses due to partial shading effect), control of DC-AC

Handbook of Artificial Intelligence Techniques in Photovoltaic Systems. https://doi.org/10.1016/B978-0-12-820641-6.00008-9
Copyright © 2022 Elsevier Ltd. All rights reserved.

inverters (by implementing controllers), optimal energy management or distributed generation systems. Most of them are related to the implementation of such techniques in electronic devices such as microcontrollers (MC), including Arduino, optimized MC (digital signal processing), microprocessors (e.g., Raspberry Pi), and reconfigurable programmable devices (PLD: Programmable Logic Device, CPLD: Complex Programmable Logic Device, and FPGA: Field Programmable Gate Array) [1–5].

Recently, more efforts have been undertaken to merge artificial intelligence (AI) techniques, Internet of things (IoT), and embedded systems to develop smart components for PV system applications, i.e., for monitoring, optimization, energy management, power sharing fault detection and control [6–9].

In the last decade, the Arduino, Raspberry Pi, and FPGA platforms received considerable attention in several application fields due to their attractive features, low-cost, reprogramming capability, parallelism, fast prototyping, flexibility, etc. Currently, researchers working in the PV area are much attracted by the application of these kind of electronic boards to address some PV system issues [10,11]. The main goal is to design smart prototypes based on embedded technologies.

This chapter focus on five photovoltaic applications: (1) smart PV monitoring systems and fault diagnosis using the IoT technique and machine learning algorithms (low-cost Arduino Mega 2560 and Raspberry Pi 4 will be employed), (2) PV system emulator using neural networks (implementation in an FPGA board), (3) fuzzy-logic-based tracking the maximum power point of a PV module using an FPGA board, (4) artificial neural network-based fault diagnosis of PV arrays (implementation in an FPGA board), and (5) deep convolutional neural networks (DCNN) for PV fault detection and classification based on thermography images (implementation in a Raspberry Pi 4 board).

8.2 Programmable electronic boards and IDE

This section aims to provide a brief description of programmable electronic boards (Arduino, Raspberry Pi, and FPGA) and programming languages employed in this chapter.

8.2.1 Arduino

Arduino is an open-source hardware and software company. It comprises a low-cost microcontroller (μ-C) with other accessories. Various hardwares have been commercialized: Arduino Uno, Arduino Mega, Nano, and others [12]. It can be programmed using C or API Arduino languages. For example, Arduino IDE 1.8.13 (Integrated Development Environment) is the open source, which makes it easy to write the code and upload it to the Arduino board [12] (see Fig. 8.1A). Fig. 8.1B shows the Arduino

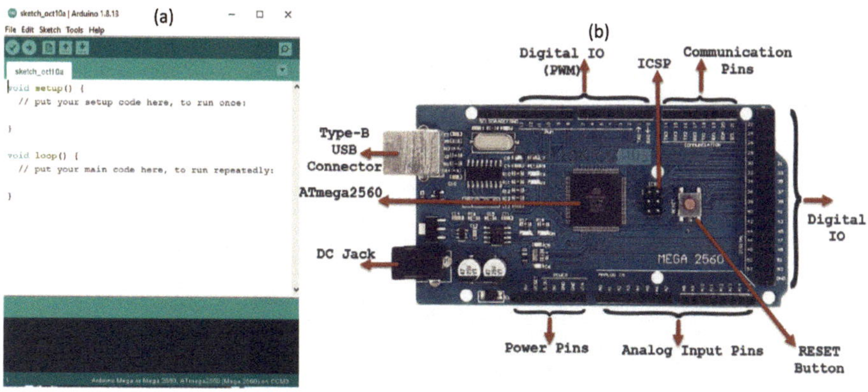

Fig. 8.1 (A) Arduino IDE 1.8.16, (B) Arduino Mega 2560 board.

Mega 2560 board; it comprises digital I/O PWM, digital I/O, USB, ATmega 2560 DC jack, analog input pins, and communication pins.

8.2.2 Raspberry Pi

A low-cost Linux and ARM-based computer on a small-circuit board sponsored by the charitable Raspberry Pi Foundation in the United Kingdom [13], Raspbian is the Debian-based Linux OS that is provided with the device [14]. The first model B of Raspberry is shown in Fig. 8.2A; it consists of the main processor CPU and other accessories such as SD slot, analog audio, Ethernet, HDMI, USB, RAM, RCA, and input/output pins. Python is the main programming language of the Raspberry Pi (see Fig. 8.2B), but it supports other programming languages such as Java, C++, etc.

8.2.3 Field programmable gate arrays

Field programmable gate arrays (FPGAs) are programmable semiconductor devices that are based on a matrix of configurable logic blocks (CLBs) connected through programmable interconnects [15]. Important characteristics of FPGAs include lower complexity, higher speed, volume designs, and programmable functions. With more technological advances, both hardware description languages, VHDL (Very High Speed Integrated Circuit (VHSIC) Hardware Description Language) and Verilog, are commonly used in programming the FPGA devices (see Fig. 8.3A). The Xilinx ISE design suite, Intel Quartus and Matlab®/Simulink environments are requested to develop different applications. The FPGA offers lower cost implementation since the functions of various components can be integrated onto the same FPGA chip as opposed to the Digital Signal Processing (DSP), which can perform only DSP-related computations.

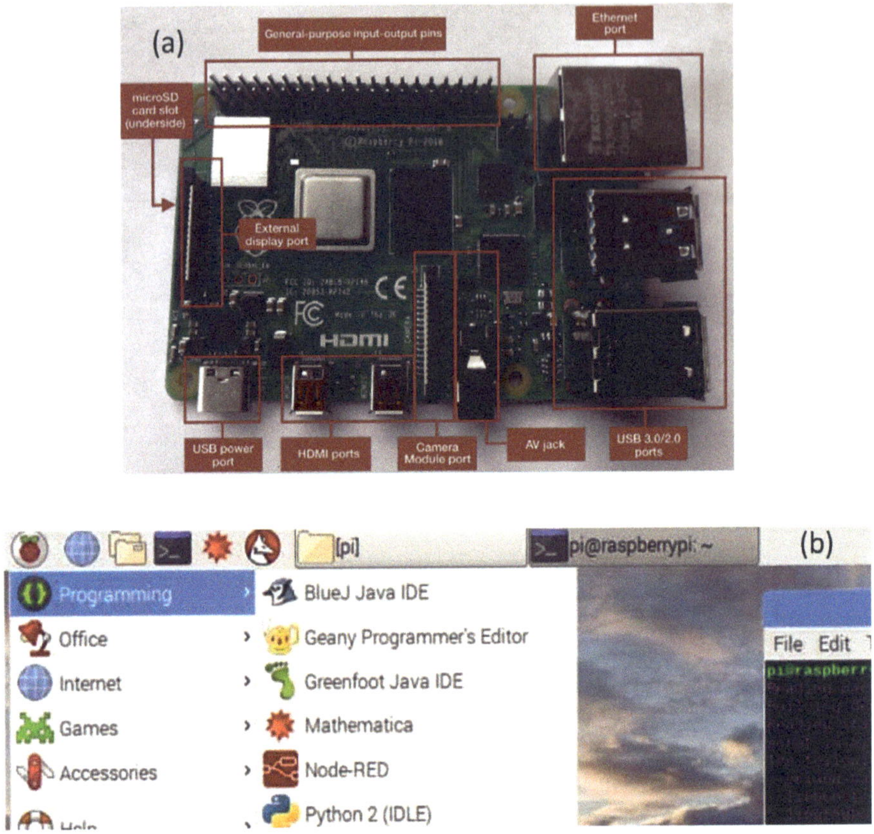

Fig. 8.2 (A) The first model B of Raspberry Pi, (B) Raspberry Pi environment (Python IDE).

8.3 Internet of things (IoT)

The Internet of things (IoT) is a very generic term for designating connected objects and can encompass many objects, many services, many situations, etc. It can be simply defined as a concept, which permits communication between different equipment using Internet or others networks, without human interaction. The clear vision of the IoT was firstly introduced in 1999 [16]. The main idea is to connect all devices, systems, and services to share data between these elements (see Fig. 8.4A).

Currently, IoT allows objects to be identified and/or controlled remotely over existing network infrastructure, making opportunities for pure integration of the physical world in computer-based systems, and the main advantages are: accuracy and efficiency improvement as well as economic benefit, and the most important advantage is the reduction of human intervention [17]. Fig. 8.4B shows Wi-Fi modules used in the Arduino board.

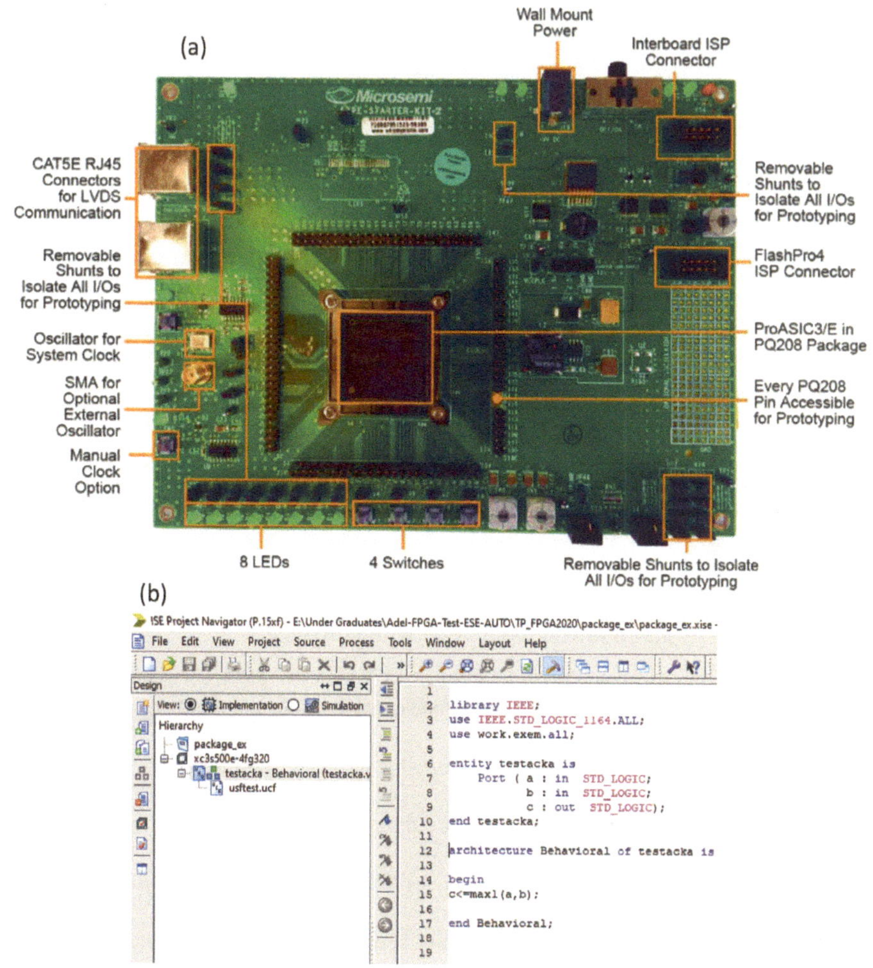

Fig. 8.3 (A) ProASIC3 Flash-based FPGA development board (Actel) and (B) Xilinx ISE design suite environment.

The IoT has been deployed in many areas: smart houses, smart city, smart grid, industrial IoT, health care, transportation, agriculture automation, building management, security and surveillance and recently in renewable energy applications including solar energy systems.

8.4 Literature review of the application of FPGA, Arduino, and Raspberry Pi for photovoltaic systems

This section presents a brief survey of the major applications of FPGA, Arduino, and Raspberry Pi in photovoltaic systems. Over the past two decades, a good number of

Fig. 8.4 (A) IoT and connected devices and (B) example of Wi-Fi modules (ESP-8266) used with Arduino boards.

contributions of electronic devices in photovoltaic systems have been attempted showing the increasing interest for these embedded technologies. Table 8.1 summarizes the most important applications of FPGA, Arduino, and Raspberry Pi devices. Focus is given on the capacity of PV systems, applications, methods, programming languages, type of electronic boards, complexity of implementation, and cost.

From Table 8.1, the following points can be highlighted:

- Many applications are related to tracking the maximum power point, PV system emulators, PV monitoring systems, PV remote sensing, and control of converters and inverters (DC-DC and DC-AC).
- Few methods are really validated in real time, while the majority are verified experimentally on a laboratory scale.
- VHDL, Python, C, and Matlab/Simulink are the most used programming languages and software.
- Methods based on machine learning techniques include a need for some skills and knowledge for their implementation.
- IoT-based remote monitoring PV systems have been used for domestic applications (small-scale PV plants).
- Smart PV monitoring systems based on the IoT technique and ML methods are the most promising topic for photovoltaic applications.
- Increase in FPGA development can help for rapid prototyping of such applications (MPPT, control, diagnosis, etc.).
- FPGA boards are mainly used for controlling power electronic devices (DC-DC and DC-AC), while few applications are related to emulator systems (e.g., generating *I-V* curves, solar irradiance prediction, etc.), Arduino and Raspberry Pi are mainly used for designing a PV monitoring system.
- Free open sources and tools, such as IoT platforms ThinkSpeak and Blynk, and hardware platforms Raspberry Pi and Android are used, as they provide a rapid development and a cost-effective solution for smart monitoring systems as well as for fault diagnosis.
- The complexity of implementation in FPGA depends mainly on the type of the algorithm developed as algorithm-based AI-techniques are relatively complex. Co-simulation (hardware in the loop, HIL) can help users to verify their method before attempting the experimental verification.
- The application of the IoT technique in this area will continue to progress.

Table 8.1 Embedded technologies for photovoltaic applications.

Ref. and year	Type of system and capacity	Application and method	Hardware language and software	Electronic board	Remarks
[18] 2009	PV module	– Simulation of a PV module – DC-DC PWM control	VHDL & ModelSim	Xilinx 4010XLPC84-3	– Real-time simulation – Relatively easy – Verified experimentally
[19] 2010	Stand-alone PV system 1 kW	– Simulation of PV system – ANN-based simulator	VHDL & ModelSim	Xilinx Virtex II XC2V1000	– Verified by simulation – Relatively complex, as the method necessity the use of ANN
[20] 2010	PV module 120 W	– Simulation of PV module – Intelligent ANN-based PV module simulator	VHDL & ModelSim	Xilinx Virtex II ProXC2v1000	– Verified by simulation – Relatively complex – Good experience on ANN is needed
[21] 2011	PV module 75 W	Control of DC-AC (DPWM)	VHDL & ModelSim	Xilinx Spartan 3E board	– Verified experimentally – Can be easily implemented
[22] 2011	PV module 120 W	Control of DC-DC Intelligent MPPT-based Fuzzy logic and genetic algorithm	VHDL, ISE Matlab/ Simulink	Xilinx Virtex II ProXC2v1000-	– Real-time simulation – Complex, it requires knowledge on genetic algorithm and fuzzy logic
[23] 2011	PV module 120 W	Control of DC-DC Intelligent MPPT-based FL	VHDL, ISE Matlab/ Simulinkl	Xilinx Virtex II ProXC2v1000	– Real-time simulation – Complex, it requires knowledge on -fuzzy logic – Good tracking accuracy

Continued

Table 8.1 Continued

Ref. and year	Type of system and capacity	Application and method	Hardware language and software	Electronic board	Remarks
[24] 2011	PV module 120 W	DC-DC control MPPT-based P&O	VHDL, ISE ModelSim	Xilinx Virtex II ProXC2v1000-	– Real-time simulation – Easy to be implemented – Verified experimentally – Acceptable accuracy
[25] 2012	PV module emulator	DC-DC, PI controller Emulator	VHDL	Intel Cyclone-III EP3C120	– Verified by simulation – Simple to be implemented
[26] 2012	PV module 120 W	DC-DC P&O algorithm	VHDL	Xilinx Virtex II ProXC2v1000-	– Verified by simulation – Simple to be implemented
[27] 2013	PV module 200 W	DC-AC SVPWM control multi-inverter	VHDL-ISE simulator Matlab/Simulink	Xilinx Spartan A3	– Online – Can be easily implemented – Acceptable results
[28] 2014	Dual-axe sun PV tracking	FL-based controller	VHDL	Altera Cyclone II EP2C35F672C6	– Verified experimentally – Relatively easy
[29] 2014	Grid-connected PV system	MPPT-P&O DC-DC (PWM) DC-AC (SPWM)	VHDL XSG ModelSim	Xilinx Virtex II ProXC2v1000-	– Co-simulation – Easy implemented
[30] 2014	PV module 150 W	MPPT Control DC-DC (PWM)	VHDL, ISE	Xilinx Virtex-5 XC5VLX50	– Verified experimentally – Relatively easy
[31] 2014	PV array	GMPPT-algorithm DC-DC (PWM)	VHDL MatlabSimulink	Intel Cyclone III— EP3C120F780C7	– Verified experimentally – Relatively complex

[32] 2016	PV module	Reflex charge control in a dual-axis solar tracking system with maximum power point tracking (MPPT)	Labview	Xilinx Spartan-3	– Verified experimentally – Easy
[33] 2016	PV string	Fault detection and diagnosis	Matlab/ Simulink XSG	Xilinx ZedBoard Zinq-7000	– Verified by simulation – Complex
[34] 2017	PV module	Monitoring ACS712 DHT11	IDE Arduino ThinkSpeak	Arduino UNO and Raspberry PI 3	– Verified experimentally – Easy and Low-cost
[10] 2017	PV module 50 W	Control dual ax tracking MPPT (PWM) FLC	VHDL	Xilink Virtex 5	– Verified experimentally – Relatively easy
[35] 2017	Grid-connected PV system	Control of DC-DC and DC AC PLL (SPWM) MPPT (PWM)	VHDL Matlab/ Simulink	Xilinx Spartan 3AN XC3S50AN	– Co-simulation – Easy
[36] 2018	PV module	Monitoring ADCs Current sensors	IDE Arduino ThinkSpeak ThingView	Arduino, Ethernet Shield, 3G modem	– Verified experimentally – Relatively easy – Low cost
[28] 2018	PV module	Control of DC-DC MPPT (PWM)	VHDL Malab/Simulinl	Xilinx Spartan 6E	– Verified by simulation – Easy
[37] 2018	PV module	Monitoring DS18B20, DhT22, current sensor, voltage sensor, and other	IDE Arduino ThinkSpeak	3G Arduino Uno Ethernet Shield W5100	– Verified experimentally – Easy – Low-cost about140.71 euro

Continued

Table 8.1 Continued

Ref. and year	Type of system and capacity	Application and method	Hardware language and software	Electronic board	Remarks
[38] 2018	Grid-connected PV system	DC-AC (SPWM)	Matlab/ Simulink XSG	Virtex-6 FPGA ML605	– Co-simulation (hardware-in-loop) – Easy
[39] 2018	PV array	Monitoring ACS712, LM35, Reference solar cell	IDE Arduino Webpage locally hosted	Arduino, Mega 2560, ESP01,	– Verified experimentally – Easy – Low cost, 75 euros
[40] 2018	PV module	Monitoring DTH22, CMP21 LM35, STC-013-000 hall effect sensor	Python Webpage application	Raspberry Pi Arduino, RFM69CW	– Verified experimentally – Relatively easy – Low-cost 99 euro
[41] 2019	PV module simulator	Current: ACS712 Voltage	IDE Arduino ThinkSpeak	Arduino Uno ESP8266 Wi-Fi module	– Verified experimentally – Easy – Low-cost
[42] 2019	PV array	Equalization algorithm	VHDL	Altera Cyclone II Altera DE2-70	– Verified experimentally – Relatively easy
[43] 2019	PV module	MPPT DC-DC (PWM) Parallel duty cycle control	VHDL	Xilinx DS610 Spartan-3A DSP	– Verified experimentally – Easy
[44] 2019	PV string	Monitoring ACS712, LM35, reference solar cell	IDE Arduino Webpage application	Arduino, Mega 2560, ESP01	– Verified experimentally – Easy – Low cost 72 euros
[45] 2019	Small PV array	Monitoring DS12820, ADS1115 LP02 Pyranometer	Python Dashboard home page	Raspberry Pi	– Verified experimentally – Easy – Low-cost 55.87 euros

[46] 2019	PV module	DS12820, DHT, voltage transducer, hall effect	Python	Raspberry Pi Arduino, RFM95 LoRa Breakout	– Verified experimentally – Easy – Low-cost 39.26 euros
[47] 2020	PV module	Intelligent method FL DC-DC (PWM)	VHDL, ISE	Xilinx Spartan 6	– Verified experimentally – Relatively easy
[48] 2020	PV module	*I-V* curves generating	VHDL, Quartus	Intel Cyclone III Altera EP3C120F780C7	– Verified by simulation – Easy
[49] 2020	Three phases grid-connected PV	*Z*-Source Inverter	VHDL, ISE Matlab/Simulik	Xilinx Virtex-5 LX50T-1C	– Verified experimentally

8.5 Real-time applications

8.5.1 *Smart PV monitoring system and fault diagnosis based on the IoT technique and machine learning algorithms*

A good number of PV monitoring systems are available in literature. Classical monitoring systems used wires to collect data [50], but recently with the advanced electronic devices technology such as sensors and microcontrollers, telecommunication, IoT, etc., researchers are much more interested in designing smart monitoring systems based on the IoT technique [51,52].

Example 8.1: PV monitoring system using an Arduino Mega 2560 board and IoT technique

Photovoltaic monitoring systems aim to monitor data, such as PV array current (I_{pv}), PV array voltage (I_{pv}), air temperature (T_a), solar irradiance (G), and others. Data stored play a very important role to supervise the PV systems online.

Fig. 8.5 shows a basic structure of a PV monitoring system based on the IoT technique. It consists basically of sensors (current ACS712, voltage B25, air temperature LM35CZ, and an LDR), a microcontroller (Arduino Mega 2560 board) and a Wi-Fi module (ESP8266-01).

In this example, to measure the global solar irradiance, a light-dependent resistor (LDR) sensor may be an alternative to a pyranometer. The LDR is calibrated in order to obtain the solar irradiance value, $G = qV_{ref}$. The q is calculated based on the formula reported in Ref. [53]. The LM35CZ is used to measure the air temperature. **Listing 8.1** presents the main code in the Arduino software IDE.

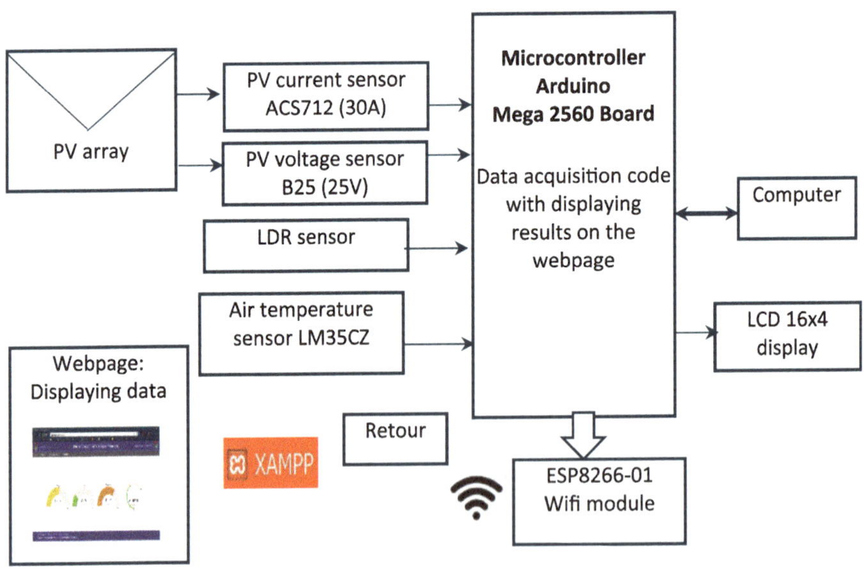

Fig. 8.5 Basic configuration used for remote PV monitoring systems using the IoT technique.

```
//Listing 8.1 main Arduino code
void loop()
{
 APV_Value = (float)analogRead(A0)*5.0/1024.0/0.2;
 APV_ValueCh = (String)(APV_Value);
 Serial.println(APV_ValueCh);
 lcd.setCursor(0, 1);
 lcd.print("V="+APV_ValueCh);
 APC_Value = (2.5-((float)analogRead(A1)*5/1024))*10;
 APC_ValueCh = (String)(APC_Value);
 Serial.println(APC_ValueCh);
 lcd.setCursor(8, 1);
 lcd.print("C="+APC_ValueCh);
 APR_Value = map(analogRead(A2),0,1023,0,1276);
 APR_ValueCh = (String)(APR_Value);
 Serial.println(APR_ValueCh);
 lcd.setCursor(0, 2);
 lcd.print("R="+APR_ValueCh);
 lcd.print(" ");
 APT_adc = analogRead(A3);
 APT_adc = analogRead(A3);
 APT_Value = ((float)APT_adc *500)/1024;
 APT_ValueCh = (String)(APT_Value);
 Serial.println(APT_ValueCh);
 lcd.setCursor(8, 2);
 lcd.print("T="+APT_ValueCh);
 Serial.println("connecting to server...");
 lcd.setCursor(0, 3);
 lcd.print("CNX To Server...");
 Serial1.println("AT+CIPSTART=\"TCP\",\"" + server + "\",80");
 if(Serial1.find("OK")){
   Serial.println("Server connection successful...");
   lcd.setCursor(0, 3);
   lcd.print("CNX Successful..");
 }
```

More details about the code implemented in the Arduino, including LCD display and connection with server using ESP8266-01 Wi-Fi module, are shown in the Appendix G (G.1 and G.2). Fig. 8.6 shows a real photo of the PV monitoring system [39].

The code implemented in the Arduino Mega 2560 board is developed under Arduino IDE environment. The procedure is summarized in the following steps:

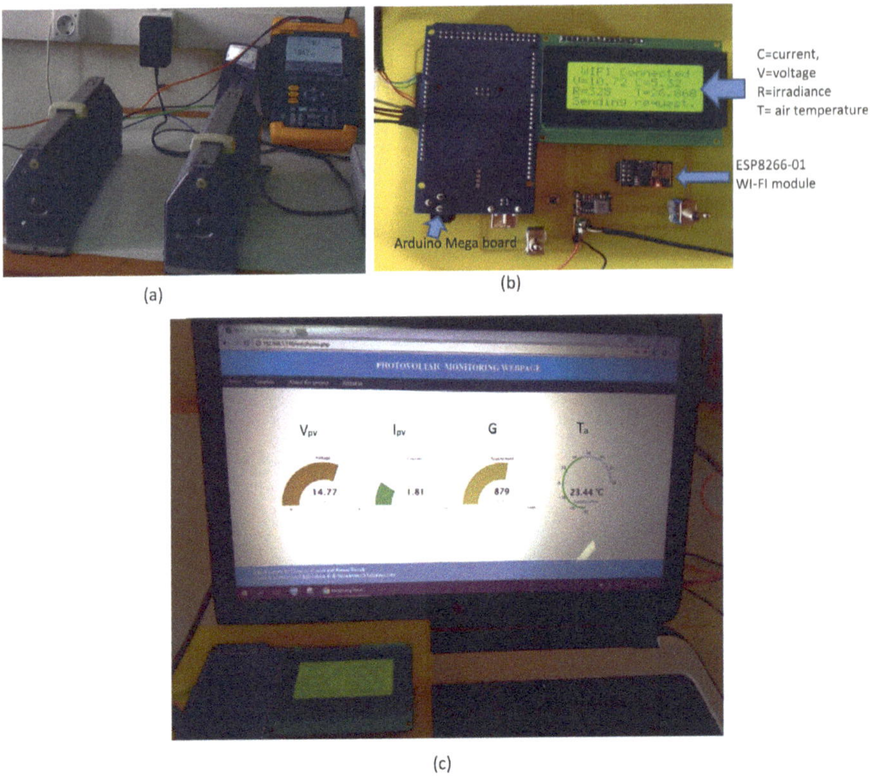

Fig. 8.6 A real photo of the PV monitoring system: (A) current and voltage measurement facility, (B) the electronic board designed based on Arduino Mega 2560 and ESP8266-01Wi-Fi module, and (C) the data uploaded I_{pv}, V_{pv}, T_{a}, and G on the webpage (photovoltaic monitoring webpage): $V_{\mathrm{pv}} = 14.77\,\mathrm{V}$, $I_{\mathrm{pv}} = 1.81\,\mathrm{A}$, $G = 879\,\mathrm{W/m^2}$, and $T_{\mathrm{a}} = 23°\mathrm{C}$.

Step #1: Initialize the LCD and the ESP8266-01.
Step #2: Test if the ESP8266-01 is ready, go to the next step, else display on the LCD an error message and retest the ESP8266-01.
Step #3: Connect the ESP8266-01 with the router.
Step #4: Read current, voltage, temperature, and solar irradiance.
Step #5: Display the results on the LCD.

- Create a TCP connection to the server.
- Prepare http request.

Step #6: Close the connection and go to **Step 4**.

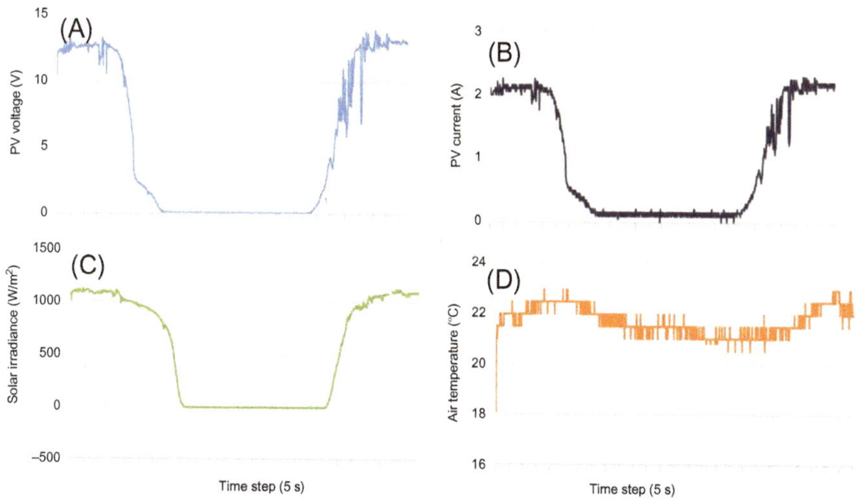

Fig. 8.7 The data monitored during a period of 24h from 12:00 18/06/2018 to 12:00 19/06/2018: (A) PV voltage, (B) PV current, (C) solar irradiance, and (D) air temperature.

The process begins with the initialization of the LCD screen and the Wi-Fi module (ESP8266-01). After the initialization step, the microcontroller waits for the ESP8266-01 to be ready. Then, it will connect automatically to the router. Fig. 8.7 shows the monitored data during a period of 24 h from 12:00 (18/06/2018) to 12:00 (19/06/2018).

Then, the monitored data will be read, saved as character strings, and displayed on the LCD screen. For data transmission, ESP8266-01 creates a TCP (transmission control protocol) connection to the server. The data are transmitted as an HTTP (HyperText Transfer Protocol) request (the message that is sent by a client to a server is what is known as an HTTP request).

To visualize the data on the website a webpage should be developed. Thus, the web development process can be divided into two parts: the client side and the server side. For the client side the webpage is designed using HTML (HyperText Markup Language), CSS (Cascading Style Sheets), and JavaScript programming languages, while the server side is handled by using PHP (Personal Home Page) programming language. The open-source XAMPP (Apache + MySQL) is used to create a local server (virtual server in a computer) in order to test the webpage before uploading it in the cloud. XAMPP is a free and open-source cross-platform web server solution stack package, which supports several programming languages. Fig. 8.8A and B show the main window of the main XAMPP control panel and the *phpMyAdmin* window for creating the database (MySQL). MySQL is the most popular database system used with PHP. Using *phpMyAdmin* a table is created which contains six columns: id, date, voltage, current, solar radiation, and cell temperature [54] (see Fig. 8.8B).

Fig. 8.9 displays the home page code of the webpage developed using the PHP language. This code permits communicating with the microcontroller and the dataset. Thus, it allows receiving the monitored data from the controller and store them in the MySQL database, as well as retrieving the data from the database to display them on the webpage.

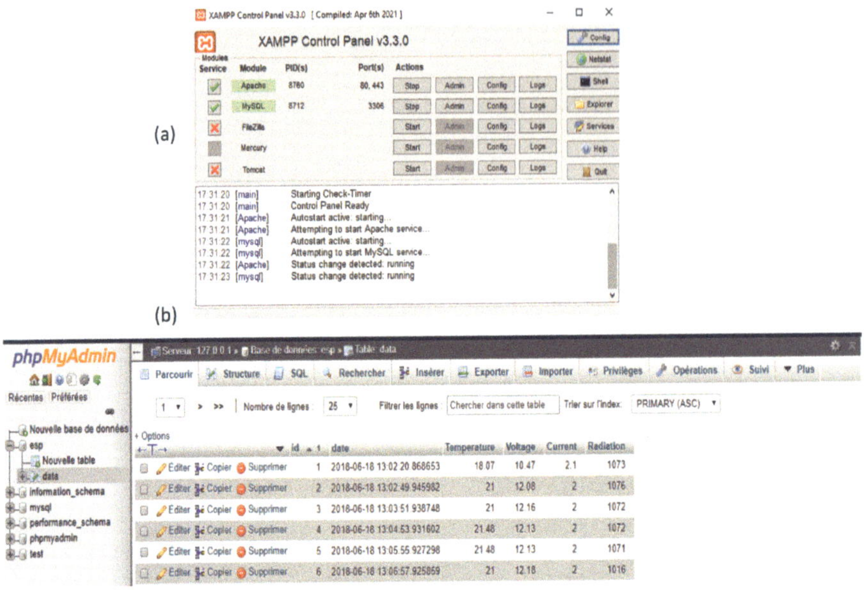

Fig. 8.8 (A) Main XAMPP control panel and (B) the *phpMyAdmin* page.

Fig. 8.9 Example of the home page code using PHP language under the Visual Studio Code (VSC) environment.

Fig. 8.10 shows the posted data on the developed webpage; also, the monitored data can be stored in an Excel file, as shown in Fig. 8.11A.

Thanks to the PLX-DAQ Excel Macro [55] (Fig. 8.11B), this tool automatically establish a communication between the microcontroller Arduino and Excel. Fig. 8.11A shows the data monitored by this tool.

Another possibility to monitor the data online (without setting up servers or developing web software), is to use the ThingSpeak platform, which is an open-source software (MathWorks). It facilitates data access, retrieval, and logging of data by

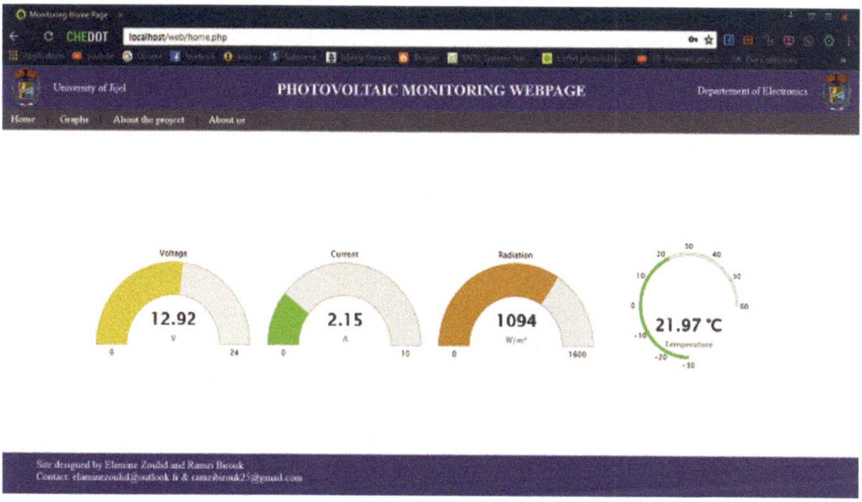

Fig. 8.10 Uploaded data on the webpage (local host) for real-time PV monitoring (voltage, current, solar irradiance, and air temperature).

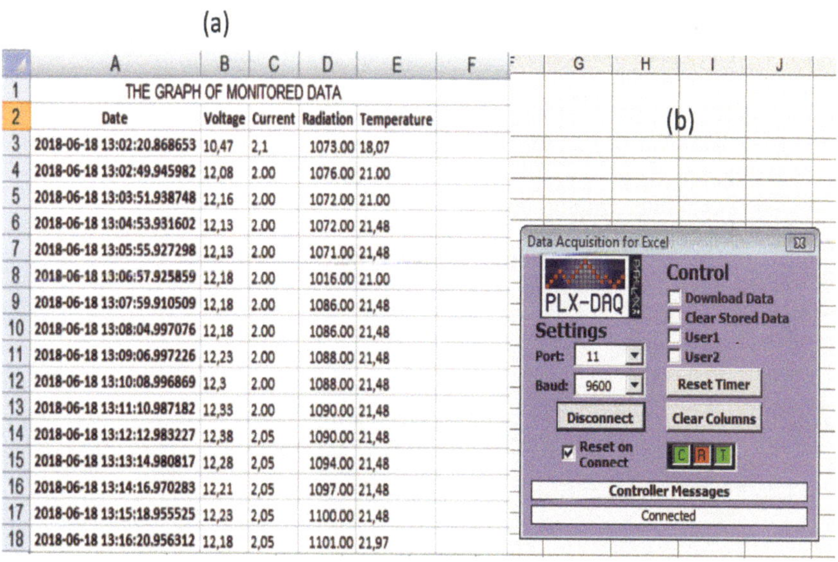

	A	B	C	D	E
1	THE GRAPH OF MONITORED DATA				
2	Date	Voltage	Current	Radiation	Temperature
3	2018-06-18 13:02:20.868653	10,47	2,1	1073.00	18,07
4	2018-06-18 13:02:49.945982	12,08	2.00	1076.00	21.00
5	2018-06-18 13:03:51.938748	12,16	2.00	1072.00	21.00
6	2018-06-18 13:04:53.931602	12,13	2.00	1072.00	21,48
7	2018-06-18 13:05:55.927298	12,13	2.00	1071.00	21,48
8	2018-06-18 13:06:57.925859	12,18	2.00	1016.00	21.00
9	2018-06-18 13:07:59.910509	12,18	2.00	1086.00	21,48
10	2018-06-18 13:08:04.997076	12,18	2.00	1086.00	21,48
11	2018-06-18 13:09:06.997226	12,23	2.00	1088.00	21,48
12	2018-06-18 13:10:08.996869	12,3	2.00	1088.00	21,48
13	2018-06-18 13:11:10.987182	12,33	2.00	1090.00	21,48
14	2018-06-18 13:12:12.983227	12,38	2,05	1090.00	21,48
15	2018-06-18 13:13:14.980817	12,28	2,05	1094.00	21,48
16	2018-06-18 13:14:16.970283	12,21	2,05	1097.00	21,48
17	2018-06-18 13:15:18.955525	12,23	2,05	1100.00	21,48
18	2018-06-18 13:16:20.956312	12,18	2,05	1101.00	21,97

Fig. 8.11 (A) Monitored data stored on an Excel file and (B) PLX-DAQ tools.

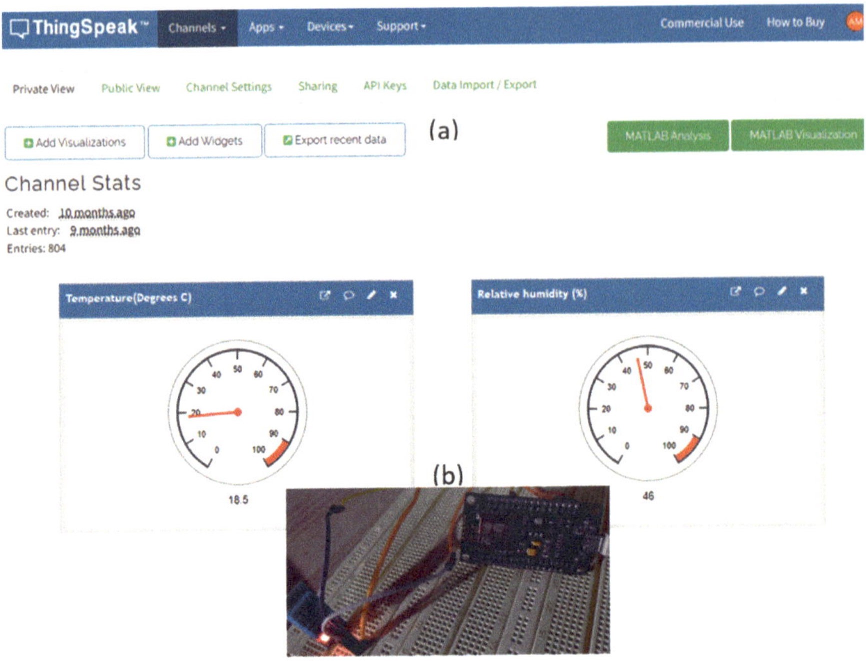

Fig. 8.12 (A) Posted values of air temperature and relative humidity in the ThingSpeak platform and (B) the electronic circuit connection: sensor with the NodeMCU ESP-8266 Wi-Fi module.

providing an API to both the devices and social network websites. So, data can be sent to ThingSpeak from your devices and create instant visualizations of live data. With Matlab analytics inside ThingSpeak, you can write and execute Matlab code to perform preprocessing, visualizations, and analyses. For example, Fig. 8.12A shows the posted air temperature and relative humidity under the ThingSpeak platform.

The data have been measured and posted on the platform using a dht22 sensor (air temperature and relative humidity) and a NodeMCU ESP8266-01 Wi-Fi module, as shown in Fig. 8.12B.

The corresponding Arduino code is shown in **Listing 8.2.**

```
#Listing 8.2: Posting data via ThingSpeak platform
#include <ESP8266WiFi.h
#include "secrets.h"
#include "ThingSpeak.h"
#include <Adafruit_Sensor.h>
#include <DHT.h>
#include <Wire.h>
#include <DHT_U.h>
float temp;
```

```
float hum;
String myStatus = "";
char ssid[] ="*****" ; // your network SSID (name)
char pass[] = "*****"; // your network password
int keyIndex = 0;  // your network key Index number (needed only for
  WEP)
WiFiClient client;

unsigned long myChannelNumber = *******;
const char * myWriteAPIKey = "***************";

#define DHTPIN D4        // Digital pin connected to the DHT sensor
#define DHTTYPE  DHT11  // DHT 11

DHT_Unified dht(DHTPIN, DHTTYPE);
uint32_t delayMS;
void setup() {
 delay(10000);
   Serial.begin(9600);
   Serial.println("CLEARDATA");
   Serial.println("LABEL,CLOCK,temperature,relative humidity");
   dht.begin();
   Wire.begin(D4);

  sensor_t sensor;
  dht.temperature().getSensor(&sensor);
  delayMS = sensor.min_delay / 1000;
     while (!Serial) {
     only
 }
 WiFi.mode(WIFI_STA);
 ThingSpeak.begin(client);
}
void loop() {
   delay(delayMS);
   sensors_event_t event;
   dht.temperature().getEvent(&event);
   if (isnan(event.temperature)) {
   }
   else {
   temp = event.temperature;

   }
   dht.humidity().getEvent(&event);
   if (isnan(event.relative_humidity)) {
       }
   else {
   hum = event.relative_humidity;
   }
   if(WiFi.status() != WL_CONNECTED){
```

```
   while(WiFi.status() != WL_CONNECTED){
     WiFi.begin(ssid, pass);
         delay(5000);
   }
  }
  ThingSpeak.setField(1, temp);
  ThingSpeak.setField(2, hum);
  ThingSpeak.setStatus(myStatus);

    int x = ThingSpeak.writeFields(myChannelNumber, myWriteAPIKey);
  if(x == 200){
    }
  else{
    }
 Serial.print("DATA,TIME,");
 Serial.print(temp);
 Serial.print(",");
 Serial.println(hum);
 delay(1000);
 }
```

Example 8.2: Smart PV monitoring and fault diagnosis system using the Arduino Mega board

Fig. 8.13 shows the structure of a PV fault detection system based on the IoT technique. It consists basically of sensors (current ACS712, voltage B25, air temperature MAX 6675 K, and solar irradiance sensor), a microcontroller (Arduino board Mega

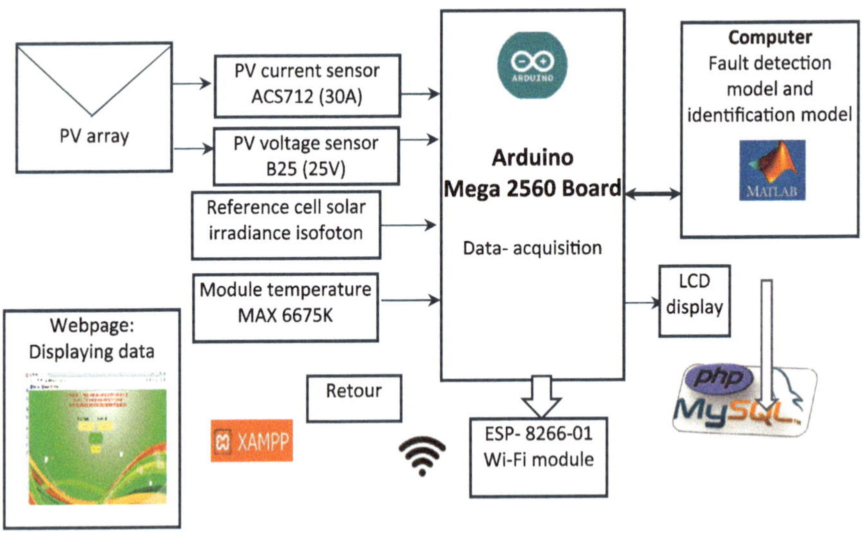

Fig. 8.13 PV remote monitoring system with a fault diagnosis method implemented in Matlab and the Arduino board.

```
Command Window
>> a = arduino
Updating server code on board Uno (/dev/tty.usbmodem1421). Please wait.

a =

  arduino with properties:

                    Port: '/dev/tty.usbmodem1421'
                   Board: 'Uno'
           AvailablePins: {'D2-D13', 'A0-A5'}
               Libraries: {'I2C', 'SPI', 'Servo'}

>>
```

Fig. 8.14 Connection between Matlab and Arduino.

2560) and a Wi-Fi module (ESP8266-01). It is similar to the configuration shown in Fig. 8.5; the new element is the fault detection and identification block. In addition, the monitoring task of a fault detection and identification code is written in the Matlab environment. To create the webpage, the same procedure is followed as shown in Example 8.1. The same tools and languages have been used (XAMPP, PHP, JavaScript, HTML, and Visual studio code environment).

The serial port is used to communicate between the Arduino and Matlab, but the Arduino support packages should be installed in Matlab. The following command line needs to be executed in Matlab:

```
>> a=Arduino,
```

So, the Matlab will detect the Arduino board, and display the following information (see Fig. 8.14).

The fault detection and identification program is implemented under Matlab, while an acquisition code is implemented in the Arduino Mega 6025 board in order to measure and communicate the data with Matlab. Then, MySQL is used to store the data and the XAMPP tool is employed to create a virtual webpage to visualize the data online. The fault detection and identification procedure is based on the electrical circuit shown in Fig. 8.15. Three PV modules are connected in parallel with voltage and current sensors, two relays K_1 and K_2, and a resistive load of 10Ω.

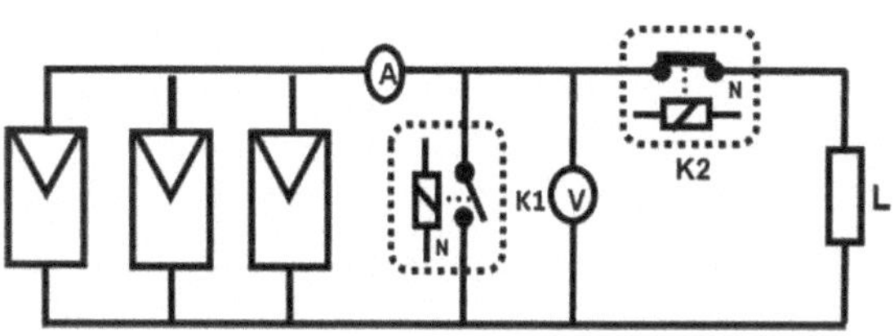

Fig. 8.15 Electrical circuit for measuring the VOC and I_{SC} using two relays K_1 and K_2.

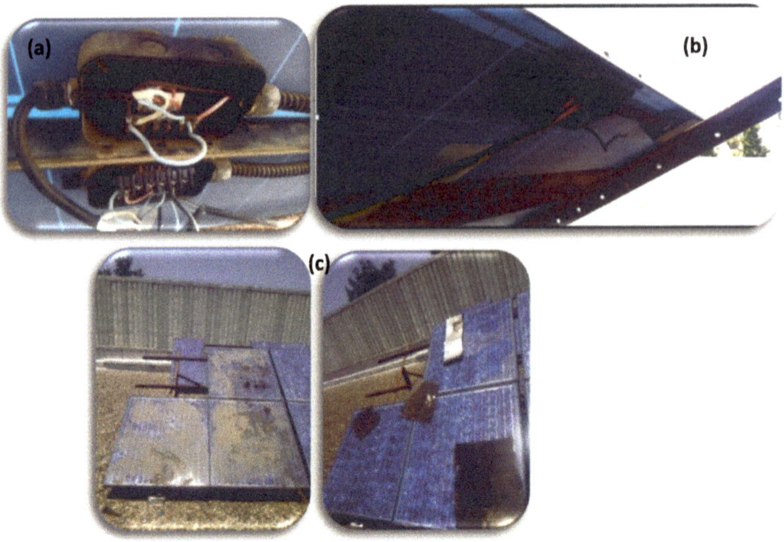

Fig. 8.16 The faults examined: (A) short-circuited PV module, (B) disconnected PV modules, and (C) other faults (e.g., shading or dust accumulation fault).

The faults considered in this example are (see Fig. 8.16): short-circuited PV module, disconnected PV modules, and other faults (e.g., shading or dust accumulation faults).

The fault detection and identification method is summarized as follows:

Define ϵp , ϵv (power and voltage thresholds), $\alpha=0.35$
Loop1 Take measurement I_{PV}^{m} , V_{PV}^{m} and calculate P_{PV}^{m} ,
Take measurement G and T_C and calculate the power $P_{\widetilde{PV}}$ $I_{\widetilde{SC}}$ and $V_{\widetilde{OC}}$
If $|P_{PV}^{m}-P_{\widetilde{PV}}|\geq\epsilon p$ then
 open the relay K_2 and measure V_{OC}^{m}
    if $|V_{OC}^{m}-V_{\widetilde{OC}}|\geq\epsilon v$ then
        If $V_{OC}^{m}>\epsilon v$ then
        print ('Shading or dust fault')
        Else
       print ('Short circuit or disconnected system')
 Else open the relay K_1 and measure I_{SC}^{m}
    if $|I_{OC}^{m}-I_{\widetilde{OC}}|<\alpha I_{SC}$ then
      print ('One PV module is disconnected')
    else print ('Two PV modules or more are disconnected')
Else print ('No fault'), loop 1
End

The code implemented in Matlab is given in the **Listing 8.3**.

```
#Listing 8.3 Matlab program for fault detection and identification
close all, clc, clear all;
ID=0;Time=5000;id3=1;a=arduino;
 for i=1:5000
      k1=0;k2=0; k3=0;k4=0;ge=0;te=0;ve=0;ce=0;
      for j=1:10
         R_i=readVoltage(a,'A10');% Read the current
         R_v=readVoltage(a,'A7');%Read the voltage
         R_t=readVoltage(a,'A3');% Read the solar irradiance
         R_g=readVoltage(a,'A6');%Read the air temprature
         k1=R_i-0.435; k2=R_v-0.435; k3=R_t-0.435; k4=R_g-0.435
         ce=ce+k1; ve=ve+k2; te=te+k3; ge=ge+k4;
      end
      R_i0=ce/10;R_v0=ve/10; R_t0=te/10; R_g0=ge/10;
% The conversion process
ID=ID+1;Time=Time-1;I_mes=(R_i0)*7;V_mes=R_v0*4.95;T_mes=R_t0*16;
  G_mes=R_g0*480;
conn = database('mes','root','');
insert(conn,'mes1',{'timee','curant','voltage','temp','rad','id'},
...{num2str(Time),num2str(I_mes),num2str(V_mes),num2str(T_mes),
  num2str(G_mes),num2str(ID)})
pause(2);
% PV module BP MSX 120W
Impp=3.56; Vmpp=33.7; ns=1;Ns=72;Voc=41.7, Isc=3.87;
ki=2.68;kv=-80; Ki=(ki*10o(-3))/Isc;Kv=(kv*10o(-3));ki=Ki; kv=Kv;
RSH0=350;T=15+273; G=G_mes;
% Call fucntion I-V model
[II1 V mm]=para5_temp_irra(G,T,Voc,Isc,Impp,Vmpp,ns,ki,kv,Ns)
I=II1;Ic=II1;Vc=V./72;Im=2.*Ic;Vm=36.*Vc;Ip=Im.*3;Vp=Vm;
 pc=max(Vp.*Ip);% Pmpp
 for i=1:length(Vp);
    if Vp(i)*Ip(i)==pc;
       Vmm=Vp(i);% Vmpp
       Imm=Ip(i);%Impp
    end
 end
 % Calculate Isc and Voc
Imin=min(abs(Ip));
for i=1:length(Vp);
    if Ip(i)>=0;
        Voc=Vp(i);
    end
end
Voc_sim=max(Voc);Isc_sim=max(II1)
```

```
%%%Fault diagnosis step
Tr=4; Thv=2; %Thresholds
p_mes=I_mes*V_mes;
dp=abs(pc-p_mes); % ABS(Pmeasured -Psimulated)
if dp<Tr
   s1=0; s2=0; s3=0; s4=0;
   disp('No faults');
  insert(conn,'mes3',{'id','delta_p','statut1','statut2','statut3','statut4'},
  ...{num2str(id3),num2str(dp),num2str(s1),num2str(s2),num2str(s3),
  num2str(s4)})
     id3=id3+1
else
%Open k2
     Voc_mes=20; dvoc=abs(Voc_mes-Voc_sim);
   if dvoc>Thv
       if Voc_mes>Thv
           s1=0; s2=0; s3=0; s4=0;
           disp('Other faults, dust, shading');
           s4=1
  insert(conn,'mes3',{'id','delta_p','statut1','statut2','statut3','statut4'},
  ...{num2str(id3),num2str(dp),num2str(s1),num2str(s2),num2str(s3),
  num2str(s4)})
         id3=id3+1;
       else
         s1=0; s2=0; s3=0; s4=0;
         disp('short-circuit, or all system disconnected')
         s3=1;   insert(conn,'mes3',{'id','delta_p','statut1','statut2',
  'statut3','statut4'},
     ...{num2str(id3),num2str(dp),num2str(s1),num2str(s2),num2str
  (s3),num2str(s4)})
        id3=id3+1;
    end
  else
%Close k1
   Isc_mes=3;disc=abs(Isc_mes-Isc_sim);
   if disc<0.35*Isc_sim;
        s1=0; s2=0; s3=0; s4=0;
        disp('one module is disconnected');
        s2=1;
insert(conn,'mes3',{'id','delta_p','statut1','statut2','statut3','statut4'},
     ...{num2str(id3),num2str(dp),num2str(s1),num2str(s2),num2str
  (s3),num2str(s4)})
     id3=id3+1;
   else
     s1=0; s2=0; s3=0; s4=0;
```

```
      disp('2 modules are disconnected');
      s1=1;
insert(conn,'mes3',{'id','delta_p','statut1','statut2','statut3','statut4'},
...{num2str(id3),num2str(dp),num2str(s1),num2str(s2),num2str(s3),
  num2str(s4)})
        id3=id3+1;
    end
  end
end
end
```

A real photo of the test facility and the designed electronic board are shown in Fig. 8.17.

The function permits plotting the *I-V* curve based on one-diode model under Matlab as presented in Appendix G.3.

The experiments have been carried out on July 7, 2019. After connecting to the server, the data from the system have been sent to the cloud via the IoT technique. The results can be displayed online on the webpage designed. To check the fault detection and identification procedure, an experimental test has been carried out. Fig. 8.18 shows an example of the monitored data as well the state of the system, displayed on the webpage designed.

From Fig. 8.18, it can be seen that the system works correctly without any fault (see Fig. 8.18A), while in another instance (see Fig. 8.18B) a fault is shown, i.e., a

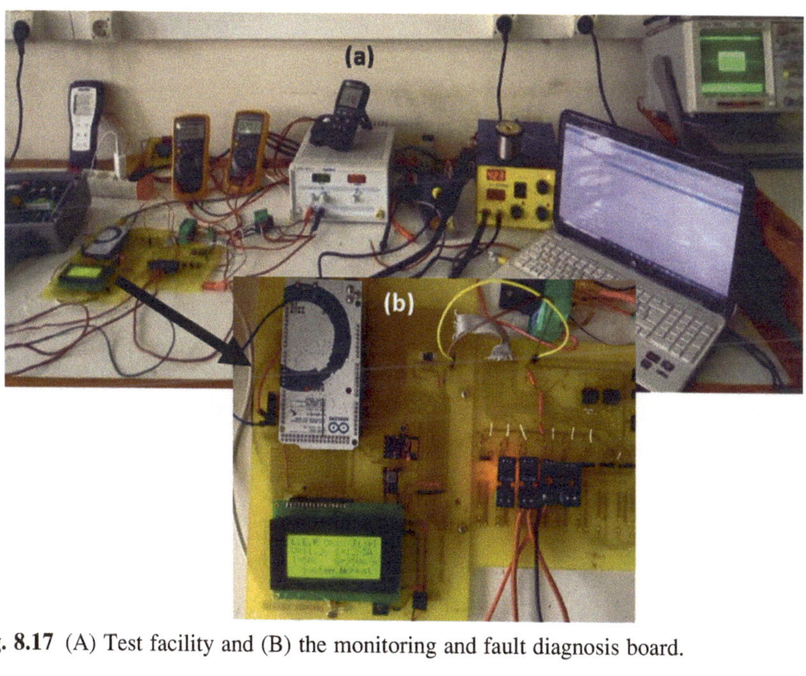

Fig. 8.17 (A) Test facility and (B) the monitoring and fault diagnosis board.

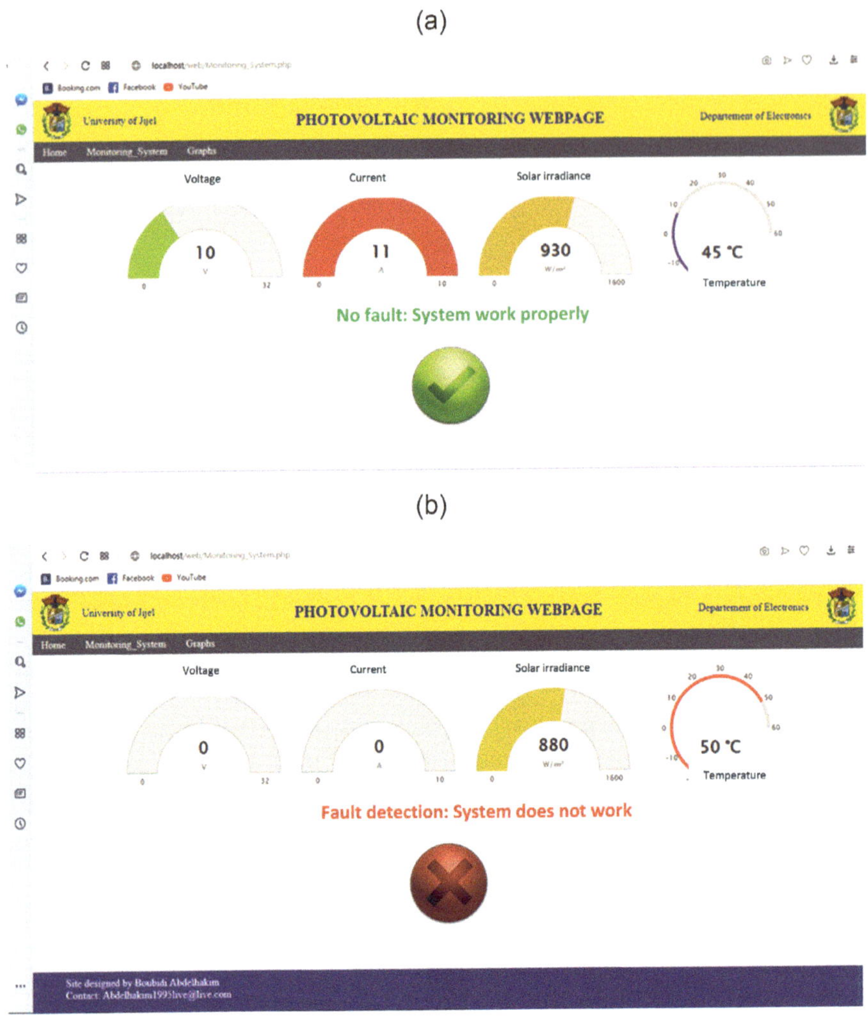

Fig. 8.18 Monitored data posted on the website: (A) normal PV system and (B) faulty PV system.

disconnected system with $V_{pv} = 0$ and $I_{pv} = 0$. This can be seen clearly on the posted PV voltage and current in Fig. 8.19A and B, respectively.

Another example of monitored data is shown in Fig. 8.20. In this case, a fault is detected and indicated on the webpage, but this fault cannot be identified by the method. The system classifies this as "other fault detection" (it could be dirty PV module, or shading effect, etc.).

In these experiments, the possibility to design smart PV monitoring systems is shown based on the IoT technique; the first prototype (**Example 8.1**) was employed and helped users to monitor their PV installations online by supervision of the most important parameters like current, voltage, air temperature, and solar irradiance.

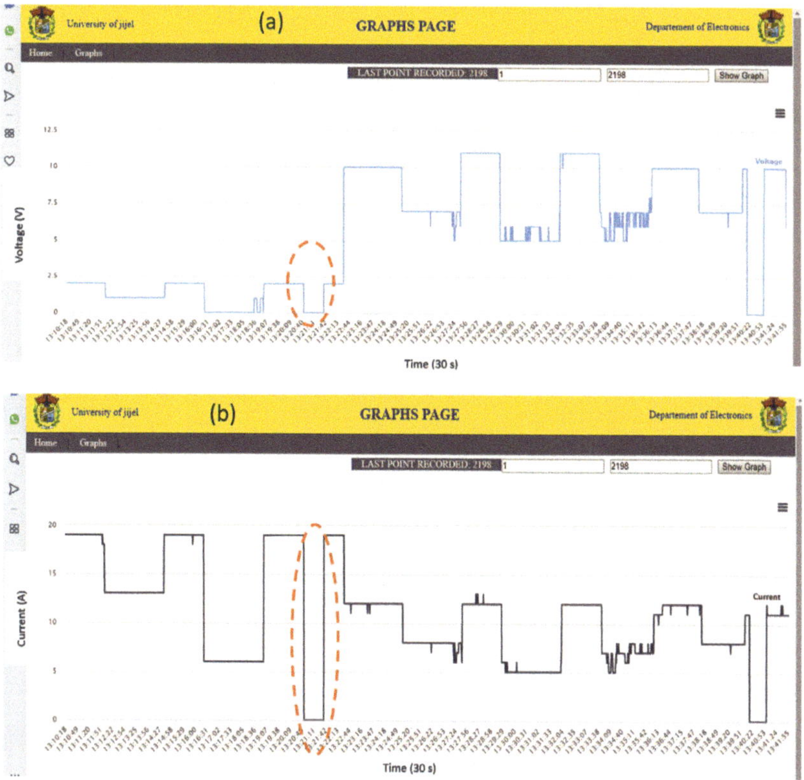

Fig. 8.19 (A) PV current and (B) PV voltage posted on the website for online monitoring.

The second prototype (**Example 8.2**) has another option, which is the possibility to detect and identify three types of faults that can occur on the PV modules (DC side). Other type of faults could be added in the procedure and integrated in the prototype.

Example 8.3: Machine learning for PV fault diagnosis using Raspberry Pi 4

The above examples show the capability of Arduino Mega 2560 board for PV monitoring and fault detection, as well as fault identification of a very small PV array. However, this device is not appropriate for the application of machine learning or deep learning methods due to its limited resources. For this reason, in this example a Raspberry Pi 4 is employed, so monitoring of data are required (I_{pv}, V_{pv}, G, and T_C: cell temperature) and these can be used to implement a PV fault detection and classification method based on machine learning algorithms. Thus, the decision tree classifier is chosen for the detection procedure (see Chapter 6, **Example 6.4**); the detection accuracy was 98%. Regarding the classification process, the random forest classifier is selected; an example is shown in Chapter 6, **Example 6.8**; the classification accuracy was 88%.

The block diagram is shown in Fig. 8.21, and it consists of a Raspberry Pi 4, a PV array, and sensors.

The algorithms (detection and classification) have been trained and tested online using Google Colab. Fig. 8.22 shows an example of the classification procedure based on the random forest algorithm.

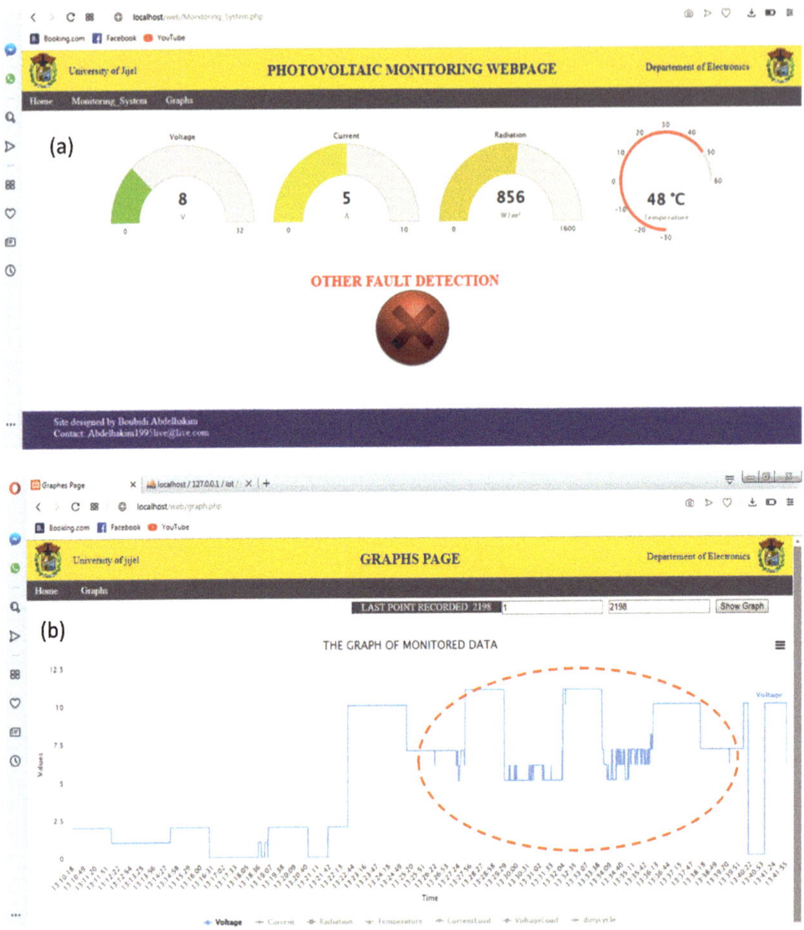

Fig. 8.20 (A) Monitored data and fault identification and (B) PV voltage posted on the website for online monitoring.

A one-diode model available in PVLib [56] is used to estimate the *I-V* curve based on the measured T_C and G (see Fig. 8.23), with a simple algorithm for parameters' extraction (I_{mp}, V_{mp}, I_{sc}, V_{oc}, P_{mp}, and FF). These features (I_{mp}/I_{sc}, V_{mp}/V_{oc}, $P_{mp}/P_{m_{p\_stc}}$, and FF) are used as input to the fault detection (input of the classifier). The binary classifier provides two value "0" or "1." If "1" is provided it means that there is a fault. Then, the fault classification is activated; this also accepts as input the extracted features with G and T_C and then predicts the class of fault (e.g., class 1 means partial shading, class 2 means dust accumulation, etc.). These algorithms are then implemented inside the Raspberry Pi 4.

Fig. 8.24 shows the real photo of the fault diagnosis prototype based on ML algorithms, IoT technique, and Raspberry Pi 4.

The website has been designed in a way to allow users to observe the data parameters monitored in real time. All data are stored using the basic management system of

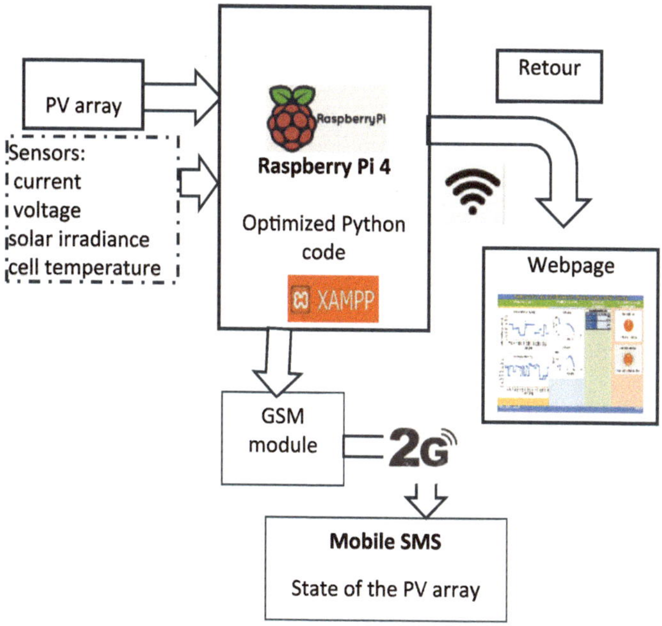

Fig. 8.21 PV remote monitoring system with a simple fault diagnosis method based on machine learning algorithms.

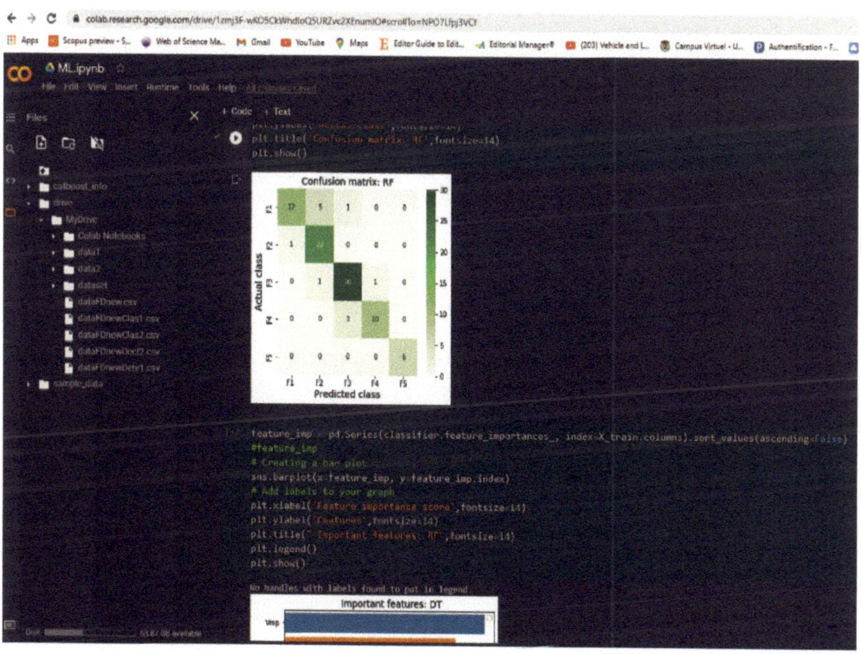

Fig. 8.22 Random forest based-fault classification algorithm implemented in Google Colab.

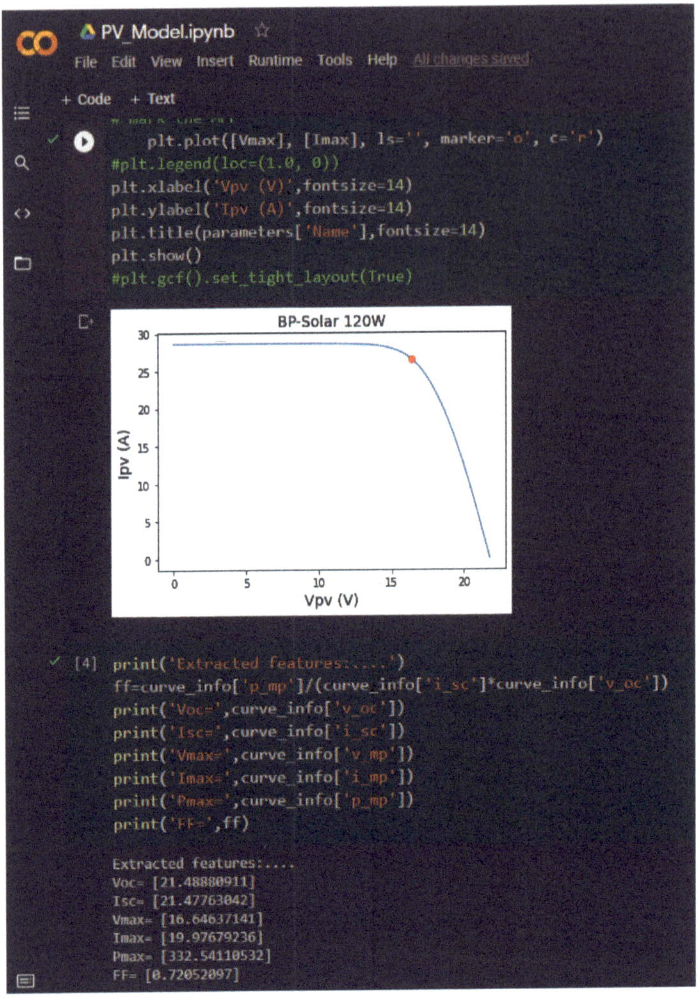

Fig. 8.23 *I-V* curve based on the one-diode model at STC conditions.

MySQL data, which is necessary for the analysis step. Fig. 8.25A shows the monitored data for online fault diagnosis of the PV array.

A mobile application is developed in order to notify the user about the state of the system. Fig. 8.25B shows a mobile SMS sent to the user indicating the state of the system and the type of fault. The system is planned to be executed one time each day.

It can be seen from this webpage, that the system has a problem, based on the plotted *I-V* and PV curves (the curves are distorted); from the circle in the right side of the webpage the type of fault is clearly indicated which in this case is partial shading abnormal condition. It is possible to detect multiple faults belonging to one of the three types of faults investigated here, but this technique is not enough for other new emerging faults. In the case of new faults, the classifier tries to find the closer type which is

Fig. 8.24 Real photo of the designed PV fault diagnosis prototype based on machine learning and the IoT technique.

not always true; thus, a large dataset of various faults is needed. For example, shading abnormal conditions can be obtained from various sources, see Fig. 8.26.

Each *I-V* curve has at least two maximum power points; in this situation, the use of machine learning methods is justified in order to make a clear decision.

8.5.2 Implementation of intelligent off-grid photovoltaic system simulator using an FPGA board

An off-grid photovoltaic system consists mainly of a PV array, batteries, charge regulators (DC-DC), and inverters (DC-AC). The system simulated is shown in Fig. 8.27.

Modeling the whole PV system can help designers and users to analyze the performance of their installations. Matlab/Simulink is among the most used environments for this purpose [42,43]; however, analyzing the performance under variable environmental conditions of such installations is relatively complex, due to the complex mathematical equations used in modeling and control of the whole systems, including a PV array, a battery, a DC-DC, and a DC-AC.

An alternative way is based on the use of data, so data-driven approaches based on machine learning (ML) including neural networks (NNs) are the most suitable in such situations [44–46].

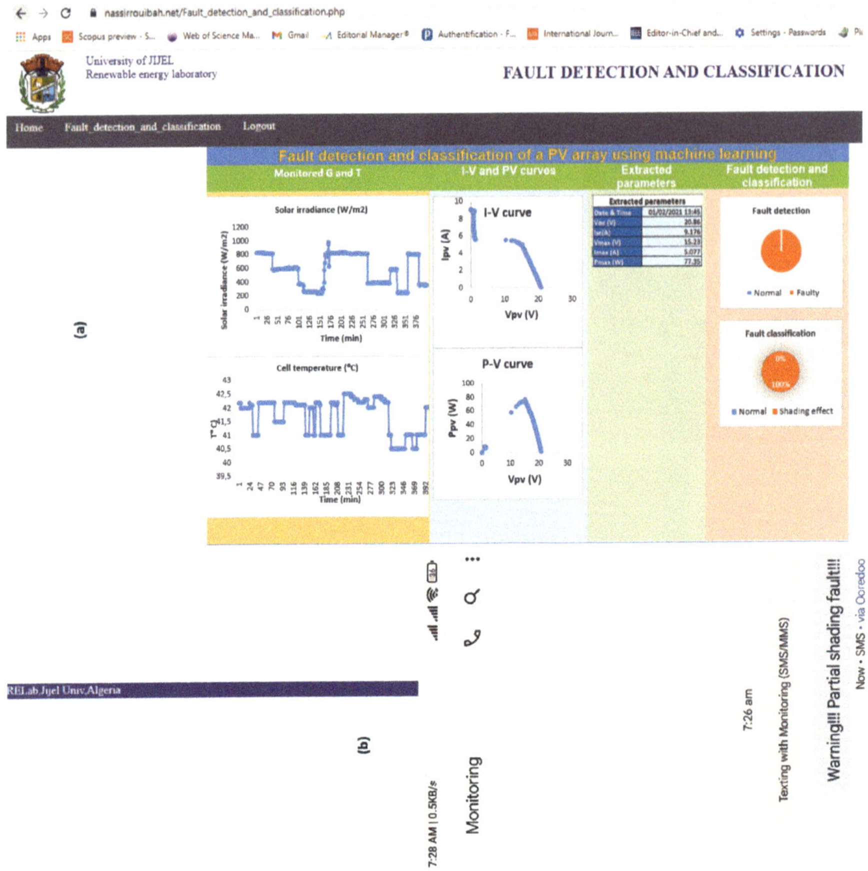

Fig. 8.25 (A) The webpage used for PV fault diagnosis, (B) Message sent to the user indicating the state of the PV system and the type of fault (Warning: shading effect fault).

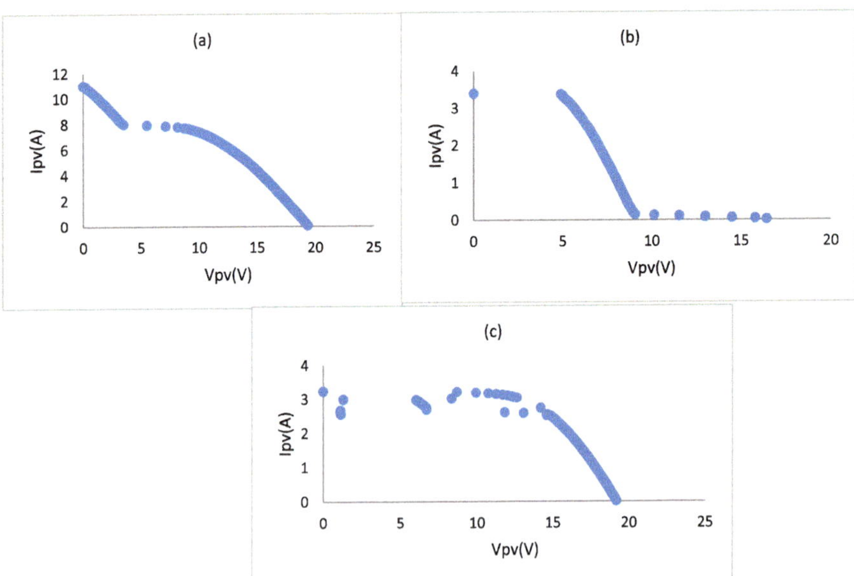

Fig. 8.26 Shading abnormal conditions, (A) artificial shading, (B) bird dropping or sand accumulation, and (C) cloud moving during the measurement of the *I-V* curve.

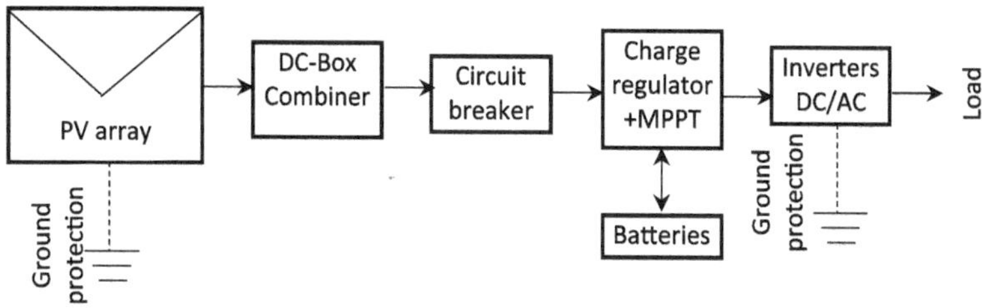

Fig. 8.27 An off-grid photovoltaic system with MPPT and protection devices.

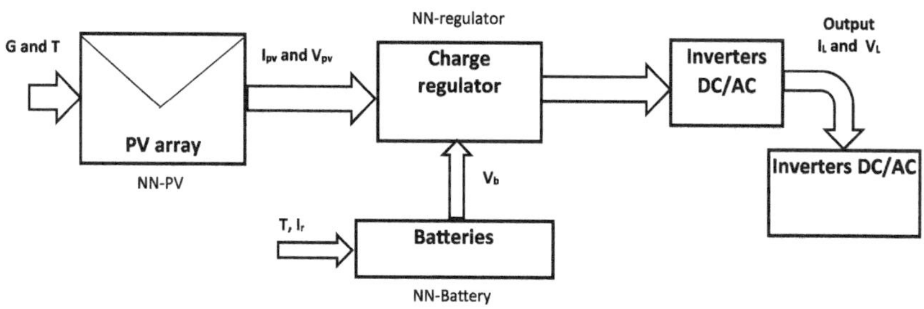

Fig. 8.28 Block diagram of the whole ANN-based off-grid PV system.

Example 8.4: ANN for modeling and simulation of a small-scale PV system

In this example, the whole off-grid PV system, shown in Fig. 8.28, is modeled based on NN. It consists of three MLP-based models [19]:

(a) MLP-PV has as the input G and T_a, and output I_{pv} and V_{pv}.
(b) MLP-battery has as the input T_a and I_r, and output is V_b,
(c) MLP-regulator is accepted as the input I_{pv}, V_{pv}, V_b and output is I_L and V_L.

where T_a is the temperature, I_r is the current from regulator, V_b is the battery voltage, and I_L and V_L are the output current and voltage of the system.

The application of the NN-based model for performance prediction of the important data issue of an off-grid-PV system is first presented, then the optimized NN-PV module, NN-battery, and NN-regulator are simulated and implemented in an FPGA device. To design different NN-based models, first it is needed to run the different NN-models separately and then use the same VHDL-NN, developed in Ref. [20].

The first step is to write the corresponding code of the neuron (basic element of the multilayer perceptron, MLP network). So, the basic element (neuron) in the MLP can be presented by the following diagram of the neuron [20]. It consists mainly of three modules: ROM, MAC, and AF (see Fig. 8.29):

(a) **ROM** is used to store the weights (W_{ij}) and bias (b_i).
(b) **MAC** is a function that allows the calculation of the following sum:

$$Y_k = \sum_{i-1}^{N} X_i W_{ij} \tag{8.1}$$

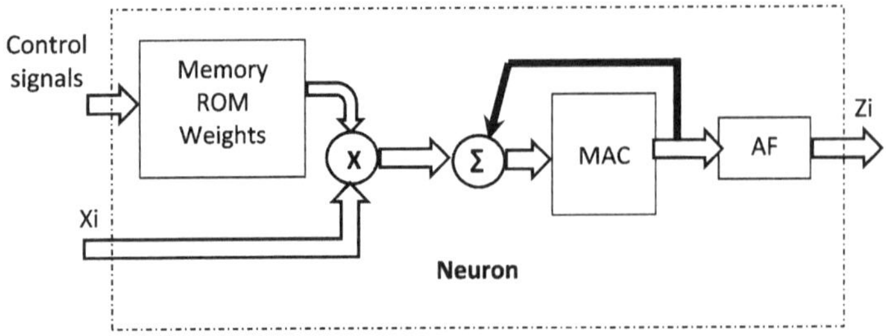

Fig. 8.29 A basic schematic representation of the elementary neuron in the MLP network structure.

(c) **AF** is the activation function; for this example, the *Tansig* function is selected which is given as

$$Z_i = \frac{2}{1+\exp(-2Y_k)} - 1 \tag{8.2}$$

It should be pointed out that the MLP-based implementation of a PV module in a reconfigurable FPGA has some benefits: (1) design a miniature intelligent PV module, (2) real-time performance evaluation, and (3) involves less computational efforts.

The VHDL codes of the above modules are given in **Listing 8.4**.

```
--Listing 8.4a: MAC function
library IEEE;
use IEEE.STD_LOGIC_1164.ALL;
use IEEE.STD_LOGIC_ARITH.ALL;
use IEEE.STD_LOGIC_SIGNED.ALL;
entity MAC is
Port ( X,Y,Z : instd_logic_vector(17 downto 0); A : out std_logic_vector(35
  downto 0); CK,CE,LOAD : in std_logic);
end MAC;
architecture Behavioral of MAC is
signal ACC:std_logic_vector(35 downto 0);
signal XY :std_logic_vector(35 downto 0);
signal Z0 :std_logic_vector(35 downto 0);
begin
Z0<=Z(17)&Z(17)&Z(17)&Z(17)&Z(17)&Z(17)&Z(17)&Z(17)&Z(17)
  &Z&'0'&X"00";
XY<=X*Y;
process(ck,ce,Z0,XY)
begin
        if falling_edge(CK)and (CE='1') then
                if (LOAD='1') then
                        ACC<= Z0;
                else
                        ACC<=ACC+XY;
```

```
            end if;
     else
                 ACC<=ACC;
     end if;
end process;
A<=ACC;
end Behavioral;
```

```
--Listing 8.4b: Activation
function
library IEEE;
use IEEE.STD_LOGIC_1164.
ALL;
use IEEE.STD_LOGIC_ARITH.
ALL;
use IEEE.STD_LOGIC_SIGNED.
ALL;
entity seg_function is
Port ( X : in
std_logic_vector(17 downto
0);
Y : out
std_logic_vector(17 downto
0);
        en_seg : in
std_logic);
end seg_function;
--Segment
architecture Behavioral of
seg_function is
constant a0:integer:=508;
constant a1:integer:=473;
constant a2:integer:=414;
constant a3:integer:=342;
constant a4:integer:=270;
constant a5:integer:=205;
constant a6:integer:=152;
constant a7:integer:=110;
constant a8:integer:=78;
constant a9:integer:=55;
constant aa:integer:=39;
constant ab:integer:=27;
constant ac:integer:=19;
constant ad:integer:=13;
constant ae:integer:=9;
constant af:integer:=5;
begin
abs_X<=abs(X);
-- when X(17)='1' else abs_X<=X;
cmp00:process(abs_X)
begin
if (abs_x>= X0)and (abs_x< X1) then
a<=CONV_std_logic_vector(a0,18);
b<=CONV_std_logic_vector(b0,18);
elsif (abs_x< X2) then
a<=CONV_std_logic_vector(a1,18);
b<=CONV_std_logic_vector(b1,18);
elsif (abs_x< X3) then
a<=CONV_std_logic_vector(a2,18);
b<=CONV_std_logic_vector(b2,18);
elsif (abs_x< X4) then
a<=CONV_std_logic_vector(a3,18);
b<=CONV_std_logic_vector(b3,18);
elsif (abs_x< X5) then
a<=CONV_std_logic_vector(a4,18);
b<=CONV_std_logic_vector(b4,18);
elsif (abs_x< X6) then
a<=CONV_std_logic_vector(a5,18);
b<=CONV_std_logic_vector(b5,18);
elsif (abs_x< X7) then
a<=CONV_std_logic_vector(a6,18);
b<=CONV_std_logic_vector(b6,18);
elsif (abs_x< X8) then
a<=CONV_std_logic_vector(a7,18);
b<=CONV_std_logic_vector(b7,18);
elsif (abs_x< X9) then
a<=CONV_std_logic_vector(a8,18);
b<=CONV_std_logic_vector(b8,18);
elsif (abs_x< XA) then
a<=CONV_std_logic_vector(a9,18);
b<=CONV_std_logic_vector(b9,18);
elsif (abs_x< XB) then
a<=CONV_std_logic_vector(aa,18);
```

```
-Bias
constant b0:integer:=0;
constant b1:integer:=7;
constant b2:integer:=29;
constant b3:integer:=70;
constant b4:integer:=124;
constant b5:integer:=185;
constant b6:integer:=244;
constant b7:integer:=299;
constant b8:integer:=347;
constant b9:integer:=386;
constant ba:integer:=416;
constant bb:integer:=441;
constant bc:integer:=459;
constant bd:integer:=473;
constant be:integer:=483;
constant bf:integer:=495;
-Interval limits
constant x0:integer:=0;
constant x1:integer:=96;
constant x2:integer:=192;
constant x3:integer:=288;
constant x4:integer:=384;
constant x5:integer:=480;
constant x6:integer:=576;
constant x7:integer:=672;
constant x8:integer:=768;
constant x9:integer:=864;
constant xa:integer:=960;
constant xb:integer:=1056;
constant xc:integer:=1152;
constant xd:integer:=1248;
constant xe:integer:=1344;
constant xf:integer:=1440;
constant x10:integer:
=1792;
signal Abs_X:std_logic
_vector(17 downto 0);
signal LATCH:std_logic
_vector(17 downto 0);
signal Y_S:std_logic
_vector(17 downto 0);
signal Abs_Y:std_logic
_vector(35 downto 0);
signal a,b:std_logic
_vector(17 downto 0);
signal bx :std_logic
_vector(26 downto 0);
```

```
b<=CONV_std_logic_vector(ba,18);
elsif (abs_x< XC) then
a<=CONV_std_logic_vector(ab,18);
b<=CONV_std_logic_vector(bb,18);
elsif (abs_x< XD) then
a<=CONV_std_logic_vector(ac,18);
b<=CONV_std_logic_vector(bc,18);
elsif (abs_x< XE) then
a<=CONV_std_logic_vector(ad,18);
b<=CONV_std_logic_vector(bd,18);
elsif (abs_x< XF) then
a<=CONV_std_logic_vector(ae,18);
b<=CONV_std_logic_vector(be,18);
elsif (abs_x< X10)then
a<=CONV_std_logic_vector(af,18);
b<=CONV_std_logic_vector(bf,18);
else
a<=(others=>'0');
b<=CONV_std_logic_vector(512,18);
end if;
end process;
bx(8 downto 0)<=X"00"&"0";
bx(26 downto 9)<=b;
ABS_Y<=a*abs_X+bx;
Sym_out:process(x,ABS_Y)
begin
if x(17)='1' then
Y_S<="00"&X"0000"-abs_Y( 26 downto
9);
else
Y_S<=abs_Y( 26 downto 9);
end if;
end process;
LATCHED_0: LATCH<=Y_S when
(en_Seg='1') else LATCH;
LATCHED_OUT: Y<=LATCH;
end Behavioral;
```

```
--Listing 8.4c: ROM functions
library IEEE;
use IEEE.STD_LOGIC_1164.ALL;
use IEEE.STD_LOGIC_ARITH.ALL;
use IEEE.STD_LOGIC_SIGNED.ALL;
entity ROM_3_X is
generic (constant x0:integer:
=10;
constant x1:integer:=20;
constant x2:integer:=20;

constant Size: Natural:=32);
port (adr: in std_logic_vector(1
downto 0);
d: out std_logic_vector(Size-1
downto 0));
end ROM_3_X;

architecture Behavioral of
ROM_3_X is
begin
process ( adr)
begin
case adr is
when "00" =>
d<=CONV_std_logic_vector(x0,
Size);
when "01" =>
d<=CONV_std_logic_vector(x1,
Size);
when "10" =>
d<=CONV_std_logic_vector(x2,
Size);
when others => d<=(others=>'0') ;
end case;
end process;
end Behavioral;

library IEEE;
use IEEE.STD_LOGIC_1164.ALL;
use IEEE.STD_LOGIC_ARITH.ALL;
use IEEE.STD_LOGIC_SIGNED.ALL;

entity ROM_7_X is
generic (constant x0:integer:
=10;
constant x1:integer:=20;
when "010" =>
d<=CONV_std_logic_vector(x2,
Size);
when "011" =>
d<=CONV_std_logic_vector(x3,
Size);
when "100" =>
d<=CONV_std_logic_vector(x4,
Size);
when "101" =>
d<=CONV_std_logic_vector(x5,
Size);
when "110" =>
d<=CONV_std_logic_vector(x6,
Size);
when others =>d<=(others=>'0') ;
end case;
end process;
end Behavioral;
___
library IEEE;
use IEEE.STD_LOGIC_1164.ALL;
use IEEE.STD_LOGIC_ARITH.ALL;
use IEEE.STD_LOGIC_SIGNED.ALL;
entity ROM_12_X is
generic (constant x0:integer:
=10;
constant x1:integer:=20;
constant x2:integer:=10;
constant x3:integer:=20;
constant x4:integer:=10;
constant x5:integer:=20;
constant x6:integer:=10;
constant x7:integer:=20;
constant x8:integer:=10;
constant x9:integer:=10;
constant xA:integer:=10;
constant xB:integer:=10;
constant Size: Natural:=32);
port(adr: in std_logic_vector(3
downto 0);
d: out std_logic_vector(Size-1
downto 0));
end ROM_12_X;
architecture Behavioral of
ROM_12_X is
```

```
constant x2:integer:=10;
constant x3:integer:=20;
constant x4:integer:=10;
constant x5:integer:=20;
constant x6:integer:=10;
constant Size: Natural:=32);
port(adr: in std_logic_vector(2
downto 0);
d: out std_logic_vector(Size-1
downto 0));
end ROM_7_X;
architecture Behavioral of
ROM_7_X is
begin
process ( adr)
begin
case adr is
when "000" =>
d<=CONV_std_logic_vector(x0,
Size);
when "001" =>
d<=CONV_std_logic_vector(x1,
Size);
```

```
begin
process ( adr)
begin
case adr is
when X"0" =>
d<=CONV_std_logic_vector(x0,
Size);
when X"1" =>
d<=CONV_std_logic_vector(x1,
Size);
when X"2" =>
d<=CONV_std_logic_vector(x2,
Size);
when X"3" =>
d<=CONV_std_logic_vector(x3,
Size);
when X"4" =>
d<=CONV_std_logic_vector(x4,
Size);
when X"5" =>
d<=CONV_std_logic_vector(x5,
Size);
when X"6" =>
d<=CONV_std_logic_vector(x6,
Size);
when X"7" =>
d<=CONV_std_logic_vector(x7,
Size);
when X"8" =>
d<=CONV_std_logic_vector(x8,
Size);
when X"9" =>
d<=CONV_std_logic_vector(x9,
Size);
when X"A" =>
d<=CONV_std_logic_vector(xA,
Size);
when X"B" =>
d<=CONV_std_logic_vector(xB,
Size);
when others => d<=(others=>'0') ;
end case;
end process;
end Behavioral;
```

A dataset of different data taken from the grid-off installation is used to train and validate the whole system including MLP-PV, MLP-battery, and MLP-regulator. The size of the dataset is 1825; 1460 samples have been used to train different MLP-based

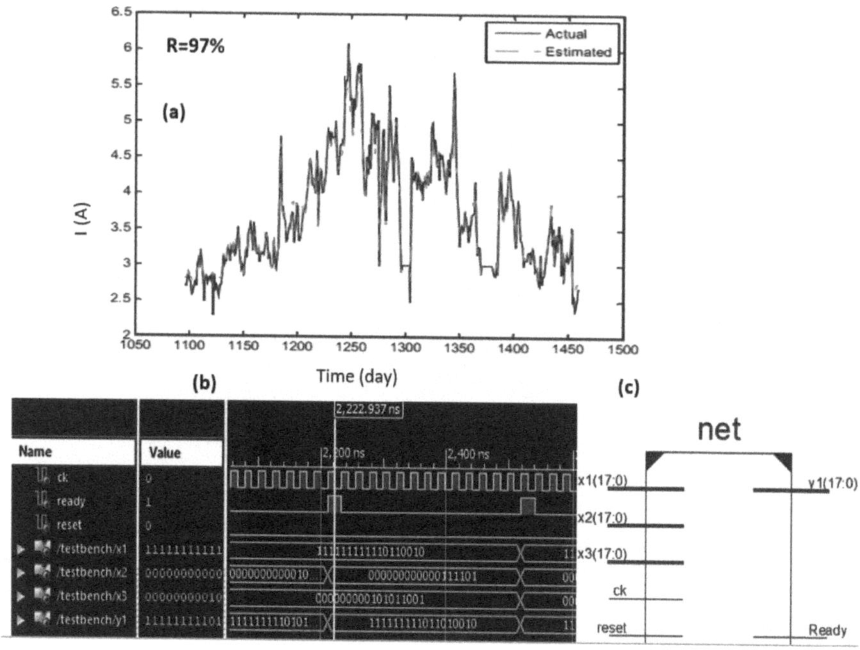

Fig. 8.30 (A) Measured versus estimated output current (I_L) based on the MLP model, (B) output current (I_L) simulation under an ISE simulator ($x1 = I_{pv}$, $x2 = V_{pv}$, $x3 = V_b$ and $y1 = I_L$), and (C) the RTL of the designed MLP-regulator.

models and 365 samples have been used to validate the MLP-based models. Fig. 8.30A shows a comparison between the measured and estimated regulator current by using the developed MLP network [8]. The correlation coefficient is 97%, which is satisfactory.

As can be seen from the results shown in Fig. 8.30A, these are very close to the ones produced by the MLP-based regulator model, which confirm the good accuracy of the developed VHDL code. Fig. 8.30A and B shows the simulated output current (I_L) under the ISE simulator and the generated register transfer level (RTL).

The last step concerns the implementation and download of the whole code in a reconfigurable FPGA device for eventual applications in this area (see Fig. 8.31). Table 8.2 shows the resources used by FPGA (Spartan 3E board).

In this example, the capability of a simple MLP neural network for behavior simulation of an off-grid-PV system (1.8 kW) is shown. It should be noted that the designed ANN-based modeling and simulation of an off-grid PV system is for a specific location and PV technology. Some future direction can be summarized as follows:

- Designing of an online ANN model for real-time verification (using a variety of datasets).
- Studying the capability generalization of this approach for large-scale PV systems.
- Improve the prediction performance by using deep learning NNs (deep convolutional neural networks).

Fig. 8.31 The VHDL-uploaded code in a Xilinx Spartan 3E FPGA.

Table 8.2 FPGA device utilization summary.

Logic utilization	Used	Available	Utilization
Total number slice registers	1083	71,680	1%
Number of used FFs	741	–	–
Number of used as latches	342	–	–
Total number of 4 LUT units	5555	71,680	7%
Number of occupied slices	2808	35,840	7%
Number of slices containing only related logic	2808	2808	100%
Number of slices containing unrelated logic	0	2808	0%
Number used logic	5347	–	–
Number used as a route-thru	208	–	–
Number of bonded IOBs	75	768	9%
IOB latches	18	–	–
Number of BUFG/BUFGCTRLS	2	32	6%
Number of DSP48s	39	80	48%
Number used as BUFGs	2	–	–
Average Fanout of Non-Clock Nets	3.07	–	–

8.5.3 Implementation of an intelligent MPPT algorithms using an FPGA board

Over the last decade, numerous maximum power point tracking (MPPT) systems based on AI techniques were developed. The main MPPT-AI techniques designed are fuzzy logic (FL) and neural networks (NN) [8].

Example 8.5: Tracking the MPPT using the FL controller

In this example, the application of FL-MPPT is described in detail as well as its implementation in a reprogrammable FPGA board. Fig. 8.32 shows the structure of the MPPT-based fuzzy logic controller (FLC).

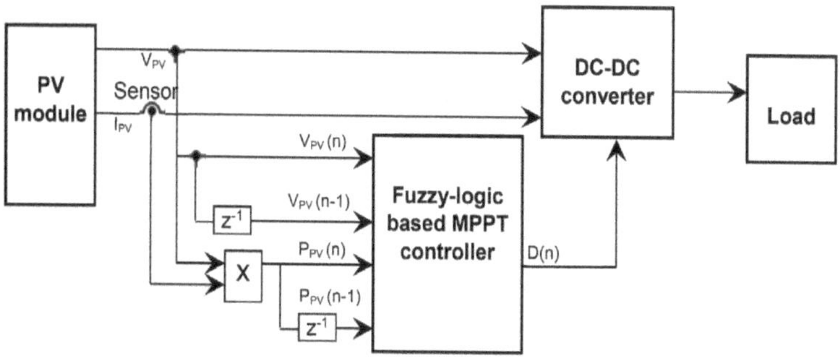

Fig. 8.32 MPPT-based fuzzy logic scheme.

It consists of a PV module, a voltage sensor, a current sensor, a DC-DC boost converter, and a resistive load. As can be seen from Fig. 8.32, the FL controller (FLC) has as input the actual measured $V_p(n)$, PV(n), the corresponding previous values of both amount $V_{pv}(n-1)$ and $P_{pv}(n-1)$, and as output the duty cycle (D), which can be used to command the DC-DC converter switch (e.g., Transistor). Mathematical equations for the calculation of error (E), the change of the error (ΔE), and D are presented in Chapter 5. Five linguistic variables (NB, NS, ZE, PS, and PB) have been adopted for input and output variables (see Chapter 5). Fig. 8.33 depicts the block diagram of the implemented MPPT-based FLC in the FPGA board. As can be seen, the MPPT includes four principal units:

- fuzzification unit, which converts a crisp input in a fuzzy term set
- a rule selector unit, which stores fuzzy rules describing how the fuzzy system performs
- an inference engine unit, which performs approximate reasoning by associating input variables with fuzzy rules, and finally
- the defuzzification unit, which converts the FLC's fuzzy output to a crisp value representing the control action

The MPPT-based fuzzy logic code is showing in **Listing 8.5**. It consists of three VHDL modules: Sequencer, Controller, and the PWM signal.

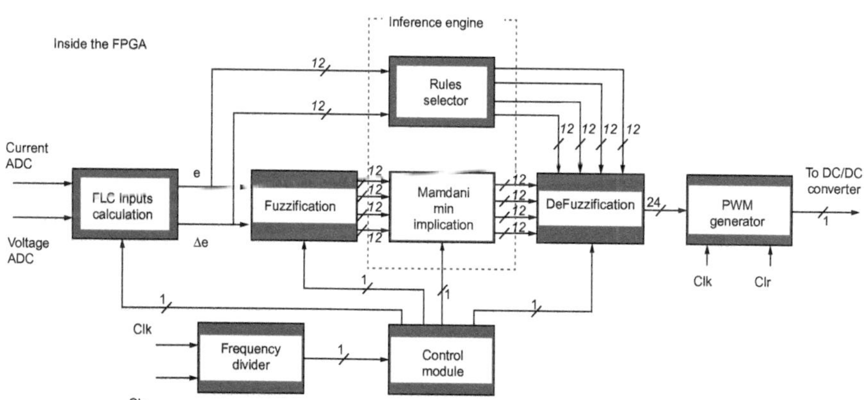

Fig. 8.33 Block diagram of the FPGA implemented MPPT-based FL.

```
--Listing 8.5a: Sequencer
library IEEE;
use IEEE.STD_LOGIC_1164.ALL;
use IEEE.STD_LOGIC_ARITH.ALL;
use IEEE.STD_LOGIC_SIGNED.ALL;
entity sequencer is
port ( clk : in std_logic; Reset : in std_logic;
       Contr_bus : out std_logic_vector(4 downto 0)
attribute FSM_EXTRACT : string;
attribute FSM_EXTRACT of sequenceur : entity is "YES";
attribute FSM_ENCODING : string;
attribute FSM_ENCODING of sequenceur : entity is "ONE-HOT";
end sequencer;
architecture Behavioral of sequencer is
type TState is
(S0,S1,S2,S3,S4,S5,S6,S7,S8,S9,S10,S11,S12,S13,S14,S15,S16,S17,S18,
S19);
signal Curente_State:TState;
signal Next_State:TState;
begin
proc_seq:process(clk,Reset)
begin
if rising_edge(clk) then
     if (Reset='1')then
           Curente_State<=S0;
     else
      Curente_State<=Next_State;
     end if;
end if;
end process;
proc_com:process(Curente_State)
begin
case Curente_State is
      when S0 =>Next_State<=S1;
      when S1 =>Next_State<=S2;
      when S2 =>Next_State<=S3;
      when S3 =>Next_State<=S4;
      when S4 =>Next_State<=S5;
      when S5 =>Next_State<=S6;
      when S6 =>Next_State<=S7;
      when S7 =>Next_State<=S8;
      when S8 =>Next_State<=S9;
      when S9 =>Next_State<=S10;
      when S10=>Next_State<=S11;
      when S11=>Next_State<=S12;
      when S12=>Next_State<=S13;
      when S13=>Next_State<=S14;
      when S14=>Next_State<=S15;
```

```
        when S15=>Next_State<=S16;
        when S16=>Next_State<=S17;
        when S17=>Next_State<=S18;
        when S18=>Next_State<=S19;
        when S19=>Next_State<=S0;
        when others=>Next_State<=S0;
end case;
end process;
with Curente_State select
Cont_bus <= "00000" when S0,
"00000" when S1,
"00000" when S2,
"00000" when S3,
"10000" when S4,
"10000" when S5,
"01000" when S7,
"01000" when S8,
"00000" when S9,
"00100" when S10,
"00100" when S11,
"00000" when S12,
"00010" when S13,
"00010" when S14,
"00000" when S15,
"00001" when S16,
"00001" when S17,
"00000" when S18,
"00000" when S19,
"——" when others;
end Behavioral;
```

```
--Listing 8.5b Controller
library IEEE;
use IEEE.STD_LOGIC_1164.ALL;
use IEEE.STD_LOGIC_ARITH.ALL;
use IEEE.STD_LOGIC_SIGNED.ALL;
entity Controleur is
port(Erreur : IN std_logic_vector(11 downto 0); CErreur : IN
std_logic_vector(11 downto 0); en_fuzzf : IN std_logic; en_inferf : IN
std_logic; en_inpf : IN std_logic; en_compf : IN std_logic; en_somaf : IN
std_logic;     en_prodaf : IN std_logic;     en_divf : IN
std_logic; clk : IN std_logic; reset : IN std_logic; Crispf : OUT
std_logic_vector(23 downto 0); reste : OUT std_logic_vector(15 downto 0);
Out_ready : OUT std_logic);
end Controleur;
```

```
architecture Behavioral of Controleur is
signal Erreur_var : std_logic_vector(11 downto 0);
 signal CErreur_var : std_logic_vector(11 downto 0);
signal res_sopf_var : std_logic_vector(23 downto 0);
signal res_sumf_var : std_logic_vector(15 downto 0);
signal en_inpf_var : std_logic;
signal en_fuzzf_var : std_logic;
signal en_inferf_var : std_logic;
signal en_compf_var : std_logic;
signal en_somaf_var : std_logic;
signal en_prodaf_var : std_logic;
signal en_divf_var : std_logic;
signal Crispf_var : std_logic_vector(23 downto 0);
signal reste_var : std_logic_vector(15 downto 0);
signal Out_ready_var : std_logic;

component FLC_S_DIV is
Port ( Errs : in std_logic_vector(11 downto 0); CErrs : in
std_logic_vector(11 downto 0); en_fuzzs : in std_logic; en_infers : in
std_logic; en_inps : in std_logic; en_comps : in std_logic; en_somas : in
std_logic; en_prodas : in std_logic; clk : in std_logic; reset : in
std_logic; res_sops : out std_logic_vector(23 downto 0); res_sumgs : out
std_logic_vector(15 downto 0) );
end component;

component divisa IS
port ( dividend: IN std_logic_VECTOR(23 downto 0); divisor: IN
std_logic_VECTOR(15 downto 0); quot: OUT std_logic_VECTOR(23 downto
0);remd:
OUT std_logic_VECTOR(15 downto 0); clk: IN std_logic; rfd: OUT
std_logic;
sclr: IN std_logic; ce: IN std_logic);
end component;
begin
FLC_S00 : FLC_S_DIV
port map(Erreur_var, CErreur_var, en_fuzzf_var, en_inferf_var,
en_inpf_var , en_compf_var, en_somaf_var, en_prodaf_var, clk, reset,
res_sopf_var, res_sumf_var);
divisa00 : divisa
port map(res_sopf_var, res_sumf_var, Crispf_var, reste_var, clk,
Out_ready_var, reset, en_divf_var);
Erreur_var <= Erreur; CErreur_var <= CErreur; en_inpf_var <= en_inpf;
en_fuzzf_var <= en_fuzzf; en_inferf_var <= en_inferf;
en_compf_var <= en_compf; en_somaf_var <= en_somaf; en_prodaf_var <=
en_prodaf; en_divf_var <= en_divf;
Crispf <= Crispf_var; reste <= reste_var; Out_ready <=
Out_ready_var;
end Behavioral;
```

```
--Listing 8.5c PWM signal
entity PWM_00 is
generic (lim_sup : integer :=1600; lim_inf : integer :=-1600);
port ( Crisp_in : IN std_logic_vector(23 downto 0);clk : IN
std_logic;reset : IN std_logic;PWM : OUT std_logic )
end PWM_00;
architecture Behavioral of PWM_00 is
signal entreel_var:std_logic_vector(23 downto 0);
signal entree2_var:std_logic_vector(23 downto 0);
begin
entreel_var<=Crisp_in;
entree2_var<= (entreel_var + conv_std_logic_vector(lim_sup,12));
process (clk,reset)
variable count : integer := 0;
variable entree : integer;
begin
entree := CONV_INTEGER(entree2_var);
if reset = '1' then
PWM <= '0';
else
if (clk'event and clk = '1') then
if count>=3201 then count := 0;
end if;
entree <= entree2_var;
case entree is
when 0 => PWM <= '0';
when 3200 => PWM <= '1';
when others =>
 if count > entree then PWM <= '0';
       else PWM <= '1';
end if;
ount := count+1;
end case;
end if;
end if;
end process;
end Behavioral;
```

Fig. 8.34 shows the RTL view of the global FL-based MPPT controller. As can be seen, different blocks are needed to implement the FLC.

Simulation results are shown in Fig. 8.35 using the ISE simulator.

As can be seen from Fig. 8.35, the controller's robustness was tested with respect to some rapid changes of irradiance. The system stabilizes after a relatively short time and seeks the maximum power transfer in all operating conditions. Fig. 8.36 shows the designed DC-DC converter for real-time verification of the implemented FL in an FPGA.

At this stage, the designed MPPT-based FLC is ready to be used. The fuzzy controller can also provide acceptable results and requires less memory space and easy implementation compared to the relatively complex P&O or IncCond algorithms.

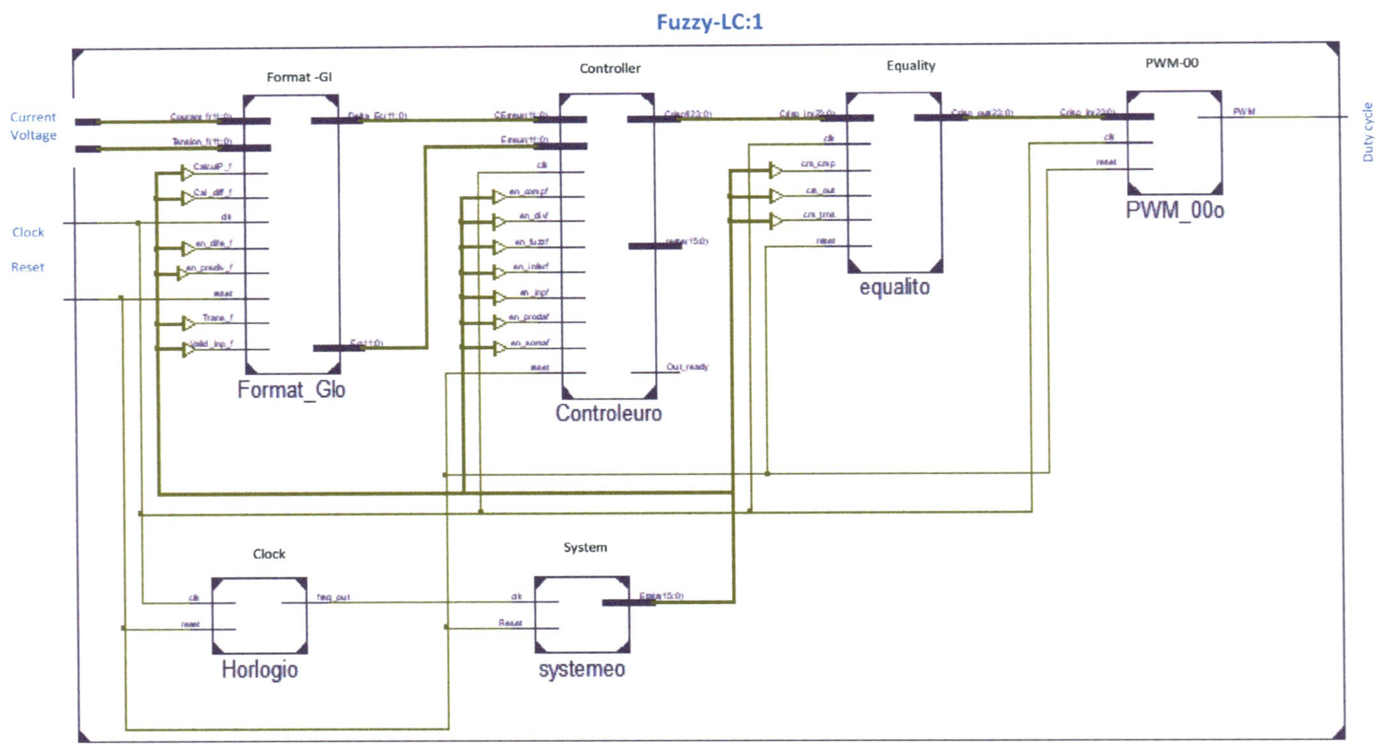

Fig. 8.34 The RTL view of the global FL-based MPPT controller.

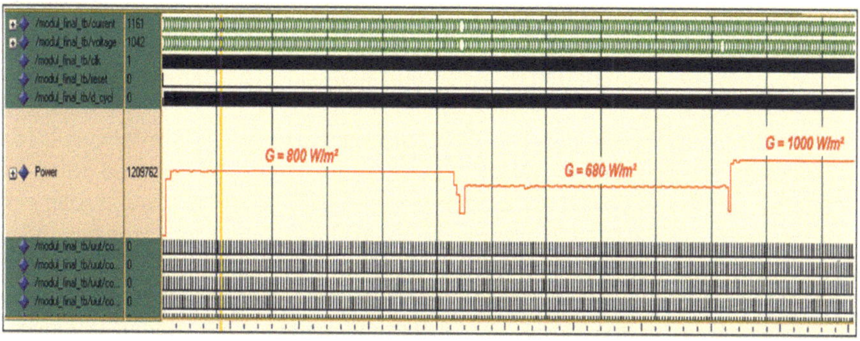

Fig. 8.35 The evolution of the output power of the solar array (W) versus time (ms) for different insolations (800, 680, and 1000 W/m^2, 25°C).

Fig. 8.36 Experimental DC-DC prototype for real-time verification of the designed MPPT-based FLC under an FPGA Spartan 3E.

The MPPT-based FL approach does not require the knowledge of the exact PV model. However, the shortcoming of fuzzy computation is to obtain the correct fuzzy rules and membership functions, which relies heavily on prior knowledge of the system and has an impact on the robustness and reliability of the method. It has been also used in the case of rapid variation of weather conditions, as reported in the literature.

8.5.4 Co-simulation of an intelligent fault diagnosis of a photovoltaic array using XSG

Photovoltaic systems are subject to a variety of failures that can involve all PV components (modules, cabling, protections, converters, and inverters), mainly due to the external operating conditions. Faults in PVS are caused by shading effects, module soiling, inverter failure, and mismatch due to variation in manufacturing or aging of PV modules.

Example 8.6: Co-simulation of a PV fault diagnosis technique using ANNs

This example is motivated by the automatic methods, which are based mainly on the analysis of the *I-V* measured curves to identify and diagnose the possible faults in a PV array. A data-acquisition system is essential in order to collect and store the *I-V* curves, and then based on the measured values of solar irradiance and solar cell temperature the simulated *I-V* curves can be plotted based, for example, on one-diode and two-diode models, Sandia model, or other improved versions.

The main idea of the method based on the *I-V* curves is to compare some significant points or features (e.g., I_{mp}, V_{mp}, I_{sc}, V_{oc}, and the number of MPPs) from the *I-V* curve to identify the type of fault. However, in some cases this method is not able to identify multi-faults and faults with similar faulty *I-V* curves. For example, Fig. 8.37 shows some faulty *I-V* curves. The main causes of faulty *I-V* curves are shown in Fig. 8.38.

Machine learning methods could be used to solve this problem, So, in this example the different steps for designing and implementing the fault detection and diagnosis method based on *I-V* curves and neural networks in an FPGA board (ZYNQ XC7Z010-1CLG400C) are described. The ZYNQ FPGA board was chosen due to its valuable embedded resources for neural networks.

First, the fault detection and diagnosis method is described in the following pseudo-code. It consists of a rules-based algorithm and an ANN particularly used for a group of faults [33].

By C_1, the reduction in the short circuit current is denoted, C_2 the reduction or an increase in the output current, V_1 the reduction in the open circuit voltage, V_2 the reduction or an increase in the output voltage, NP: number of MPPs in the *I-V* curve, and Th_1 and Th_2 are thresholds defined empirically [33].

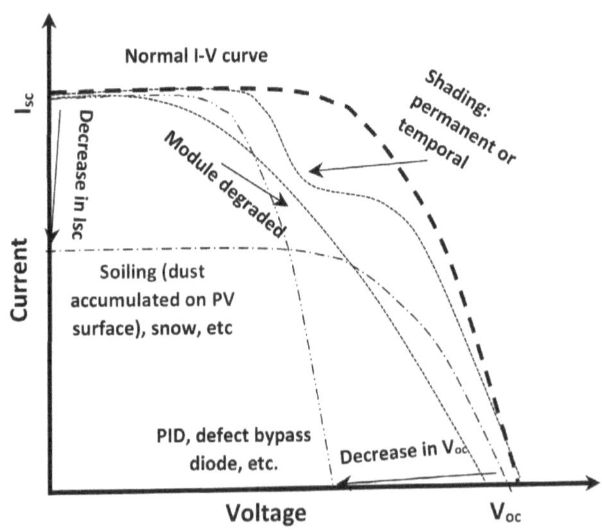

Fig. 8.37 *I-V* curves (Normal: solid line, faulty: dotted line).

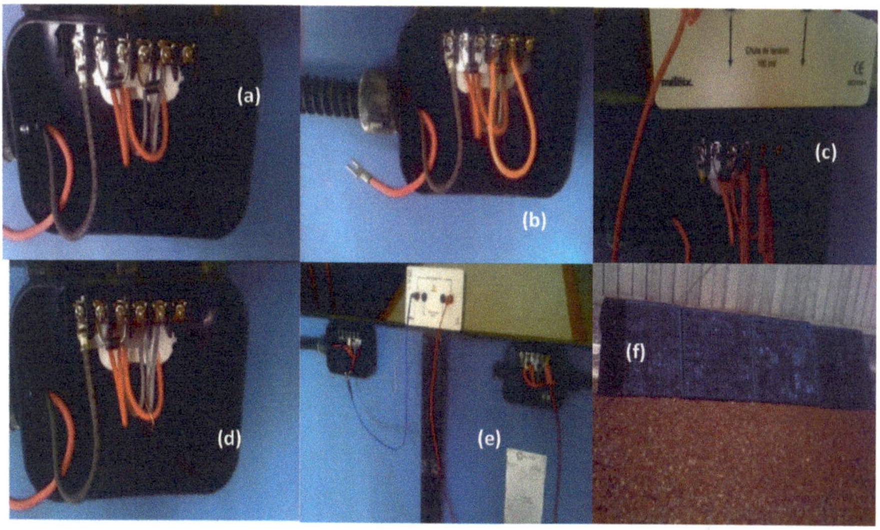

Fig. 8.38 The investigated faults: (A) shunted diode, (B) short-circuit PV module, (C) damaged diode, (D) open-circuit PV modules, (E) connection fault, and (F) partial shading and other combination of faults.

```
Parameters initialization: C1, C2, Th1 and Th2
Calculate of Voc, Isc, Imp and Vmp.
Compare the measured Pm with the calculated Pcal to detect the
fault.
If Pmeas <> Pcalc then scan for the number of MPPs (Np) in the I-V
curve
If Np>1 then,
        If C2>Th1 then
                If V2> Th2 then
                        Write(' Type of fault: Shadow with connection
faults')
                Else
                        Write('Type of fault: Partial shading with
bypass diode faults')
                Endif
        Else
                Write(' Type of fault: Partial shading with bypass
diode faults')
        Endif
Else
        If C1=Isc_meas then
                Write('Type of fault: Open circuited PV module fault')
        Else
                If C1>Th1 then
                        Write('Type of fault: Bypass diode faults');
                Else
```

(Continued)

```
                    If C2>Th1 then
                         If V1>Th2 then
                              Write('group of faults: call for
ANN classification procedure);
                         Else
                              Write('connection fault);
                         Endif
                    Else
                              Write('group of faults: Call for
ANN classification procedure );
                    Endif
               Endif
          Endif
Endif
Endif
End.
```

The faults considered are: open-circuited PV modules (F4), connection fault (F5), partial shadow (bypass diodes work correctly) (F6), partial shadow (faulty bypass diode) (F7), partial shadow with connection fault (F8), and a group of faults including: F1, F2, F3, and F5.

The ANN classification procedure is used to classify the group of four faults that cannot be distinguished by the rules-based algorithm. The well-known MLP network is used; it consists of three neurons in the input layer corresponding to the ratio between the measured and the simulated values: the open-circuit voltage (Rv_{oc}), the maximum power point current (Ri_{mp}), and the maximum power point voltage (Rv_{mp}). Four neurons in the output layer correspond to the fault class (F_i). The number of hidden layers and neurons within the hidden layers are determined during the training process. To do this, an experimental dataset of *I-V* curves (for a string of four PV modules) was collected and used to train and test the MLP classifier. Different faults have been created on the PV string to form the database [33].

After several experiments, the optimal configuration consists of three inputs, two hidden layers within five nodes in the first hidden layer and seven neurons in the second hidden layer, and the output layer has four nodes corresponding to the classified faults (binary classifier) (see Fig. 8.39).

Simulation results indicate that the minimum MSE achieved during the training and test processes are 0.008 and 0.009, respectively. The classification confusion matrix reveals that the correct and the false classification rates obtained during the test process are 90.3% and 97%, respectively [33].

The Xilinx block library provided by Simulink has a large number of building blocks, to generate a variety of functions, such as adders, multipliers, registers, and interfaces between real data and binary encoded data. Manual adjustment of the parameters of each block is essential. For example, the "Gateway in" block, which was used for the binary encoding of real numbers requires the specification of the number of bits used for the binary encoding and the position of the comma, as well

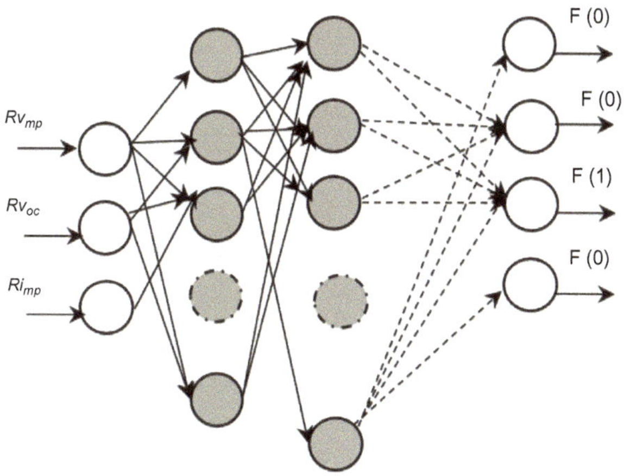

Fig. 8.39 ANN-based fault classification ($3 \times 5 \times 7 \times 4$).

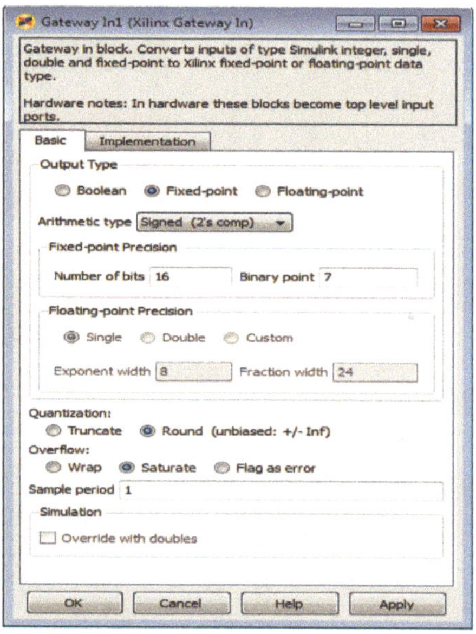

Fig. 8.40 Gateway in block of XSG.

as the choice of the type of encoding (encoding fixed point or floating point coding) as shown in Fig. 8.40.

The next step concerns the implementation of the designed method in the ZYBO (ZYNQ XC7Z010-1CLG400C) FPGA board, using the Xilinx System Generator, for co-simulation and real-time implementation. Fig. 8.41 presents the architecture of the whole system developed under XSG. It consists mainly of five blocks:

- parameter extraction of the measured *I-V* characteristic
- modeling the *I-V* characteristic using translation equations

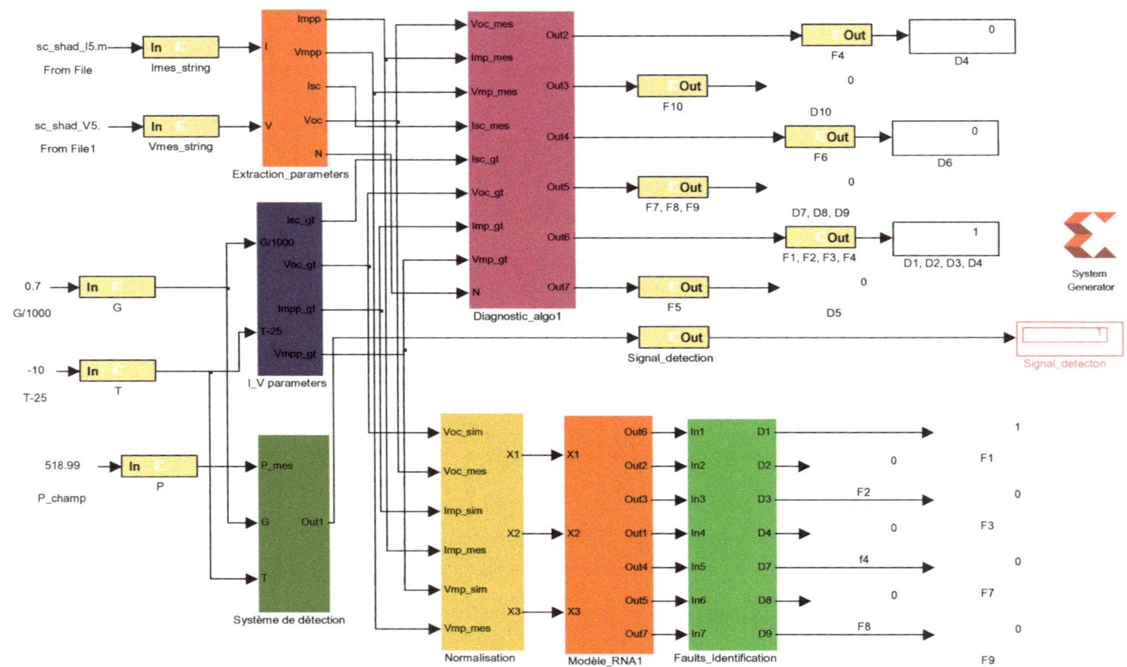

Fig. 8.41 Fault detection and diagnoses system implemented under XSG.

- fault detection procedure
- rule-based algorithm for fault diagnosis
- ANN-based algorithm for fault diagnosis

In these architectures, different building blocks are used such as: "CMulte," "Constant," "Mcode," and others. More details on the modeling of the design stages of this system can be found in Appendix H.

The data used by these blocks such as solar irradiance, temperature, current, and voltage are converted to binary data using the "Gateway in" block; in this application, it is preferred to use the fixed point binary encoding of 16 bits distributed as follows: 7 bits for the digits after the decimal point, 8 bits for the whole part, and 1 bit for the sign. The data presented at the output of these blocks must be reconverted from binary to decimal using the "Gateway out" block.

Fig. 8.42 shows an example of the *I-V* model based on XSG.

Fig. 8.43 displays the XSG configuration of the basic element of neuronal network (neuron).

Once the detection and diagnostic method is designed by Xilinx blocks, a block named "System generator" allows these blocks to be converted in a circuit in the form of a confined IP (Intellectual Property) package for hardware implementation such as the one indicated in Fig. 8.44, which reports the generated Netlist Hdl code. This is then used to generate the BitStream without moving to VHDL.

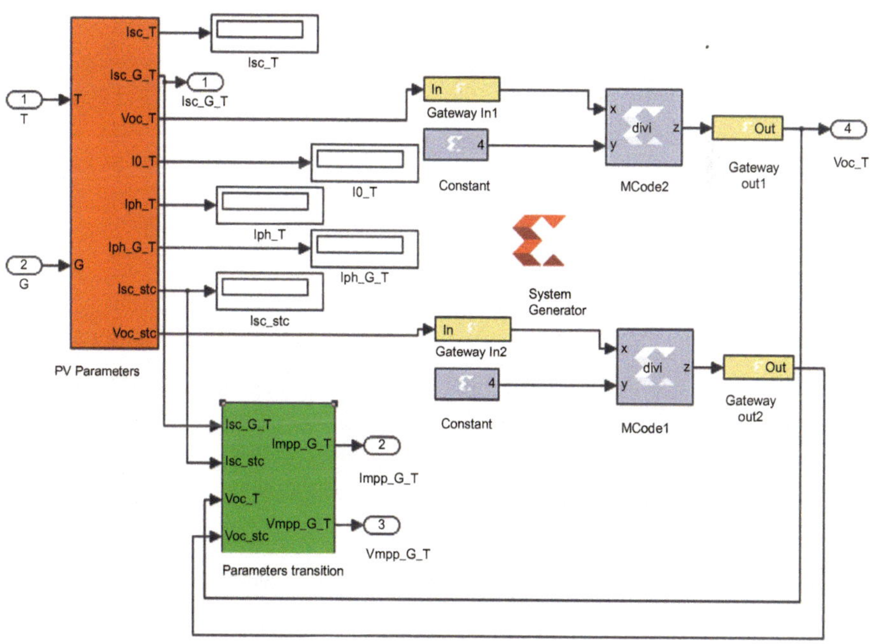

Fig. 8.42 *I-V* model-based developed under XSG.

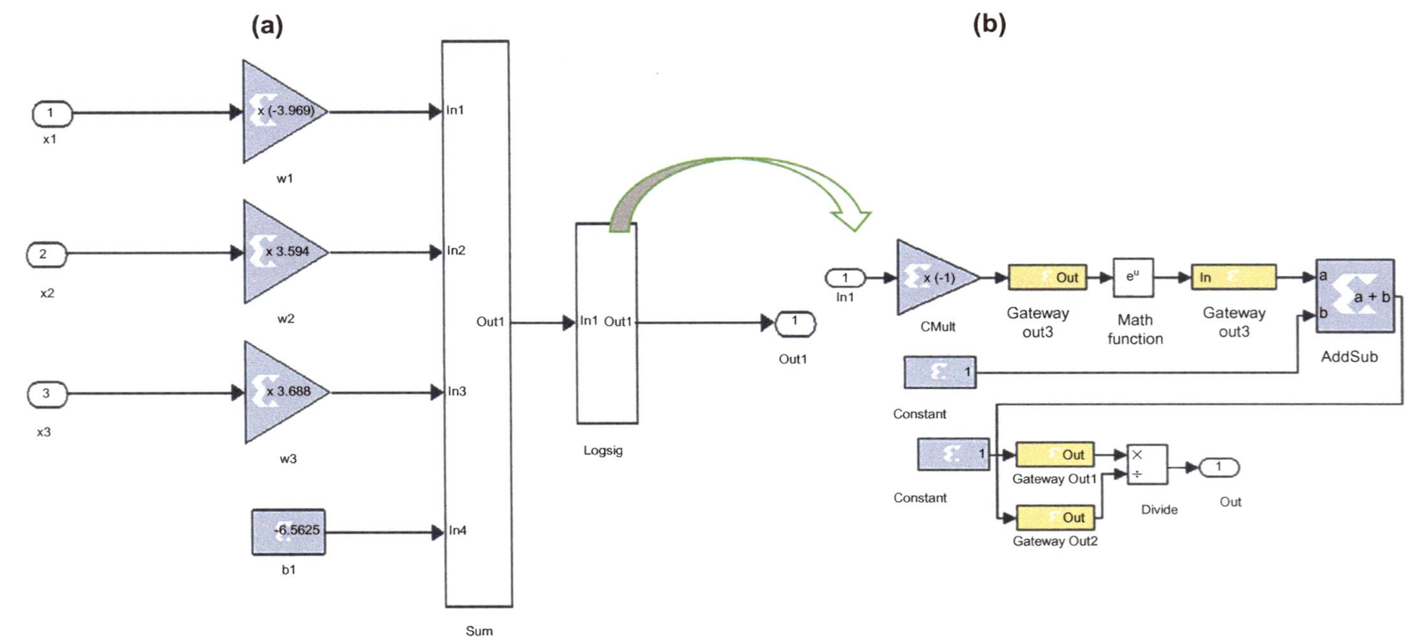

Fig. 8.43 (A) XSG of the basic neuron and (B) XSG of the log-sigmoidal function.

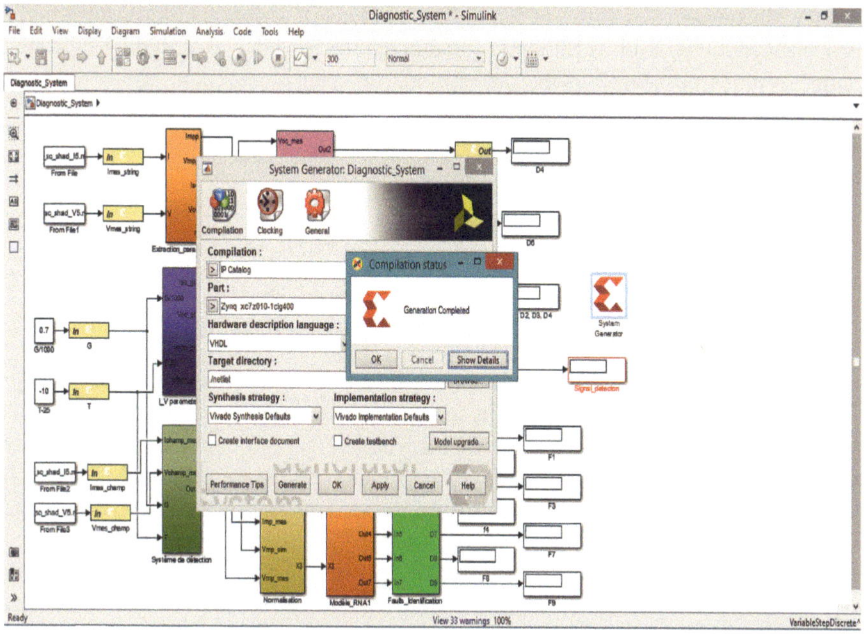

Fig. 8.44 Generated Netlist code of the whole fault detection and diagnosis procedure.

The generated code using XSG Vivado is shown in Fig. 8.45A, while Fig. 8.45B shows the generated BitStream in the FPGA board. Thus, the code is ready to be used and the final step is to verify experimentally the designed fault diagnosis method.

In this example, the possibility of the implementation of an intelligent fault diagnosis method for a PV array in the FPGA board is demonstrated. The FPGA device

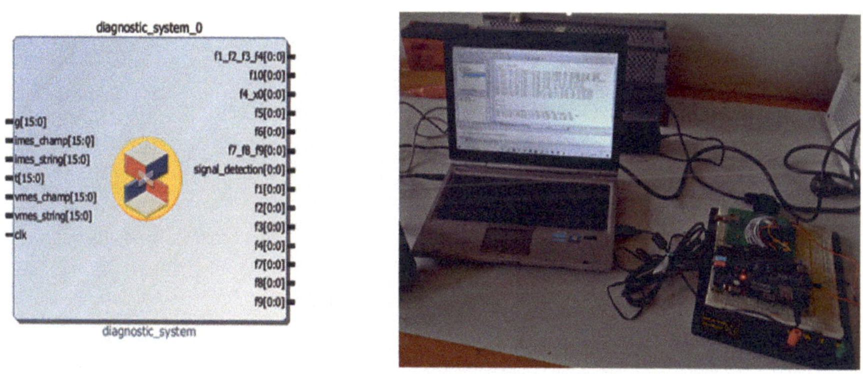

(a) (b)

Fig. 8.45 (A) IP package exported by Vivado and (B) generating the BitStream and programming the Zyng Zybo 7 FPGA board.

with the advantageous features of flexibility, parallelism of computations, robustness, and concurrent operation is the most suitable choice for rapid prototyping of digital control systems compared to the conventional microprocessor and microcontroller devices. As perspectives in this area:

- Online validation of the implemented method for real-time applications.
- Apply other DL algorithms to increase the classification accuracy.
- Study the possibility generalization of the method for large-scale PV plants.

8.5.5 Implementation of PV fault diagnosis techniques using thermography images and Raspberry Pi 4

Example 8.7: PV fault diagnosis based on thermography images and DCNN

This example aims to implement a semi-automatic fault detection and diagnosis method using a thermal camera and Raspberry Pi 4. Fig. 8.46A shows the test facility available at the UDES Tipaza, employed to prepare the dataset (thermal images). It consists of a FLIR t640 camera, and a PV array of 5 PV modules. A thermal image of the PV array is shown in Fig. 8.46B.

The DCNN structure shown in Fig. 8.47 consists of one average pooling layer with a ReLU activation function, four Max-pooling layers, four convolution filters with a ReLU activation function, one dropout layer, one flatten layer, three dense

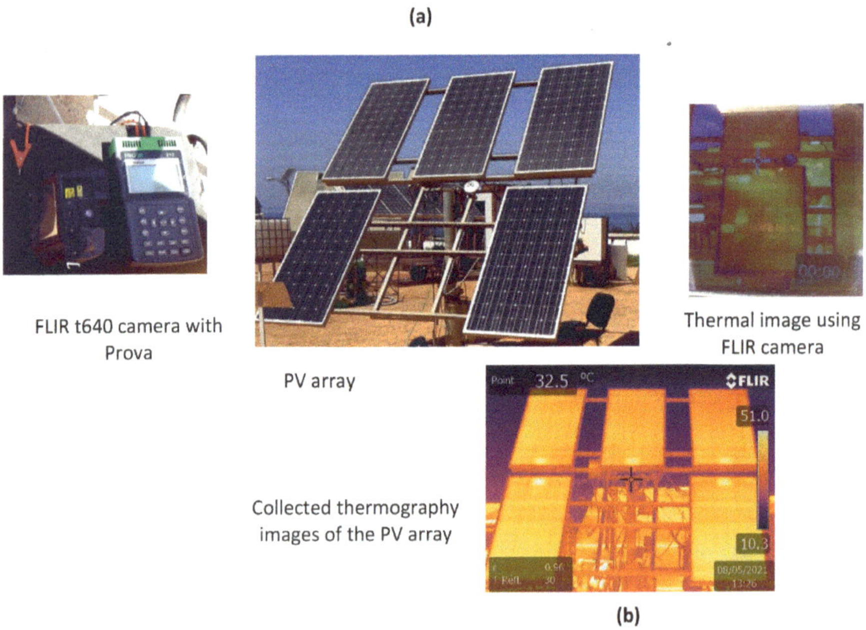

Fig. 8.46 (A) Test facility at the UDES research center, Tipaza and (B) thermography image.

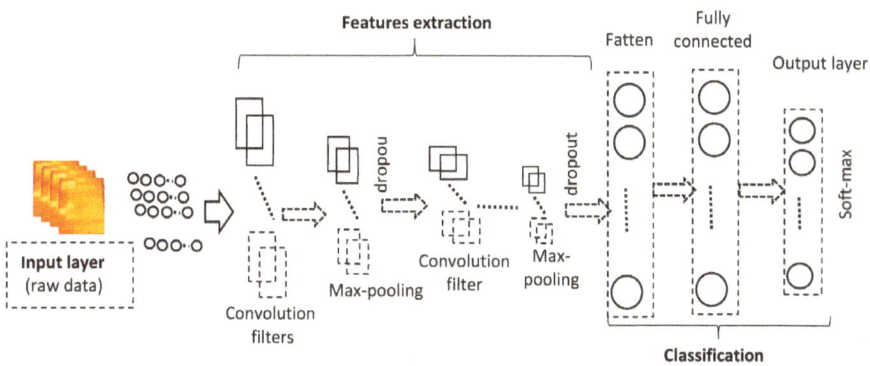

Fig. 8.47 The DCNN structure employed for fault detection and classification based on PV thermography images.

layers with a ReLU activation function, and one dense layer with Softmax activation function.

The faults investigated in this example are: F1: dust accumulation on PV modules, F2: partial shading (abnormal conditions), and F3: short-circuited diode. The implemented code in the Raspberry Pi 4 is shown in **Listing 6.8** of Chapter 6. Fig. 8.48 shows the DCNN code developed in Google Colab software.

The dataset consists of 479 images (see, for example, Fig. 8.49) used for developing the DCNN model and the accuracy is 73% as shown in Chapter 6, Section 6.8.3.

The following code **Listing 8.6** permits saving and converting the DCNN model in the TFlite model (optimized model) in order to be implemented in the Raspberry Pi 4.

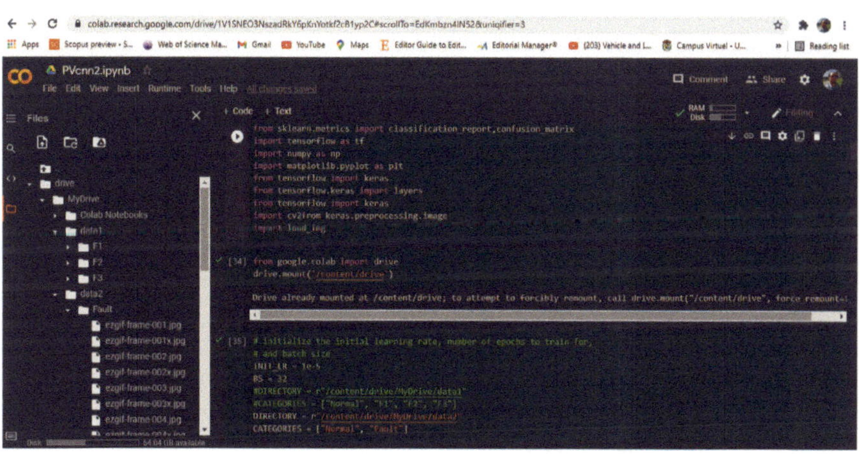

Fig. 8.48 DCNN code developed under Python using Google Colab.

```
#Listing 8.6 Save and convert the model
tf.saved_model.save(model,'/content')
converter=tf.lite.TFLiteConverter.from_saved_model('/content')
tflite_model=converter.convert()
with open('model.tflite','wb') as f:
f.write(tflite_model)
```

Fig. 8.49 PV thermography images: (A) example of used image in the training of the model and (B) augmented images (this function is used to increase the number of images in the dataset with various rotations).

The global block diagram of the fault diagnosis method is shown in Fig. 8.50. It comprises a thermal image as input of the raspberry, Raspberry Pi device, including the optimized DCNN code, Firebase, the webpage for displaying the image and results mobile application designed for notifying the users if any fault has happened.

The steps of the method are as follows:

Step 1: Train and test the DCNN model under Google Colab.
Step 2: Call the optimized model on Raspberry with tf.lite.Interpreter().
Step 3: Collect a new image and feed the model via USB to Raspberry Pi 4.
Step 4: Predict the image with the model and get the result.
Step 5: Use the Pyrebase library (which is a simple Python wrapper for the Firebase API) to make connection with the database.
Step 6: Save the result on the Firebase database.
Step 7: Send an SMS or email, if there is a problem, via a GSM module.

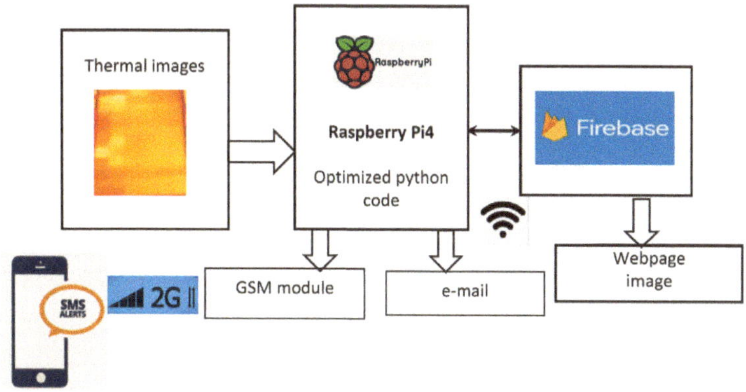

Fig. 8.50 Block diagram of the fault detection and classification using Raspberry Pi 4 and thermal images.

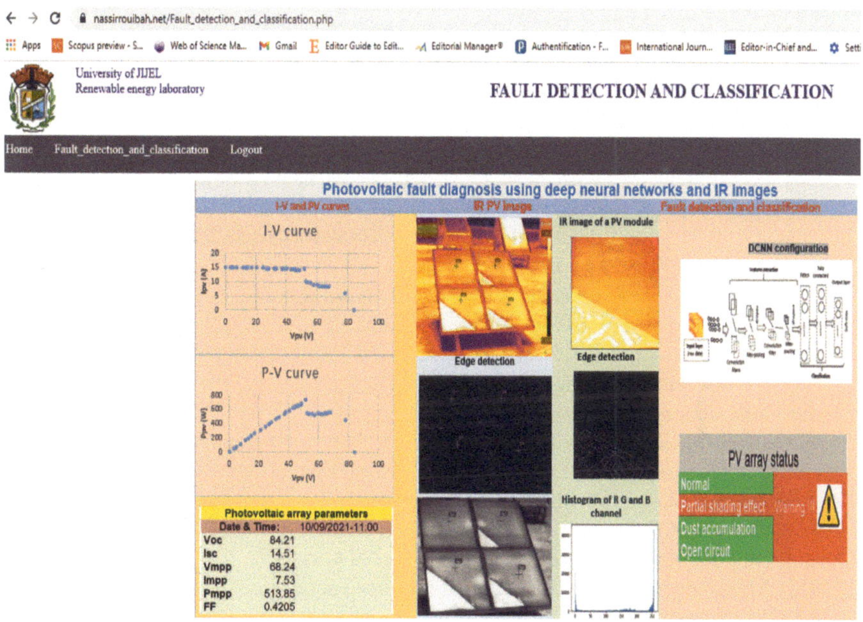

Fig. 8.51 Webpage of the fault detection and classification method using PV thermography images.

Basic codes on how to generate SMSs and sending emails are shown in Appendix I. Fig. 8.51 shows the graphical interface of the fault detection and classification method using PV thermography images.

8.6 Summary

With technological advances like the IoT technique, embedded systems, and artificial intelligence techniques (machine learning and deep learning), we are in the new era of modern photovoltaic systems. In this chapter, seven examples of the PV real-time application are described and presented. Three electronic programmable boards (Arduino, Raspberry Pi, and FPGA) have been used to implement various algorithms starting from simple algorithms to advanced algorithms, such as machine learning and deep learning.

This part of the book can help readers to understand how to implement and test some algorithms mainly used for photovoltaic applications, such as simple PV monitoring system, smart PV monitoring system, tracking the maximum power-based fuzzy-logic, fault diagnosis-based machine learning, fault diagnosis based on thermal images and deep learning, and PV array embedded emulator.

References

[1] E. Koutroulis, K. Kalaitzakis, N.C. Voulgaris, Development of a microcontroller-based, photovoltaic maximum power point tracking control system, IEEE Trans. Power Electron. 16 (1) (2001) 46–54.

[2] R. Akkaya, A.A. Kulaksiz, A microcontroller-based stand-alone photovoltaic power system for residential appliances, Appl. Energy 78 (4) (2004) 419–431.

[3] L. Khemissi, B. Khiari, R. Andoulsi, A. Cherif, Low cost and high efficiency of single phase photovoltaic system based on microcontroller, Sol. Energy 86 (5) (2012) 1129–1141.

[4] S.M. Azharuddin, M. Vysakh, H.V. Thakur, B. Nishant, T.S. Babu, K. Muralidhar, D. Paul, B. Jacob, K. Balasubramanian, N. Rajasekar, A near accurate solar PV emulator using dSPACE controller for real-time control, Energy Procedia 61 (2014) 2640–2648.

[5] M. Fuentes, M. Vivar, J.M. Burgos, J. Aguilera, J.A. Vacas, Design of an accurate, low-cost autonomous data logger for PV system monitoring using Arduino™ that complies with IEC standards, Sol. Energy Mater. Sol. Cells 130 (2014) 529–543.

[6] A. Mellit, S. Kalogirou, Artificial intelligence and internet of things to improve efficacy of diagnosis and remote sensing of solar photovoltaic systems: challenges, recommendations and future directions, Renew. Sust. Energ. Rev. 143 (2021) 110889.

[7] A. Youssef, M. El-Telbany, A. Zekry, The role of artificial intelligence in photo-voltaic systems design and control: a review, Renew. Sust. Energ. Rev. 78 (2017) 72–79.

[8] A. Mellit, S.A. Kalogirou, MPPT-based artificial intelligence techniques for photovoltaic systems and its implementation into field programmable gate array chips: review of current status and future perspectives, Energy 70 (2014) 1–21.

[9] M.D. Mudaliar, N. Sivakumar, IoT based real time energy monitoring system using raspberry Pi, Internet Things 12 (2020), 100292.

[10] H.H. Gad, A.Y. Haikal, H.A. Ali, New design of the PV panel control system using FPGA-based MPSoC, Sol. Energy 146 (2017) 243–256.

[11] R.I. Pereira, I.M. Dupont, P.C. Carvalho, S.C. Jucá, IoT embedded linux system based on Raspberry Pi applied to real-time cloud monitoring of a decentralized photovoltaic plant, Measurement 114 (2018) 286–297.

[12] https://www.arduino.cc/.

[13] https://www.raspberrypi.org.

[14] N.P. Petrikowski, Getting to Know the Raspberry Pi, The Rosen Publishing Group, 2014.

[15] Xilinx, What Is a FPGA? http://www.xilinx.com/fpga/. Accessed 5 November 2015.

[16] N. Gershenfeld, When Things Start to Think, Macmillan, 1999.

[17] C. Tran, S. Misra, The technical foundations of IoT, IEEE Wirel. Commun. 26 (3) (2019) 8.

[18] E. Koutroulis, K. Kalaitzakis, V. Tzitzilonis, Development of an FPGA-based system for real-time simulation of photovoltaic modules, Microelectron. J. 40 (7) (2009) 1094–1102.
[19] A. Mellit, H. Mekki, A. Messai, H. Salhi, FPGA-based implementation of an intelligent simulator for stand-alone photovoltaic system, Expert Syst. Appl. 37 (8) (2010) 6036–6051.
[20] H. Mekki, A. Mellit, S.A. Kalogirou, A. Messai, G. Furlan, FPGA-based implementation of a real time photovoltaic module simulator, Prog. Photovolt. Res. Appl. 18 (2) (2010) 115–127.
[21] S.N. Singh, S. Mishra, FPGA implementation of DPWM utility/DG interfaced solar (PV) power converter for green home power supply, Arch. Control Sci. 21 (4) (2011) 461–469.
[22] A. Messai, A. Mellit, A. Guessoum, S.A. Kalogirou, Maximum power point tracking using a GA optimized fuzzy logic controller and its FPGA implementation, Sol. Energy 85 (2) (2011) 265–277.
[23] A. Messai, A. Mellit, A.M. Pavan, A. Guessoum, H. Mekki, FPGA-based implementation of a fuzzy controller (MPPT) for photovoltaic module, Energy Convers. Manag. 52 (7) (2011) 2695–2704.
[24] A. Mellit, H. Rezzouk, A. Messai, B. Medjahed, FPGA-based real time implementation of MPPT-controller for photovoltaic systems, Renew. Energy 36 (5) (2011) 1652–1661.
[25] D. Ickilli, H. Can, K.S. Parlak, Development of a FPGA-based photovoltaic panel emulator based on a DC/DC converter, in: 2012 38th IEEE Photovoltaic Specialists Conference, IEEE, 2012, pp. 001417–001421.
[26] N. Chettibi, A. Mellit, M. Drif, FPGA-based implementation of IncCond algorithm for photovoltaic applications, in: 2012 24th International Conference on Microelectronics (ICM), IEEE, 2012, pp. 1–4.
[27] M.V. Rajkumar, P.S. Manoharan, FPGA based multilevel cascaded inverters with SVPWM algorithm for photovoltaic system, Sol. Energy 87 (2013) 229–245.
[28] A. Benzekri, A. Azrar, FPGA-based design process of a fuzzy logic controller for a dual-axis sun tracking system, Arab. J. Sci. Eng. 39 (8) (2014) 6109–6123.
[29] N. Chettibi, A. Mellit, Fpga-based real time simulation and control of grid-connected photovoltaic systems, Simul. Model. Pract. Theory 43 (2014) 34–53.
[30] F. Chekired, A. Mellit, S.A. Kalogirou, C. Larbes, Intelligent maximum power point trackers for photovoltaic applications using FPGA chip: a comparative study, Sol. Energy 101 (2014) 83–99.
[31] K.S. Parlak, FPGA based new MPPT (maximum power point tracking) method for PV (photovoltaic) array system operating partially shaded conditions, Energy 68 (2014) 399–410.
[32] J.-H. Chen, H.-T. Yau, L. Jin-Han, Implementation of FPGA-based charge control for a self-sufficient solar tracking power supply system, Appl. Sci. 6 (2) (2016) 41.
[33] W. Chine, A. Mellit, V. Lughi, A. Malek, G. Sulligoi, A.M. Pavan, A novel fault diagnosis technique for photovoltaic systems based on artificial neural networks, Renew. Energy 90 (2016) 501–512.
[34] S.M. Patil, M. Vijayalashmi, R. Tapaskar, IoT based solar energy monitoring system, in: 2017 International Conference on Energy, Communication, Data Analytics and Soft Computing, 2017, pp. 1574–1579.
[35] N.K. Adarsh, et al., Simulation and implementation of FPGA controlled distributed solar generation for residential network, in: 2017 IEEE PES Asia-Pacific Power and Energy Engineering Conference (APPEEC), IEEE, 2017.
[36] A. Lopez-Vargas, M. Fuentes, M. Vivar, On the application of IoT for real-time monitoring of small stand-alone PV systems: results from a new smart datalogger, in: IEEE 7th

World Conference on Photovoltaic Energy Conversion (WCPEC) (A Joint Conference of 45th IEEE PVSC, 28th PVSEC & 34th EU PVSEC), 2018, pp. 605–607.

[37] A. López-Vargas, M. Fuentes, M. Vivar, IoT application for real-time monitoring of solar home systems based on Arduino™ with 3G connectivity, IEEE Sensors J. 19 (2) (2018) 679–691.

[38] S. Jena, G. Panda, R. Peesapati, FPGA-based implementation for improved control scheme of grid-connected PV system with 3-phase 3-level NPC-VSI, Int J Circuit Theory Appl 46 (4) (2018) 942–964.

[39] A. Hamied, A. Mellit, M.A. Zoulid, R. Birouk, IoT-based experimental prototype for monitoring of photovoltaic arrays, in: IEEE 2018 International Conference on Applied Smart Systems, 2018, pp. 1–6.

[40] J.M. Paredes-Parra, A. Mateo-Aroca, G. Silvente-Niñirola, M.C. Bueso, Á. Molina-García, PV module monitoring system based on low-cost solutions: wireless raspberry application and assessment, Energies 11 (11) (2018) 3051.

[41] M.S. Khan, H. Sharma, A. Haque, IoT enabled real-time energy monitoring for photovoltaic systems, in: IEEE International Conference on Machine Learning, Big Data, Cloud and Parallel Computing, 2019, pp. 323–327.

[42] L. Bouselham, B. Hajji, A. Mellit, A. Rabhi, K. Kassmi, Hardware implementation of new irradiance equalization algorithm for reconfigurable PV architecture on a FPGA platform, in: 2019 International Conference on Wireless Technologies, Embedded and Intelligent Systems (WITS), IEEE, 2019, pp. 1–8.

[43] J.A. Prathap, T.S. Anandhi, A novel parallel duty cycle control algorithm for photovoltaic voltage regulator system using FPGA, Microprocess. Microsyst. 65 (2019) 107–120.

[44] A. Hamied, A. Boubidi, N. Rouibah, W. Chine, A. Mellit, IoT-based smart photovoltaic arrays for remote sensing and fault identification, in: Smart Energy Empowerment in Smart and Resilient Cities, International Conference in Artificial Intelligence in Renewable Energetic Systems, Springer, Cham, 2019, pp. 478–486.

[45] I.M. Dupont, P.C. Carvalho, S.C. Juca, J.S. Neto, Novel methodology for detecting nonideal operating conditions for grid-connected photovoltaic plants using internet of things architecture, Energy Convers. Manag. 200 (2019) 112078.

[46] J.M. Paredes-Parra, A.J. Garcia-Sanchez, A. Mateo-Aroca, A. Molina-Garcia, An alternative internet-of-things solution based on LoRa for PV power plants: data monitoring and management, Energies 12 (5) (2019) 881.

[47] A. Ilyas, M.R. Khan, M. Ayyub, FPGA based real-time implementation of fuzzy logic controller for maximum power point tracking of solar photovoltaic system, Optik 213 (2020) 164668.

[48] K.Ş. Parlak, Obtaining electrical characteristics of a PV module by FPGA based experimental system, Int. J. Hydrog. Energy 45 (2020) 33128–33135.

[49] S. Selvakumar, V.K. Kannan, FPGA based PV grid-connected inverter using partition input union output dual impedance Qzsi, Microprocess. Microsyst. 75 (2020) 103055.

[50] S.R. Madeti, S.N. Singh, Monitoring system for photovoltaic plants: a review, Renew. Sust. Energ. Rev. 67 (2017) 1180–1207.

[51] M.M. Rahman, J. Selvaraj, N.A. Rahim, M. Hasanuzzaman, Global modern monitoring systems for PV based power generation: a review, Renew. Sust. Energ. Rev. 82 (2018) 4142–4158.

[52] P.T. Le, H.L. Tsai, T.H. Lam, A wireless visualization monitoring, evaluation system for commercial photovoltaic modules solely in MATLAB/Simulink environment, Sol. Energy 140 (2016) 1–11.

[53] R.C. de Barros, J.M.S. Callegari, D. do Carmo Mendonca, W.C.S. Amorim, M.P. Silva, H. A. Pereira, Low-cost solar irradiance meter using LDR sensors, in: 2018 13th IEEE International Conference on Industry Applications (INDUSCON), IEEE, 2018, pp. 72–79.
[54] https://www.apachefriends.org/index.html.
[55] https://www.parallax.com/package/plx-daq/.
[56] PVLIB. https://pvlib-python.readthedocs.io/en/stable/singlediode.html.

Appendices

Appendix A: PV module modeling

The relationship between current (I) and voltage (V) of a PV module is given by the following equation:

$$I = I_{\mathrm{ph}} - I_0\left(\exp\left(\frac{V+IR_s}{n_sV_t}\right)-1\right)-\left(\frac{V+IR_s}{R_{\mathrm{sh}}}\right) \tag{A.1}$$

where I_{ph}, I_0, R_s, R_{sh}, and n_s represent the photocurrent of the cell, the saturation current of the diode, the series resistance, the parallel resistance, and the number of cells in the module, respectively, and V_t is the thermal voltage given by

$$V_t = \frac{AKT}{q} \tag{A.2}$$

where A, K, T, and q are the ideality factor of the diode, Boltzmann constant, cell temperature, and the charge of the electron, respectively.

Three equations can be determined by replacing the three particular points of the characteristic I-V in Eq. (A.1).

At $I = I_{\mathrm{sc}}$ and $V = 0$ by substituting these values in Eq. (A.1):

$$I_{\mathrm{sc}} = I_{\mathrm{ph}} - I_0 e^{\frac{I_{\mathrm{sc}}R_s}{nsVt}} - \frac{I_{\mathrm{sc}}R_s}{R_{\mathrm{sh}}} \tag{A.3}$$

At MPP, $I = I_{\mathrm{mpp}}$ and $V = V_{\mathrm{mpp}}$;olpi by substituting these values in Eq. (A.1):

$$I_{\mathrm{mpp}} = I_{\mathrm{ph}} - I_0 e^{\frac{V_{\mathrm{mpp}}+I_{\mathrm{mpp}}R_s}{nsVt}} - \frac{V_{\mathrm{mpp}}+I_{\mathrm{mpp}}R_s}{R_{\mathrm{sh}}} \tag{A.4}$$

At $I = 0$ and $V = V_{\mathrm{oc}}$, by substituting these values in Eq. (A.1):

$$I_{\mathrm{oc}} = 0 = I_{\mathrm{ph}} - I_0 e^{\frac{V_{\mathrm{oc}}}{nsVt}} - \frac{V_{\mathrm{oc}}}{R_{\mathrm{sh}}} \tag{A.5}$$

At MPP, the derivative of the power compared to the tension is null, so

$$\left.\frac{dP}{dV}\right|_{\substack{V=V\text{mpp}\\ I=I\text{mpp}}}=0 \tag{A.6}$$

The derivative of the voltage with respect to the current for the short-circuit condition equal to the parallel resistance:

$$\left.\frac{dI}{dV}\right|_{I=Isc}=-\frac{1}{R_{\text{sh}}} \tag{A.7}$$

From Eqs. (A.3), (A.5), the photocurrent can be expressed by

$$I_{\text{ph}}=I_0 e^{\frac{V_{\text{oc}}}{nsVt}}+\frac{V_{\text{oc}}}{R_{\text{sh}}} \tag{A.8}$$

By inserting Eq. (A.8) into Eq. (A.3):

$$I_{\text{sc}}=I_0\left(e^{\frac{V_{\text{oc}}}{nsVt}}-e^{\frac{I_{\text{sc}}R_s}{nsVt}}\right)+\frac{V_{\text{oc}}-I_{\text{sc}}R_s}{R_{\text{sh}}} \tag{A.9}$$

The term in the parentheses has been simplified by neglecting the second exponential before the first, then Eq. (A.9) becomes

$$I_{\text{sc}}=I_0 e^{\frac{V_{\text{oc}}}{nsVt}}+\frac{V_{\text{oc}}-I_{\text{sc}}R_s}{R_{\text{sh}}} \tag{A.10}$$

From Eq. (A.10), the saturation current of the diode can be expressed by

$$I_0=\left(I_{\text{sc}}-\frac{V_{\text{oc}}-I_{sc}R^{sc}s}{R_{\text{sh}}}\right)e^{\frac{-V_{\text{oc}}}{n_sV_t}} \tag{A.11}$$

By replacing Eqs. (A.8), (A.11) in Eq. (A.4): do you mean Eq. (A.8), etc.?

$$I_{\text{mpp}}=I_{sc}-^{sc}\frac{V_{\text{mpp}}+I_{\text{mpp}}R_s-I_{\text{sc}}R_s}{R_{\text{sh}}}-\left(I_{\text{sc}}-\frac{V_{\text{oc}}-I_{\text{sc}}R_s}{R_{\text{sh}}}\right)e^{\frac{V_{\text{mpp}}+I_{\text{mpp}}R_s-V_{\text{oc}}}{n_sV_t}} \tag{A.12}$$

The derivative of the power with respect to the voltage at the PPM can be rewritten as follows:

$$\left.\frac{dP}{dV}\right|_{\substack{V=V\text{mpp}\\ I=\text{Impp}}}=\frac{d(IV)}{dV}=I+\frac{dI}{dV}V \tag{A.13}$$

To find the derivative of the power with respect to the voltage at the point MPP, one would have to find the derivative of Eq. (A.12) with respect to the voltage, the latter is a transcendent equation, and it requires a numerical method to express. Eq. (A.12) can be written in the following form:

$$I = f(I, V) \tag{A.14}$$

where $f(I,V)$ is the second member of Eq. (A.12). By deriving the Eq. (A.14):

$$dI = dI \frac{\partial f(I,V)}{\partial I} + dV \frac{\partial f(I,V)}{\partial V} \tag{A.15}$$

And the derivative of the current with respect to the voltage is given by

$$\frac{dI}{dV} = \frac{\frac{\partial}{\partial V} f(I,V)}{1 - \frac{\partial}{\partial I} f(I,V)} \tag{A.16}$$

From Eqs. (A.15), (A.16), it follows that

$$\frac{dP}{dV} = I_{\mathrm{mpp}} + \frac{V_{\mathrm{mpp}} \frac{\partial}{\partial V} f(I,V)}{1 - \frac{\partial}{\partial I} f(I,V)} \tag{A.17}$$

From the above equations, we can write

$$\left.\frac{dP}{dV}\right|_{I=\mathrm{Impp}} = I_{\mathrm{mpp}} + V_{\mathrm{mpp}} \frac{-\frac{(I_{\mathrm{sc}} R_{\mathrm{sh}} - V_{\mathrm{oc}} + I_{\mathrm{sc}} R_s) e^{\frac{V_{\mathrm{mpp}} + I_{\mathrm{mpp}} R_s - V_{\mathrm{oc}}}{n_s V_t}}}{n_s V_t R_{sh}} - \frac{1}{R_{\mathrm{sh}}}}{1 + \frac{(I_{\mathrm{sc}} R_{\mathrm{sh}} - V_{\mathrm{oc}} + I_{\mathrm{sc}} R_s) e^{\frac{V_{\mathrm{mpp}} + I_{\mathrm{mpp}} R_s - V_{\mathrm{oc}}}{n_s V_t}}}{n_s V_t R_{\mathrm{sh}}} + \frac{R_s}{R_{\mathrm{sh}}}} \tag{A.18}$$

From Eqs. (A.7), (A.17), (A.18), it follows that

$$\left.-\frac{1}{R_{\mathrm{sh}}}\right|_{I=Isc} = \frac{-\frac{(I_{\mathrm{sc}} R_{\mathrm{sh}} - V_{\mathrm{oc}} + I_{\mathrm{sc}} R_s) e^{\frac{V_{\mathrm{mpp}} + I_{\mathrm{mpp}} R_s - V_{\mathrm{oc}}}{n_s V_t}}}{n_s V_t R_{\mathrm{sh}}} - \frac{1}{R_{\mathrm{sh}}}}{1 + \frac{(I_{\mathrm{sc}} R_{\mathrm{sh}} - V_{\mathrm{oc}} + I_{\mathrm{sc}} R_s) e^{\frac{V_{\mathrm{mpp}} + I_{\mathrm{mpp}} R_s - V_{\mathrm{oc}}}{n_s V_t}}}{n_s V_t R_{\mathrm{sh}}} + \frac{R_s}{R_{\mathrm{sh}}}} \tag{A.19}$$

Appendix B: Machine learning and deep learning functions

Table B.1 summarizes the major functions used for solving photovoltaic problems including optimization, forecasting, classification, control, fault detection, and diagnosis.

Table B.1 Functions used in ML and DL algorithms.

Algorithm and type of learning	Function/class	Libraries, platform and frameworks	Type of problem	Photovoltaic applications (example)
LR and LoR Supervised	LinearRegression LogisticRegression	Scikit-learn 0.23.2	Regression	Output PV prediction
SVM Supervised	NuVR, LinearSVM NuSVC, LinearSVC	Scikit-learn 0.23.2	Regression and Binary and multi-class classification	Prediction and fault classification
k-NN Supervised	KNeighborsRegressor KNeighborsClassifier	Scikit-learn 0.23.2	Regression and classification	Detection and classification of faults
DT Supervised	DecisionTreeRegressor DecisionTreeClassifier	Scikit-learn 0.23.2	Regression and classification	Energy scheduling, power forecasting
RF Supervised	RandomForestClassifier RandomForestRegressor	Scikit-learn 0.23.2	Regression and classification	Maximum power tracking,
NB Supervised	GaussianNB	Scikit-learn 0.23.2	Regression and classification	Energy prediction and fault classification
NNs Supervised	MLPClassifier and MLPregressor	Scikit-learn 0.23.2	Regression and classification	Power forecasting
K-Means Unsupervised	Kmeans	Scikit-learn	Clustering	MPPT, PV degradation.
FCM Unsupervised	Fuzz FCM	Skfuzzy.cmeans fcmeans	Clustering	PV fault diagnosis and control
PCA Unsupervised	decomposition.PCA	Scikit-learn	Dimensionality reduction	Shading default identification, PV cells classification
HC Unsupervised	AgglomerativeClustering	Scikit-learn	Clustering	Clustering of PV power data
CNN Supervised	Sequential, Conv2D,Con1D	Keras and TensorFlow platform	Classification	Fault diagnosis, degradation
LSTM and GRU Supervised	Sequential, LSTM, GRU	Keras and TensorFlow platform	Time series prediction	Power forecasting
BiLSTM and BiGRU Supervised	Sequential, LSTM, GRU	Keras and TensorFlow platform	Time series prediction	Power forecasting

Appendix C: Error metrics

The common error metrics are summarized below:

Error (e) is the difference between measured (x_i) and predicted value (y_i) given as follows:

$$e = (x_i - y_i) \tag{C.1}$$

Standard deviation (σ) measures the dispersion of a dataset relative to its mean:

$$\sigma = \sqrt{\sum_{i=1}^{n} \frac{(x_i - \overline{x})^2}{n-1}} \tag{C.2}$$

Standard error of the mean (SEM) measured how much discrepancy there is likely to be in a sample's mean compared to the population mean.

$$\text{SEM} = \frac{\sigma}{\sqrt{n}} \tag{C.3}$$

Mean absolute error (MAE) is the average of the absolute values of the deviation.

$$\text{MAE} = \frac{1}{n}\sum_{i=1}^{n} |x_i - y_i| \tag{C.4}$$

Mean relative percent error (MAPE) this is where mean absolute percentage error comes in as this is a percentage error and hence gives a good idea of the relative error. Lower the value of the better is the forecast.

$$\text{MAPE} = \frac{100\%}{n}\sum_{i=1}^{n} \left|\frac{x_i - y_i}{x_i}\right| \tag{C.5}$$

Weighted MAPE (wMAPE), it is a variant of MAPE in which errors are weighted by values of actuals.

$$\text{wMAPE} = \frac{\sum_{i=1}^{n} |x_i - y_i|}{\sum_{i=1}^{n} |x_i|} \tag{C.6}$$

Correlation coefficient (r) determines how measured and forecasted data are correlated, given by this expression.

$$r = \frac{\sum_{i=1}^{n} ((x_i - \overline{x})(y_i - \overline{y}))}{\sqrt{\sum_{i=1}^{n} (x_i - \overline{x}_i)^2 \sum_{i=1}^{n} (y_i - y_i)^2}} \tag{C.7}$$

The mean squared error (MSE) is the average of the square of the forecast error. As the square of the errors are taken, the effect is that larger errors have more weight on the score.

$$\mathrm{MSE} = \frac{1}{n}\left(\sum_{i=1}^{n}(x_i - y_i)^2\right) \tag{C.8}$$

The root mean squared error (RMSE) is

$$\mathrm{RMSE} = \sqrt{\frac{1}{n}\left(\sum_{i=1}^{n}(x_i - y_i)^2\right)} \tag{C.9}$$

Appendix D: Confusion matrix

Precision, recall, accuracy, and F1 score are the most commonly used error metrics in classification problems. In the case of two classes, the above metrics can be expressed as

$$\text{Accuracy} = \frac{(\text{TP} + \text{TN})}{(\text{TP} + \text{FP} + \text{TN} + \text{FN})} \tag{D.1}$$

$$\text{Precision} = \frac{\text{TP}}{(\text{TP} + \text{FP})} \tag{D.2}$$

$$\text{Recall} = \frac{\text{TP}}{(\text{TP} + \text{FN})} \tag{D.3}$$

$$\text{F1} - \text{score} = \frac{2(\text{Precision*Recall})}{(\text{Recall} + \text{Precision})} \tag{D.4}$$

$$\text{Misclassified rate} = 1 - \text{accuracy} \tag{D.5}$$

where TP is the number of true positive, TN is the number of true negative, FP is the number of fault positive, and FN is the number of fault negative. x_i and y_i are the measured and forecasted values, respectively, and $\bar{x}$ and $\bar{y}$ are the average values of the measured and the forecasted data, respectively.

To explain how the convolution matrix (CM) works, let us consider the following CM with 2×2 dimension (two classes: Normal and Fault):

Actual class	Normal	145 (TN)	5 (FP)
	Fault	10 (FN)	100 (TP)
		Normal	Fault
	Predicted class		

In total, 260 samples are present (145 + 10 + 100 + 5). The diagonal elements CM (1,1) =145, CM (2,2) = 100 are correctly classified, while the elements CM (1,2) and CM (2,1) are incorrectly classified. The accuracy is (145 + 100)/260 = 94.23%. The precision is 100/(100 + 5) =95.23%. The recall is 100/(100 + 10) =90.90%. The F1 score is 93.01% and the misclassified rate is (10 + 5)/260 = 0.057. For example, if the element CM (1,2) =5 is considered, this one is incorrectly classified in "**Fault**" class and the element CM (2,1) is also incorrectly classified in "**Normal**" class.

In the case of a CM of $N \times N$, the above metrics can be written as follows:

$$\text{Accuracy} = \frac{\sum_i \text{CM}(i,i)}{\sum_i \sum_j \text{CM}(i,j)} \tag{D.6}$$

$$\text{Precision}_i = \frac{\text{CM}(i,i)}{\sum_j \text{CM}(j,i)} \tag{D.7}$$

$$\text{Sensitivity}_i = \frac{\text{CM}(i,i)}{\sum_j \text{CM}(i,j)} \tag{D.8}$$

$$\text{F1} - \text{score} = 2\frac{(\text{sensitivity} \times \text{Recall})}{(\text{sensitivity} + \text{Precision})} \tag{D.9}$$

$$\text{Misclassified rate} = 1 - \text{accuracy} \tag{D.10}$$

Appendix E: Evaluation error metrics in Python

Error metrics and statistical test functions employed to evaluate models are reported in Table E.1.

Table E.1 Error metrics.

Error name	Function	Library	Type (regression classification and clustering)
Accuracy score	accuracy_score	sklearn	Classification
Precision	metrics.precision_score	Sklearn	Classification
recall	metrics.recall_score	Sklearn	Classification
F1_score	metrics.f1_score	*Sklearn*	Classification
F_beta	metrics.fbeta_score	*Sklearn*	Classification
MAE	mean_absolute_error	*Sklearn*	Regression
MSE	mean_squared_error	*Sklearn*	Regression
RMSE	mean_squared_log_error	*Sklearn*	Regression
MAPE	mean_absolute_percentage_error	*Sklearn*	Regression
R^2	r2_score	*Sklearn*	Regression
r	Sqrt(r2_score)	*Sklearn*	Regression
Std	statistics.stdev	*statistics*	Regression
Var	statistics.variance	*statistics*	Regression
Confusion Matrix	confusion_matrix	*Sklearn*	Classification

Appendix F: *K*-fold cross validation

K-fold cross validation (CV) is where a given dataset is split into a K number of sections/folds where each fold is used as a testing set at some point. In fact, it is a resampling procedure used to evaluate machine learning models on a limited data sample.

Let's take the scenario of 10-fold cross validation ($K = 10$). Here, the dataset is split into 10 folds. In the first iteration, the first fold is used to test the model and the rest are used to train the model. In the second iteration, second fold is used as the testing set, while the rest serve as the training set. This process is repeated until each fold of the 10 folds has been used as the testing set (see Fig. F.1).

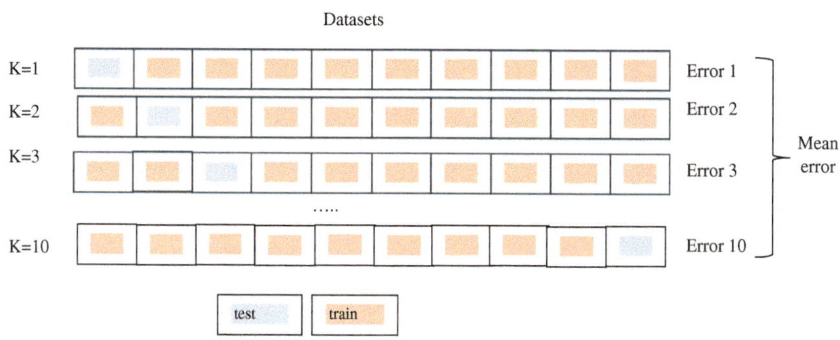

Fig. F.1 K-fold cross-validation method.

Appendix G

G.1. Arduino code for reading sensor connected to port A0

```
void setup() {
Serial.begin(9600);          // the bigger number the better
Serial.println("CLEARDATA"); //clears up any data left
from previous projects
Serial.println("LABEL,Clock,Time,Data");
//Always write LABEL, so excel knows the next things
will be the names of the columns (instead of Acolumn
you could write Time for instance)
Serial.println("RESETTIMER"); //resets timer to 0
}
void loop() {
Serial.print("DATA,TIME,TIMER,");
//Writes the time in the first column A, the elapsed
time in column B, and the data in C.
Serial.print(analogRead(A0));//Read channel A0
Serial.println(""); //be sure to add println to the
last command so it knows to go into the next row on
the second run
delay(100); //Add a 100mS delay
}
```

Listing G.2: Displaying results on the LCD and connecting to the sever

```
#include <LiquidCrystal.h>
const int rs = 43, en = 45, d4 = 47, d5 = 49, d6 = 51, d7 = 53;
LiquidCrystal lcd(rs, en, d4, d5, d6, d7);
 // #include "SoftwareSerial.h"
 // SoftwareSerial Serial1(8,12); // RX, TX
  String server = "192.168.1.118";
  float APV_Value = 0.0;
  float APC_Value = 0.0;
  int APR_Value = 0;
  int APT_adc = 0;
  float APT_Value = 0.0;
  String get_request = "";
  String APV_ValueCh = "";
  String APC_ValueCh = "";
  String APR_ValueCh = "";
  String APT_ValueCh = "";
void setup()
{
  Serial.begin(115200); // serial port used for debugging
```

```
  Serial1.begin(115200);
  lcd.begin(16, 4);
  lcd.print("System");
  lcd.setCursor(0, 1);
  lcd.print("  starting...");
  delay(2000);
  AT_send: Serial1.println("AT");
  if(Serial1.find("OK")){
    Serial.println("AT OK");
    lcd.setCursor(0,0);
    lcd.print("ESP is ready...");
    lcd.setCursor(0, 1);
    lcd.print("       ");
}
else{
   Serial.println("Error on ESP8266...");
   lcd.setCursor(0,0);
   lcd.print("Error on ESP...");
   lcd.setCursor(0, 1);
   lcd.print("       ");
   //make a reset to ESP module
   goto AT_send;
}
   Serial1.println("AT+CWMODE=1");
   if(Serial1.find("OK")){
   Serial.println("ESP8266 on STATION mode...");
   lcd.setCursor(0, 0);
   lcd.print(" Station mode ");
   lcd.setCursor(0, 1);
   lcd.print("  Enabled  ");
   }
  wifi_con:  Serial1.println("AT+CWJAP=\"wireless_home\",\"LER012
  _357951\"");
   Serial.println("Connecting to WIFI...");
   lcd.setCursor(0, 0);
   lcd.print(" Connecting to ");
   lcd.setCursor(0, 1);
   lcd.print("  WIFI...  ");
   Serial1.setTimcout(10000);
   if(Serial1.find("OK")){
   Serial.println("WIFI Connected ");
   lcd.setCursor(0, 0);
   lcd.print(" WIFI Connected ");
   lcd.setCursor(0, 1);
   lcd.print("       ");
   }
   else{
   Serial.println("Unable to connect...");
```

```
    Serial.println("Please, verify your AP ...");
    lcd.setCursor(0, 0);
    lcd.print("Unable to connect");
    lcd.setCursor(0, 1);
    lcd.print("Verify your AP ..");
    goto wifi_con;
    }
    Serial1.setTimeout(2000);
    Serial1.println("AT+CIPMUX=0");
    if(Serial1.find("OK")){
    Serial.println("Single connections enabled");
    }
}
   void loop()
   {
   APV_Value = (float)analogRead(A0)*5.0/1024.0/0.2;
   APV_ValueCh = (String)(APV_Value);
   Serial.println(APV_ValueCh);
   lcd.setCursor(0, 1);
   lcd.print("V="+APV_ValueCh);
   APC_Value = (2.5-((float)analogRead(A1)*5/1024))*10;
   APC_ValueCh = (String)(APC_Value);
   Serial.println(APC_ValueCh);
   lcd.setCursor(8, 1);
   lcd.print("C="+APC_ValueCh);
   APR_Value = map(analogRead(A2),0,1023,0,1276);
   APR_ValueCh = (String)(APR_Value);
   Serial.println(APR_ValueCh);
   lcd.setCursor(0, 2);
   lcd.print("R="+APR_ValueCh);
   lcd.print(" ");
   APT_adc = analogRead(A3);
   APT_adc = analogRead(A3);
   APT_Value = ((float)APT_adc *500)/1024;
   APT_ValueCh = (String)(APT_Value);
   Serial.println(APT_ValueCh);
   lcd.setCursor(8, 2);
   lcd.print("T="+APT_ValueCh);
   Serial.println("connecting to server...");
   lcd.setCursor(0, 3);
   lcd.print("CNX To Server...");
   Serial1.println("AT+CIPSTART=\"TCP\",\"" + server + "\",80");
   if(Serial1.find("OK")){
     Serial.println("Server connection successful...");
     lcd.setCursor(0, 3);
     lcd.print("CNX Successful..");
   }
   else{
```

```
    lcd.setCursor(0, 3);
    Serial.println("CNX Failed...");
}
    get_request= "GET /web/Get_Data.php?V=";
    get_request=get_request+APV_ValueCh;
    get_request=get_request+"&C=";
    get_request=get_request+APC_ValueCh;
    get_request=get_request+"&R=";
    get_request=get_request+APR_ValueCh;
    get_request=get_request+"&T=";
    get_request=get_request+APT_ValueCh;
    get_request=get_request+" HTTP/1.1\r\n";
    get_request=get_request+"Host: 192.168.1.118\r\n\r\n";
    Serial1.print("AT+CIPSEND=");
    Serial1.println(get_request.length());
    if(Serial1.find(">")) {
      Serial.println("Sending request. . .");
      lcd.setCursor(0, 3);
      lcd.print("Sending request.");
      Serial1.println(get_request);
    }
    delay(1500);
     if(Serial1.find("OK")) {
      lcd.setCursor(0, 3);
      lcd.print("Request Sent....");
      Serial.println("HTTP request sent...:");
      Serial.println();
     }
   Serial1.println("AT+CIPCLOSE");
   delay(1500);
  if( Serial1.find("OK")){
    Serial.println("Connection Closed...");
    lcd.setCursor(0, 3);
    lcd.print("CNX Closed......");
    delay(1500);
    //delay(295000);
  }
}
```

Listing G.3: One-diode model for *I-V* curve estimation

```
function[II1Vmm]=para5_temp_irra(G,T,Voc,Isc,Impp,Vmpp,ns,ki,kv,Ns)
%[II1 V output Rs Rsh a IO Iph mm]
K=1.3854*10o(-23); q=1.6*10o(-19);Vt=(K*T)/q;
%Voc=Voc/Ns; Vmpp=Vmpp/Ns;Ns=1
x001=(((Voc-Vmpp)/Impp)/Ns)
x002=((Vmpp/1)/(Isc-Impp))
%x001=0.33;x002=135;
```

```
x0=[x001;
    x002;
    (1/ns)*(Vmpp+Impp*x001-Voc)*(1/log((Impp-Isc+((Vmpp
+(Impp*x001)-(Isc*x001))/x002))/(-(Isc-((Voc-(Isc*x001))/
x002))))))];
   options=optimset('Display','iter'); % Option to display output
%
  [x,fval,exitflag,output] = fsolve(@myfun_stc,x0,options,Voc,
Isc,Impp,Vmpp,ns)
x = fsolve(@(x) myfun_stc(x,Voc,Isc,Impp,Vmpp,ns),x0)
a=(q*(x(3)/Ns))/(K*T);mm=a;
%ns=72;
I0=(Isc-((Voc-(Isc*x(1))))/x(2))*exp(-Voc/(ns*x(3)));
  Iph=I0*exp(Voc/(ns*x(3)))+(Voc/x(2));
  Rs=x(1);Rsh=x(2);Vt=x(3);
  [Isc1 Iph1 Voc1]=para_irra(G,Isc,Iph,ns,Rs,Rsh,Vt,I0);
  [Isc Iph Voc I0]=para_tempirra(T,Voc1,Isc1,Iph1,ns,Rs,Rsh,Vt,
  ki,kv);
% Rsh1=Rsh;
% RSH0=350; %pour si
% Rsh=Rsh1+(RSH0-Rsh1)*exp(-5.5*(G/1000));
  Iph
  I0
% Isc=Isc1;
% Voc=Voc1;
% Iph=Iph1;
% I0=I01;
%tracé de la courbe i_v et p_v
  I1(1)=Isc;
V=0:Voc+5;%45
II1=[];
for k=1:length(V);
    I1(1)=Isc;
    for j=1:10;
    df(j)=-1-I0*(x(1)/(ns*x(3)))*exp((V(k)+(I1(j)*x(1)))/(ns*x
(3)))-(x(1)/x(2));
    f(j)=Iph-I1(j)-I0*(exp((V(k)+(I1(j)*x(1)))/(ns*x(3))))-((V(k)
+(I1(j)*x(1)))/x(2));
    I1(j+1)=I1(j)-(f(j)/df(j));
    end
    II1=[II1 I1(j+1)];p1(k)=V(k)*II1(k);
end
%pm=max(p1);
```

```
function[Iscl Iph Voc]=para_irra(G,Isc,Iph,ns,Rs,Rsh,Vt,I0)
x(1)=Rs;
x(2)=Rsh;
x(3)=Vt
Iscl=Isc*(G/1000);
Iph1=Iph*(G/1000);
Iph=Iph1;
Voc(1)=0;
for mm=1:20;
df_dvoc(mm)=1-(ns*x(3))*(-1/((Iph*x(3))-Voc(mm)));
    f(mm)=Voc(mm)-(ns*x(3))*log(((Iph*x(2))-Voc(mm))/(I0*x(2)));
  Voc(mm+1)=Voc(mm)-(f(mm)/df_dvoc(mm));
end
Voc=Voc(mm);
```

```
function[Iscl  Iph1  Voc1  I01]=para_tempirra(T,Voc,Isc,Iph,ns,Rs,
Rsh,Vt,ki,kv)
x(1)=Rs;
x(2)=Rsh;
x(3)=Vt
Iscl=Isc*(1+((ki/1)*(T-298)));
Voc1=Voc+(kv*(T-298));
I01=(Iscl-((Voc1-(Iscl*x(1)))/x(2)))*exp(-Voc1/(ns*x(3)));
Iph1=I01*exp(Voc1/(ns*x(3)))+(Voc1/x(2));
```

Appendix H: XSG models

H.1. I-V model block based on XSG

The translation method is based on the following equations (Fig. H.1):

$$I(G,T) = I(G_0,T_0)\frac{G}{G_0}(1 + k_i(T - T_0)) \tag{H.1}$$

$$V(G,T) = V(G_0,T_0) - R_s(I(G,T) - (G_0,T_0)) + \frac{KT}{q}\ln\left(\frac{G}{G_0}\right) + K_v(T - T_0) \tag{H.2}$$

H.2. Extraction parameters block on XSG

See Fig. H.2.

H.3. NN model block based on XSG

See Fig. H.3.

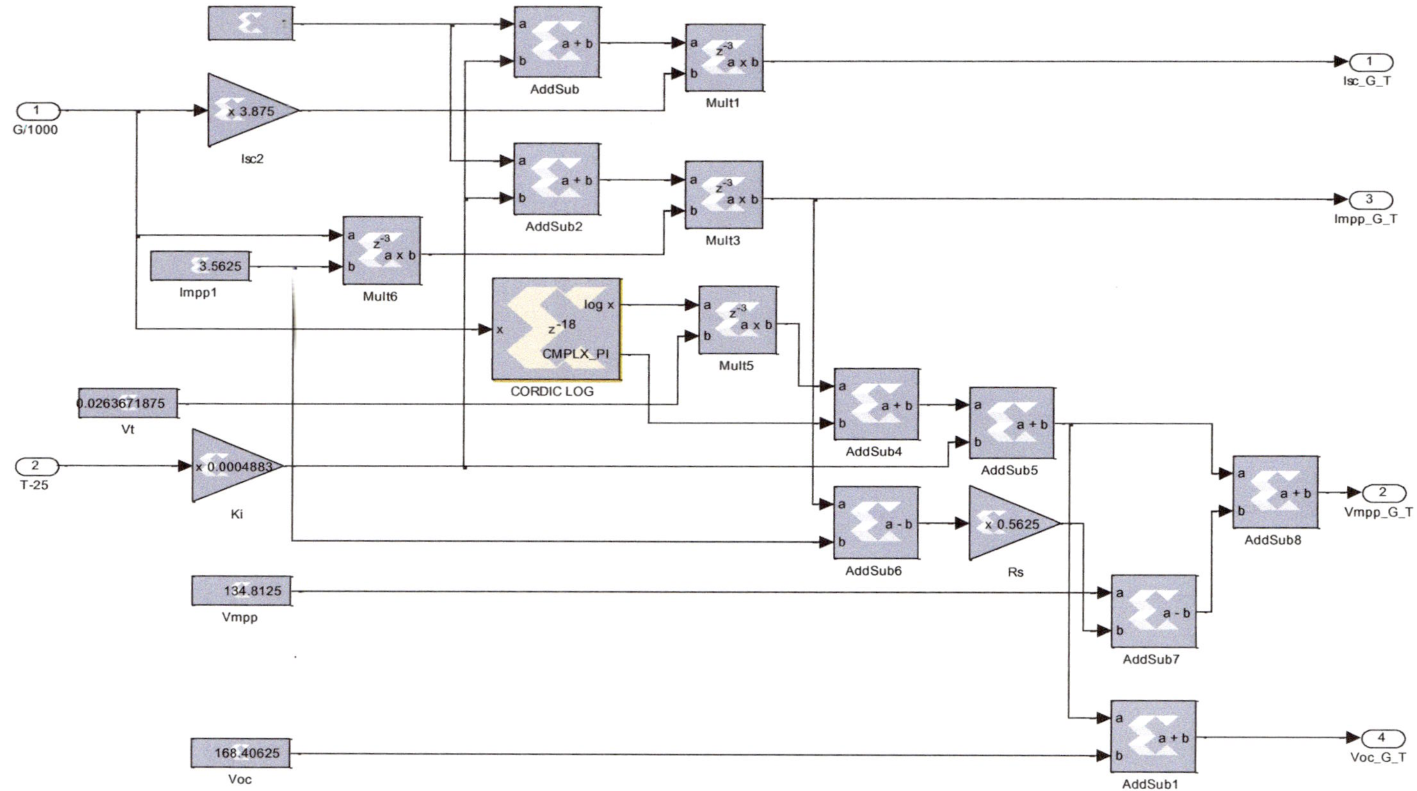

Fig. H.1 PV array model by the translation method of the characteristic *I-V*.

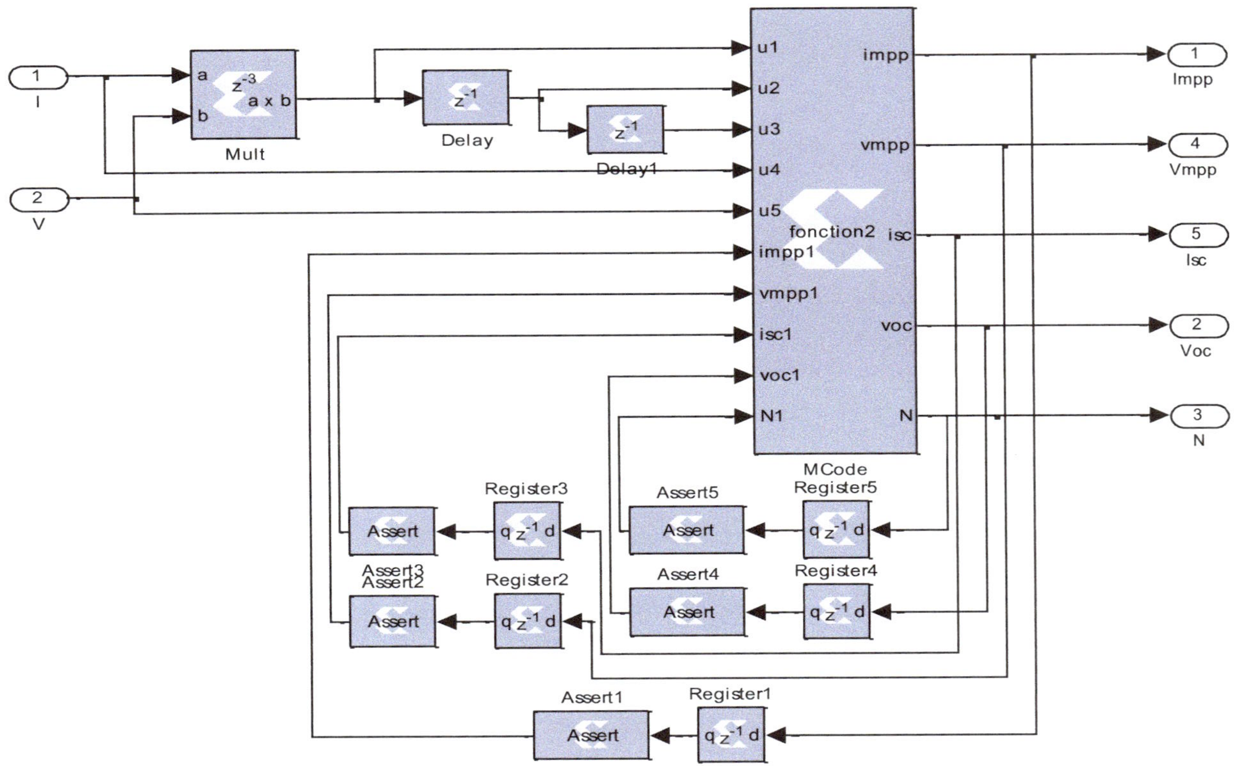

Fig. H.2 Parameters extracted block of the *I-V* curve using XSG.

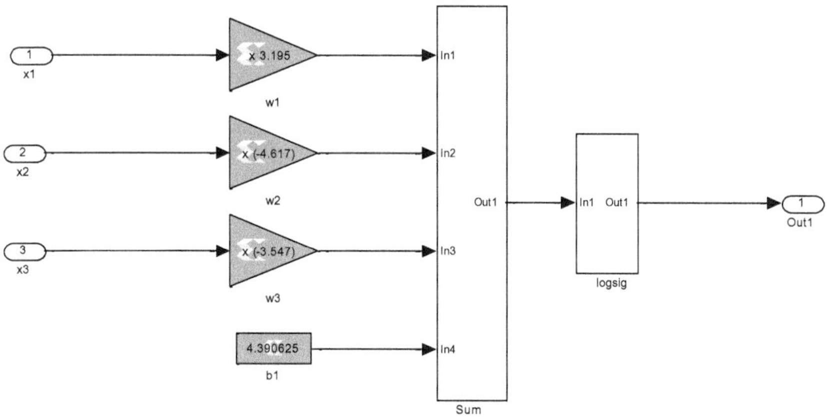

Fig. H.3 ANN block using XSG.

Appendix I

I.1. Generate SMS and sending email

```
Step 1 insert 2G activated SIM card into SIM800L GSM module
Step 2 Confirm that the hardware is connected, type the command
line: sudo raspi-config and accept serial port (Enable)
Step 3 Open the serial port and installing the screen tool, type
the following command
AT                    The device recognizes the SIM800L module
OK                    Response
AT+CFUN=1          Set phone functionality
OK
AT+CMGF=1
OK
AT+CMGS="+213*******"
>Type here your text message
OK                    Response
acting the raspberry pi to sim8ool gsm module to send, receive sms,
delete sms etc.
*****************************************************************
Import serial
Import time
from sim800l import SIM800L
sim800l=SIM800L('/dev/serial0')
receiveNum='+213551998982'
sim8001=serial.Serial(port='dev/serial0',          baudrate=9600,
parity=serial.PARITY_NONE,stopbits=serial.STOPBITS_ONE,
bytesize=serial.EIGHTBITS,timeout=1)
sms='Fault'
time.sleep(1)
sim8001.write('AT+CMGS=1\n')
print sim8001.read(24)
time.sleep(1)
cmd1='AT+CMGS=\"+str(receiverNum+'\n')
sim8001.write(cmd)
print sim8001.read(24)
time.sleep(1)
sim8001.write(str(sms))
sim8001.write(chr(26))
print sim8001.read(24)
```

I.2. Sending email

```
import smtplib
def email_alert(examp_str=""):
    server = smtplib.SMTP_SSL('smtp.gmail.com', port=587)
    print("Server Started...")
    server.login('adelmellit2013@gmail.com','*********')
    print("Server Login Successful")
    msg    ="""From:adelmellit2013@gmail.com\nTo:user@gmail.com
    \nSubject: Fault detected...\n
    """
    msg = msg+"Number = "+examp_str
    try:
      server.sendmail(' adelmellit2013@gmail.com ','user@gmail.
      com',msg)
    except:
      return server
    print("Sending Message...")
    server.quit()
    print("Quit Server")
for ii in range(0,60):
    if ii==40:
      server=email_alert(str(ii))
    else:
      continue
```

Index

Note: Page numbers followed by *f* indicate figures, *t* indicate tables, and *b* indicate boxes.

R

FSC
www.fsc.org
MIX
Papier | Fördert
gute Waldnutzung
FSC® C083411

Kolibri 360 GmbH
Ferdinand-Jühlke-Straße 7
99095 Erfurt, Deutschland
produktsicherheit@kolibri360.de